THE DEFENSE INDUSTRIAL BASE

To my parents

The Defense Industrial Base

Strategies for a Changing World

NAYANTARA D. HENSEL
National Defense University, USA

LONDON AND NEW YORK

First published 2015 by Ashgate Publishing

2 Park Square, Milton Park, Abingdon, Oxfordshire OX14 4RN
711 Third Avenue, New York, NY 10017

Routledge is an imprint of the Taylor & Francis Group, an informa business

First issued in paperback 2017

British Library Cataloguing in Publication Data
A catalogue record for this book is available from the British Library

The Library of Congress has cataloged the printed edition as follows:
Hensel, Nayantara D.
The defense industrial base : strategies for a changing world / by Nayantara D. Hensel.
pages cm
Includes bibliographical references and index.
ISBN 978-1-4094-3104-6 (hbk) -- 1. Defense industries--United States. 2. United States. Department of Defense--Appropriations and expenditures. 3. United States--Armed Forces--Appropriations and expenditures. 4. United States--Military policy. I. Title.
HD9743.U6H46 2015
338.4'735500973--dc23

2014043307

ISBN 978-1-4094-3104-6 (hbk)
ISBN 978-1-138-55985-1 (pbk)

Contents

List of Figures

List of Tables

List of Tables

About the Author

Dr Nayantara Hensel has served as the Chief Economist for the Department of the Navy. She has provided economic guidance on the defense industrial base, the financial health of defense contractors, commodities and strategic materials, the federal budget, interest rates, unemployment, exchange rates, and inflation, as well as trends in the broader economy and in the defense sector. Dr Hensel was awarded the Joint Civilian Service Commendation Medal in 2013. She received her B.A., M.A., and Ph.D. from Harvard University, where she graduated magna cum laude and Phi Beta Kappa and specialized in finance and economics. She has taught at Harvard University, the Stern School of Business at New York University (NYU), the Naval Postgraduate School, and National Defense University, and has also served as the Pentagon Scholar in Residence. Dr Hensel also previously served as Senior Manager and chief economist for Ernst & Young's litigation advisory group, Managing Economist for the New York City office of the Law and Economics Consulting Group (LECG), and an economist in the economic consulting arm of Marsh & McLennan.

Dr Hensel has written over 50 articles and research reports. She has published in journals, such as the *International Journal of Managerial Finance*, the *Review of Financial Economics*, *Business Economics*, the *European Financial Management Journal*, the *Journal of Financial Transformation*, and *Harvard Business School Working Knowledge*. She is on the Research Advisory Board for Defense Acquisition University, has served on the Board of Directors for the National Association of Business Economists (NABE) and the National Economists Club (NEC), and has served as the Chair of the NABE Industry Survey and the Chair of the NABE Outlook Survey. Dr Hensel is an elected member of NBEIC and the Conference of Business Economists, both of which are composed of the top economists in the US, and is a member of the Harvard University Industrial Economists Group. She has given seminars on the defense industrial base and the economy at a number of government agencies, academic institutions, and conferences. She has appeared on CNBC, PBS, Fox Business TV, Voice of America, and Reuters TV, and has also conducted a variety of radio interviews on ABC, CBS, NPR, Wall Street Journal Radio, Bloomberg Radio, and AP.

About the Author

Introduction

The contemporary global defense sector is facing a series of challenges from a variety of different sources, and the development of successful strategies designed to address these challenges in the current environment is necessary in order to enable the global defense sector to continue to effectively contribute to the maintenance of global stability and security. Many countries, including the US, are facing significant fiscal constraints regarding defense spending due to the strain on federal budgets and the need to limit rising government debts and deficits. Moreover, although economic growth has shown improvement from the financial crisis in 2008–2009, the economic recovery has been slow and uneven. This also has implications for the demands on federal budgets, which, in turn, limits the growth of defense budgets or even their sustainment at current levels. Moreover, countries, such as the US, are experiencing a shift in defense priorities, which impacts the demand for particular types of equipment. For example, as the US withdraws from Iraq and Afghanistan and expands its presence in the Asia-Pacific region, there is less demand for mine-resistant vehicles and greater demand for certain types of vessels and aircraft. Nevertheless, the challenge of balancing between fiscal constraints and the need for particular types of equipment in terms of modernization, replacement, and maintenance due to shifting defense priorities requires the development of new strategies. Indeed, as new challenges emerge in the Middle East (Syria, Iraq, etc.) and in the Ukraine, defense priorities may further shift, leading to changes in demand for equipment in the global defense industrial base.

The current decisions that are made in the defense sector will have a long-term impact on the defense industrial base and, potentially, on the ability of countries to flexibly adjust to shifting defense priorities and equipment demand in future years. In the US, a significant fraction of the defense industrial workforce will retire in the coming years and, if there is little demand for certain types of equipment, there will be no incentive for defense contractors to hire younger workers and for intergenerational skillset transfer to occur. As a result, if defense priorities were to shift in the future and the demand for certain types of equipment increases, it could be difficult for defense industries to adapt to the changes.

The global defense industrial base is a network of linkages across countries. Slow economic growth and rising debt have put strain on federal budgets and defense budgets in many countries which are facing similar challenges regarding tradeoffs between fiscal constraints and the need for modernizing equipment as a result of changing defense priorities. Due to the substantial involvement of many countries in global trade in the defense sector and other sectors, slow demand for exports can hurt defense contractors that are dependent on overseas markets. This, in turn, can impact the sustainability of the defense industrial base. Moreover, many overseas countries are trying to develop their own defense industrial bases in an effort to purchase equipment which is less expensive, but which meets their needs.

In response to these uncertainties in demand for equipment and services due to fiscal constraints and shifting defense priorities, defense contractors must develop new strategies or revisit older ones in an effort to mitigate risk. Investment in innovation of new designs or in expanding production facilities could represent a loss if there is a significant reduction in orders. Moreover, smaller firms could have difficulty accessing capital markets in an environment of slow economic growth. Following the end of the Cold War, there was substantial consolidation in the US defense sector due to excess capacity in the industry. Did this consolidation lead to greater efficiencies and preservation of the defense industrial base? Should defense contractors consolidate in the coming years? Would forming alliances in an effort to leverage research and development (R&D) rather than merging lead to less permanence in integration, but provide more flexibility? If defense contractors expanded into the civilian market and away from the defense sector, could they produce equipment which could be developed using common designs and skillsets that could also be used in the defense sector? This could provide opportunities to continue to innovate, preserve skillsets, and maintain profitability. Should the defense contractors expand into overseas markets to provide civilian and military equipment? The chapters in this book seek to provide perspectives on these issues.

With the recent shift in defense priorities and the current fiscal constraints, how will the US and global defense industrial bases evolve? This book examines the interaction between the economic environment, budget constraints, and shifting defense priorities on the global defense industrial base. It discusses the challenges facing:

a. the US Department of Defense (DoD) and the services (the Army, the Navy and Marines, and the Air Force);
b. the US defense contractors and the US defense industrial base in certain regions;
c. European countries and their defense contractors; and
d. developing countries and their defense industrial bases in Asia, the Middle East, etc.

Chapter 1 examines the impact of the US budgetary environment and shifting defense priorities on defense spending in order to identify and evaluate the tradeoffs facing DoD. It evaluates the growth of the overall defense budget in the context of the larger federal budget and the importance of defense spending as a share of GDP relative to other expenditures. It then assesses recent defense budget proposals in the context of fiscal constraints and shifting defense priorities. It compares the growth in the various categories of the overall defense budget – operations and maintenance; military personnel; procurement; research; development; testing and evaluation (RDT&E); and military construction – as well as the growth in areas of the budget by service. The chapter discusses the impact of sequestration and the Budget Control Act on the services and on various budget categories. The level and volatility of DoD monthly outlays by budget category and by service are also evaluated. Finally, the chapter highlights the tradeoffs for the services in terms of reducing current capacity in order to protect and further expand modernization programs versus focusing on the sustainment of current capabilities, while sacrificing future technologies. It provides examples of these tradeoffs between procurement of new equipment, operations and maintenance, and RDT&E.

Chapter 2 focuses on the US defense industrial base and the impact of fiscal constraints and shifting defense priorities on the demand for various types of equipment, which has implications for the types of strategies which defense contractors are developing to meet these challenges. It first discusses the areas of transformation in the US defense industrial base in recent years, providing historical and contemporary examples of growth and atrophy. Areas of the defense industrial base experiencing growth in recent years include unmanned aerial vehicles (UAVs), cybersecurity, etc., while areas that have experienced less rapid growth include the shipbuilding segment. The chapter then assesses the impact of growth and contraction in areas of the defense industrial base on various regions of the US, as well as the trends in employment over time. This section discusses the employment trends in age distribution, as well as in various categories of equipment production, which impact the type of skillsets and the ability to intergenerationally transfer skillsets, based on DoD demand for products. It also provides examples of the impact of reductions in defense spending and sequestration on various types of equipment, which, in turn, impacts firm employment and regional production lines. The chapter highlights the tradeoffs in demand for different types of equipment in the defense industrial base due to budgetary constraints and the shift in defense priorities. It also provides examples of the challenges facing the Army, Navy, Air Force, and Coast Guard, and the resulting impact on the defense industrial base. The chapter then discusses the strategies for defense contractors in combating defense industrial base pressures. One strategy involves the expansion of defense contractors into the civilian sector in an effort to mitigate the risk of uncertainty in DoD demand and fiscal constraints, as well as to leverage investment in developing new technologies. A second strategy involves the expansion of international sales by defense contractors. Finally, the chapter discusses the US firms in the context of the global defense industry and discusses the trends in the top global defense firms by country and by defense revenues. It emphasizes that, while expansion into international markets can be helpful in diversifying risk, many overseas countries are also experiencing fiscal constraints.

Chapter 3 provides perspectives regarding the impact of consolidation in the defense industrial base as a strategy for defense contractors. It examines the causes and consequences of the wave of defense mergers in the US during the 1990s and evaluates the impact of consolidation in improving cost efficiencies from rationalization of plant capacity, pooling of knowledge sets, etc. The chapter evaluates the cost efficiencies following consolidation by defense contractors (as acquirers or targets), as well as by weapons system category. It explores which of the services benefited from the cost efficiencies and discusses the improvements in cost efficiencies (or lack thereof) in particular mergers. Would consolidation in the current environment be a possible strategy for defense contractors? An evaluation of the historical consolidation wave in the defense sector following the end of the Cold War provides some insights.

Chapter 4 assesses the degree to which alliances could be a strategy for defense contractors to sustain the defense industrial base. With the shrinkage of defense budgets in the US and in Europe, alliance formations could result in more cost-effective methods for investing in innovation, as well as in providing stronger bids in overseas competition as defense contractors expand into international sales. The chapter examines the costs and benefits of consolidations versus alliances and assesses a variety of case studies to provide perspectives on the evolution and outcome of various alliances. These case studies involve alliances which:

a. spur other alliances between competitors, ultimately creating new products;
b. promote national defense strategies among different countries;
c. focus on new product creation;
d. share R&D costs;
e. develop interoperable equipment between allied nations; and
f. enhance firm strategies in an era of fiscal constraints.

Chapter 5 focuses on the relationship between global defense industries and the US defense industrial base by providing a case study of the tanker competition in the US between Boeing and Northrop Grumman/EADS. It first discusses a historical taxonomy of strategies which often limit the role of foreign manufacturers in the US defense industrial base, including the difficulties faced by foreign entrants in merging with domestic competitors, as well as the enactment of tariffs or quotas to protect domestic industries and the strengthening of the Buy America legislation. It then discusses in detail the 2008 competition between Boeing and Northrop Grumman/EADS to supply the US Air Force (USAF) with a new fleet of aerial refueling tankers. The case study evaluates the strategies of the traditional incumbent (Boeing) as it attempted to maintain its historical role in the defense industrial base, as well as the strategies of the alliance between a domestic firm (Northrop Grumman) and a foreign firm (EADS). Moreover, it highlights the impact of the tanker proposals on various regions and the resulting role of Congressional representatives in determining the outcome of the tanker competition. Finally, the chapter discusses the third and final competition for the tankers, which began in the fall of 2009. It provides perspectives on the interaction of DoD, defense contractors, and Congressional representatives in an environment of increasing globalization of the defense sector and of the need to replace aging equipment.

Chapter 6 focuses on the role of the US in the international defense market. This builds upon the earlier discussion in Chapter 2 regarding the strategy of US defense contractors in attempting to minimize the risks associated with uncertain demand and fiscal constraints in the US by emphasizing further expansion into overseas markets. Moreover, it builds on the perspectives in Chapter 5 regarding the openness of the US defense market to foreign suppliers. The chapter first explores the degree to which the US is engaged in the international defense market by assessing the tendency for the US to import military equipment and related civilian equipment from foreign suppliers in various categories. It further compares the number and value of contracts awarded by DoD to US suppliers relative to foreign suppliers in various equipment categories and, to the degree that DoD awards contracts to foreign suppliers, it discusses which countries are more likely to receive contracts from the US. The chapter then assesses the role of US defense industrial firms in exporting defense equipment, especially through foreign military sales (FMS) and provides examples of the role of the US in significant exports to overseas countries. The second half of the chapter examines the role of the US in international trade in the aerospace sector as a case study, which also provides insights into the trends for US contractors in expanding into overseas markets through exports of military and civilian aerospace equipment. The chapter provides examples of the challenges and opportunities facing US military sales of aircraft as exemplified by overseas sales of the F-35, as well as US civilian sales of aircraft, through a discussion of the sales to the Middle East. Finally, it explores US imports of aerospace products, including civilian aircraft.

Chapter 7 focuses on the global defense market and the impact of economic growth, shifting defense priorities, and fiscal constraints on global defense trade. It first examines military expenditures in a global context by discussing economic growth and government debt for the countries with the highest military expenditures, as well as the importance of these military expenditures as a share of GDP. The chapter then explores the trends in trade patterns for defense equipment globally by assessing economic growth and government debt for the countries which are the top exporters of defense equipment, as well as the countries which are the top importers, since fiscal constraints and economic growth in importing and exporting countries can influence growth in the global defense trade. It explores the global trade market in ships and aircraft as a case study, and assesses the trends among the top importing and exporting countries in ships and aircraft in recent years, as well as the sustainability of market share by the leading countries in these markets. The next section of the chapter evaluates the strategies employed by certain countries as they attempt to preserve their respective defense industrial bases in the wake of global trade. Some of these strategies include: (a) countries re-selling used defense equipment to other countries to sustain the operations and maintenance skillsets of their defense contractors; and (b) countries ordering upgrades of existing equipment to preserve skillsets and/or to replace another type of equipment because they cannot order newer models due to budgetary constraints. The similarities of the challenges facing the US and European countries, as well as their defense contractors, is evident in the case study on the European defense industrial base and its fiscal constraints. The case study evaluates the budget limitations in many of the European countries and the impact of budget constraints on the closure of production facilities and employment, which is similar to the issues discussed in Chapters 1 and 2 in relation to the US defense industrial base. It also examines the tendency toward alliances between European defense contractors, which reinforces many of the motivations for alliances, discussed in Chapter 4. Finally, the chapter examines the conflict between fiscal constraints (which reduce defense spending) and emerging defense threats (which expand the need to increase defense spending) by examining the potential impact of the Ukrainian crisis on the European, as well as the global, defense industrial bases.

Chapter 8 continues to examine global trade by focusing on a detailed analysis of global arms deliveries in the overseas defense market beyond the US and Europe. Global arms deliveries involve a variety of different types of equipment, including: missiles, aircraft, ships, submarines, armored cars, artillery, and tanks. The chapter assesses the trends in overall global arms deliveries by supplier countries in recent years and examines the sustainability of leading suppliers over time as market leaders. Moreover, it also focuses on the trends in arms deliveries to developing countries by supplier countries, as well as discusses the developing countries which are the leading recipients of global arms deliveries. The sustainability of supplier countries in terms of the quantity and type of weapons delivered to developing countries is evaluated in detail. The next sections of the chapter evaluate global arms deliveries to various regions in detail. The section on Asia highlights the importance of shifting defense priorities for the Asian countries in response to increasing concerns over China's potential expansion of control in the East and South China Seas and discusses the types of equipment that is needed by countries, such as Taiwan, South Korea, Indonesia, and Japan. Similarly, it examines India's expansion of its domestic defense industrial base, as well as its demand for upgrading ships and aircraft, in response to perceived threats from China and Pakistan. The section also examines in detail the trends in defense suppliers by type and quantity of weapons delivered to Asian nations. The section on the Middle East/Near East discusses the leading countries in that region in military expenditures and in arms deliveries, as well as the types of equipment needed and the development of indigenous defense industrial bases. It evaluates the trends in arms deliveries by supplier countries to this region and compares the trends by type and quantity of weapons delivered to Middle Eastern/Near Eastern countries. Finally, the last two sections of the chapter address global arms deliveries to Latin America and Africa, respectively. These sections examine the trends in global arms deliveries by supplier countries, as well as by quantity and type of weapons that are delivered to these regions.

In short, this book examines the challenges and opportunities facing the contemporary US defense industrial base, as well as the global defense industrial base. There are many similarities between the current challenges facing nations in terms of handling tradeoffs due to shifting defense priorities and fiscal constraints.

While these tradeoffs emphasize the procurement of the types of equipment that are currently deemed to be necessary and/or desirable in response to the present-day security and fiscal environment, the impact of these tradeoffs can, however, lead to the atrophy of neglected, but potentially important areas of the global defense industrial base. This could, in turn, become problematic if these neglected sectors of the defense industrial base were to be re-emphasized in the future in response to unanticipated national security challenges. Developing strategies designed to sustain production facilities, employment, and skillsets, while simultaneously responding to the current national security challenges and fiscal realities, is vital to maintaining the global defense industrial base. Consequently, the role of this book is to provide the reader with contemporary perspectives concerning the types of strategies that could be developed by defense industrial firms and governments to sustain a robust global defense industrial base and, thereby, help to ensure the maintenance of global peace, stability, and prosperity.[1]

1 The opinions, conclusions, and/or recommendations expressed or implied within this book are solely those of the author and do not necessarily represent the views of the United States government, the US Department of Defense, or any other US government agency or institution.

Chapter 1

The Evolution and Challenges of the US Defense Budget and the Tradeoffs Impacting the Defense Industrial Base

Economic growth in the United States, as measured by Gross Domestic Product (GDP), has varied over the past 50 years and has impacted federal spending, as well as defense and non-defense spending as a share of GDP. Nevertheless, as more government programs were developed and expanded (such as Social Security and Medicare), spending in those areas steadily increased as a share of GDP. Military expenditures, on the other hand, despite procurement and development of new types of equipment, as well as expenses from operations and maintenance and the rising costs of military healthcare and retirement, declined as a share of GDP. Within the context of this overall trend, military expenditures fluctuated over various periods (the Korean War, the Vietnam War, the Cold War, and the War on Terror) due to the assessment of global challenges and opportunities, shifting defense priorities, and the types of equipment and manpower that were necessary.

The current fiscal constraints and shifting defense priorities provide a variety of tradeoffs in the budget. This chapter will first examine the historical trends in the defense budget relative to the broader federal budget and then will discuss the DoD budgets in recent years. It will provide perspective on the recent defense budgets in terms of:

a. challenges due to fiscal constraints and sequestration;
b. challenges due to evolving defense strategies and priorities; and
c. challenges within various budget categories and by service (Army, Navy, Air Force, and Marine Corps).

The chapter examines the recent developments in DoD budgets and in DoD monthly expenditures in various categories and by service. In addition, it discusses the impact of sequestration on budgets and outlays for DoD and for the services. Finally, it concludes with a discussion of the difficulties and tradeoffs faced by each of the services in various areas of the budget, especially the difficult choice of modernizing equipment or sustaining current capabilities at the risk of sacrificing research and development for future technologies.

Historical Trends in Defense Budgets in the Context of the Federal Budget

The historical trends in US government expenditures, defense spending, and other areas of the budget provides perspectives on the broader federal budgetary demands in the context of economic growth.

Figure 1.1 below shows US government expenditures as a share of GDP over the past 50 years, as well as Social Security/Medicare and defense spending as a share of GDP. The figure indicates that government expenditures, as well as Social Security and Medicare, have increased as a share of GDP over the past 50 years, but defense spending as a share of GDP has declined over that same period, especially relative to the Korean War and the Vietnam War.

Government expenditures were about 17.2% of GDP in 1948, following World War II. By 1952, they were over a quarter of GDP and remained in that range, reaching about 30% of GDP in 1968, during the height of the Vietnam War.

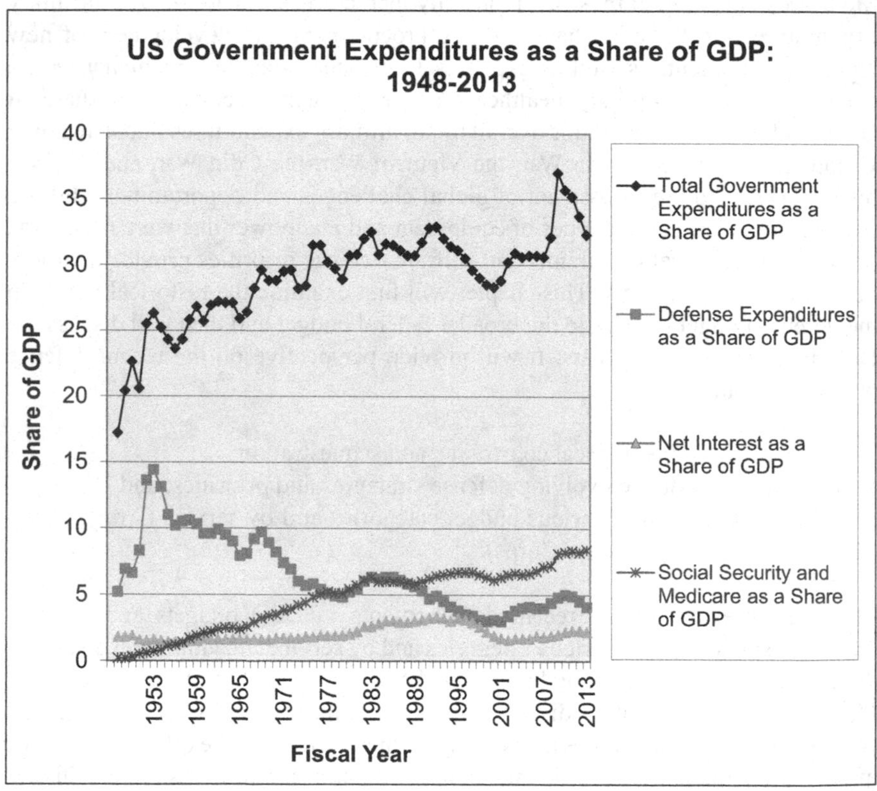

Figure 1.1 US Government Expenditures as a Share of GDP: 1948–2013

Source: The President's Budget FY 2015, Table 15.5. See http://www.whitehouse.gov/omb/budget

Government expenditures as a share of GDP continued to trend upward, reaching 37% in 2009, dropping to 35–36% in 2010 and 2011, and then dropping to 33.7% of GDP in 2012 and 32.3% in 2013.

Medicare and Social Security as a share of GDP have steadily risen over the past 50 years. Although the Medicare program was not implemented until 1965, the Social Security program had increased from 1.1% of GDP in 1955 to 2.1% in 1960. By 1967, Social Security and Medicare had risen to almost 3% of GDP. Between 1967 and 1976, they rose to between 3% and 5% of GDP. During the period from 1982–1990, they ranged between 5.8% and 6.3% of GDP and they ranged from 6.2% to 6.7% between 1991 and 2001. Most recently, between 2009 and 2013, they were between 8.1% and 8.4% of GDP.

Defense expenditures as a share of GDP, on the other hand, trended downwards over time. Although they were only 5.2% of GDP in 1948 at the end of World War II, they peaked out at 14.4% of GDP in 1953, during the Korean War. Between 1956 and 1963, they ranged from between 9.5% and 10.6% of GDP. By 1970, defense expenditures had dropped to 8.2% of GDP, and, as the Vietnam War wound down, fell to 5.8% in 1975, and then as low as 4.8% in 1979. Beginning in 1982 and throughout the 1980's, defense spending as a share of GDP rose to a range of 5.6%-6.3% of GDP. Indeed, in 1989, they were at 5.6% of GDP, but, as the Cold War ended, defense spending continued to drop, reaching 3.8% in 1995. By 2001, defense spending was 3% of GDP, but, with the beginning of the War on Terror following 9/11, it increased to 5% of GDP between 2009 and 2011, but then fell to 4.5% in 2012 and 4.1% in 2013.

The trends in defense spending as a share of GDP suggest that the importance of defense priorities have historically impacted defense spending to a greater degree than the trends in the overall government expenditures as a share of GDP and the federal budget. Nevertheless, both forces have played a valuable role in defense spending in recent years and will likely do so in future years. Moreover, the volatility of defense spending as a share of GDP relative to the increasing trend of Social Security and Medicare as a share of GDP highlights the tradeoffs between defense and other areas of the budget, as well as the challenges faced by DoD when it needs to expand again, following earlier contractions.

Recent DoD Budgets and Challenges Faced by Budget Cuts and Shifting Defense Priorities

The most recent developments in the defense budget indicate the likelihood for changes in subsequent years, due to the need for the defense budget to contribute to restraining the rising federal budget, as well as to handle the shift in defense priorities, away from Iraq and Afghanistan, and toward a balance between expanding our military presence in the Asia-Pacific region, as well as maintaining the US military presence in the Middle East.

In November, 2013, Secretary of Defense Hagel announced the six priorities which had been developed from the Strategic Choices and Management Review, which had ended in July 2013. These priorities are likely to impact the size and importance of various categories of the defense budget in the upcoming years. These six focus areas included:

a. balancing between capability and capacity;
b. institutional reform;
c. re-evaluation of the force planning construct (which includes the assumptions establishing the planning of organization of troops, acquisition of equipment, and training of the military);
d. protecting investment in newly developing military capabilities;
e. compensation and personnel policy; and
f. preparing for a prolonged military readiness challenge.

Secretary Hagel noted the importance of DoD protecting investments in "emerging military capabilities, especially space, cyber, special operations forces, and intelligence, surveillance and reconnaissance …"[1]

Secretary Hagel also observed that "the military would have to come to grips with much smaller budgets and reset expectations as the nation moves 'off on a perpetual war footing.'" On the other hand, he also noted, in his speech at the Center for Strategic and International Studies in November 2013, that "even after a retrenchment, the United States alone will still account for nearly 40% of all military spending in the world." Secretary Hagel suggested that there will be a move toward "a smaller, modern, and capable force" relative to "a larger force with older equipment" and that the conventional land forces will continue to shrink, while Special Operations and weapons with new technologies will be expanded. Finally, he suggested that DoD "needs to rely more heavily on allies, and should play a supporting role, not a leading role, in US foreign policy."[2]

As defense priorities evolve, DoD is also faced with fiscal constraints. The Budget Control Act, passed in August 2011, required $487 billion in defense cuts over the next 10 years, portions of which were reflected in the subsequent budgets. Due to the Super Committee's lack of success in developing a blueprint to reduce federal budget deficits by $1.2 trillion in November 2011, the Budget Control Act called for sequestration to begin in January 2013.[3] It was deferred and partially took effect on March 1, 2013. Department of Defense, under full sequestration, would bear an additional $500 billion of the discretionary spending cuts over the next 10 years. Under full sequestration, DoD would have to cut $52 billion in spending each year through FY 2021.

---

1 Weisgerber, 2013b.
2 Whitlock, 2013.
3 McGarry and Taborek, 2013.

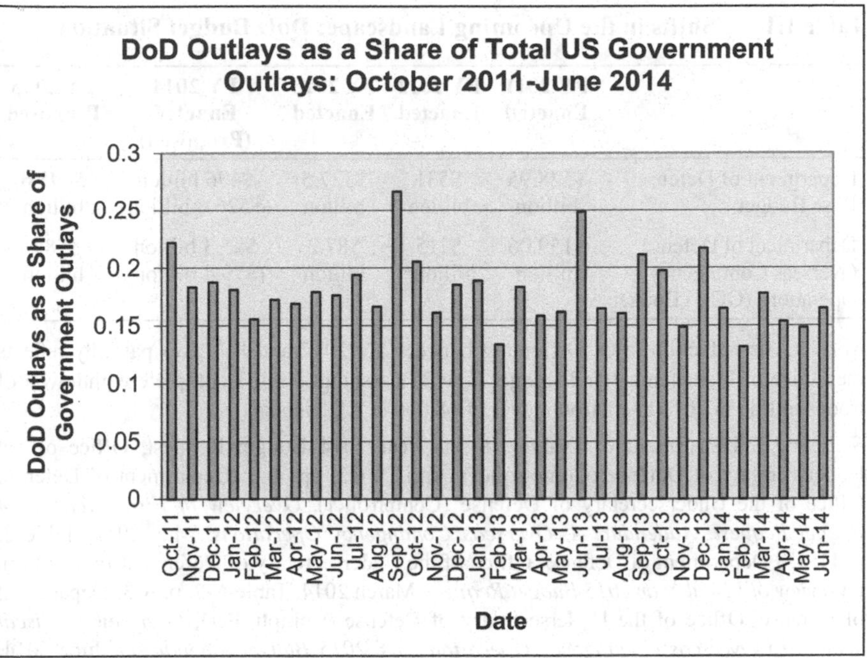

Figure 1.2 DoD Outlays as a Share of Total US Government Outlays: October 2011–June 2014

Source: Department of the Treasury, Bureau of the Fiscal Service, *Monthly Treasury Statement of Receipts and Outlays of the United States Government*, October, 2011–June, 2014, p. 2 and p. 8.

The share of monthly DoD outlays relative to US government outlays in FY 2012, FY 2013, and the first nine months of FY 2014 is illustrated in Figure 1.2 and suggests the impact of upcoming fiscal constraints and shifting defense priorities. DoD outlays peaked in September 2012 – the end of the fiscal year for FY 2012 – at 27% of US government outlays. During FY 2012, DoD outlays varied between 15.5% and 27% of US government outlays. In FY 2013, DoD outlays peaked as a share of federal government outlays at 24.9% in June, 2013. During FY 2013, DoD outlays varied between 13.3% and 24.9%. During the first nine months of 2014, DoD outlays varied between 13.1% in February 2014 and 21.7% in December 2013 and ranged between 14% and 18% in six of the nine months.

Defense expenditures were affected by the uncertainties in FY 2013 regarding sequestration. Indeed, defense expenditures were as low as 13% of federal outlays by February 2013 and remained between 15% and 18% throughout the rest of the year, except for June, 2013 (24.9%) and September, 2013 (21.1%). This was a slightly lower range than in FY 2012. The sequestration cuts, which were mandated by the Budget Control Act, resulted in the need for DoD to

Table 1.1 Shifts in the Upcoming Landscape: DoD Budget Situation

	FY 2011 Enacted	FY 2012 Enacted	FY 2013 Enacted	FY 2014 Enacted (Proposed)	FY 2015 Proposed
Department of Defense Base Budget	$528.95 billion	$531 billion	$527.5 billion	$496 billion ($526.6 billion)	$495.6 billion
Department of Defense Overseas Contingency Operations (OCO) Budget	$159.06 billion	$115 billion*	$87.2 billion	$85.3 billion ($79.4 billion)	$58.6 billion

Note: * The reduction in OCO funding between FY 2012 and FY 2013 partially reflects the Office of Management and Budget (OMB)'s assumption of a complete withdrawal of troops in Iraq in 2013 and a troop level of 68,000 in Afghanistan.

Source: US Department of Defense Fiscal Year 2014 Budget Request, Office of the Undersecretary of Defense (Comptroller), April, 2013, pp. 1–3; Department of Defense, Office of the Undersecretary of Defense (Comptroller), *Overview of Fiscal Year 2014 Budget Request: Addendum A, Overseas Contingency Operations*, May 2013, Table 2, p. 11; US Department of Defense, Office of the Undersecretary of Defense (Comptroller), *Overview of Fiscal Year 2015 Budget Request,* March 2014, Table A-7, p. A-3: Department of Defense, Office of the Undersecretary of Defense (Comptroller), *Overview of Fiscal Year 2015 Overseas Contingency Operations: FY 2015 Budget Amendment*, June 2014, Table 2, p. 12.

cut $37 billion between March and September, 2013 and included a reduction in training.[4] Similarly, in the first nine months of 2014, defense expenditures exceeded 18% of federal outlays in two of the nine months and were in the 14%–16% range over half of the time.

Table 1.1 shows the Department of Defense base budget over the last few years, ranging between $495.6 billion, which was proposed for FY 2015, and $531 billion, which was enacted for FY 2012. The Department of Defense Overseas Contingency Operation (OCO) budget, on the other hand, has fallen over the past few years with the withdrawal of troops from Iraq and Afghanistan. Not surprisingly, with the shift in defense priorities, the OCO budget has fallen from $159.06 billion in 2011 to $115 billion in 2012, and then to $87.2 billion in 2013 and $85 billion in 2014. For the FY 2015 OCO proposal, DoD requested $58.6 billion.

The FY 2014 budget included an emphasis on the pivot to the Asia-Pacific region and the Middle East and focused on:

a. reducing ground forces with the withdrawal of Army troops from Iraq and Afghanistan;

4 Weisgerber, 2013b.

b. a renewed focus on air and naval power, as the presence in the Pacific and the Middle East potentially expands; and
c. a continued ability to conduct and expand cyber warfare.

The proposed budget also emphasized longer-term plans which focused on the concept of reversibility, in which the size of the ground forces would be reduced, but manpower and equipment/platforms could be quickly scaled up.

The key focus of the FY 2015 budget proposal was the need to handle the tradeoffs between capabilities, capacity, and readiness. The budget proposal attempted to focus on improving efficiencies, and streamlining the force size, while continuing to maintain readiness. The FY 2015 budget request was linked to the strategy of the 2014 Quadrennial Defense Review (QDR). The 2014 QDR emphasized the importance of institutional reforms to support several key initiatives: "to protect the homeland, to deter and defeat threats to the United States and to mitigate the effects of potential attacks and natural disasters; build security globally, to preserve regional stability, deter adversaries, support allies and partners, and cooperate with others to address common security challenges; and project power and win decisively, to defeat aggression, disrupt and destroy terrorist networks, and provide humanitarian assistance and disaster relief."[5]

Unfortunately, however, FY 2013, FY 2014, and FY 2015 programs have been and/or can be adversely affected by partial sequestration in their given year, as well as by ripple effects from partial sequestration in the prior year. These effects have included training cutbacks, civilian furloughs, deferral of equipment and facility maintenance, reductions to energy conservation investments, contract inefficiencies, and curtailed deployments.[6] If sequestration continues in the subsequent years, it can lead to negative spillover effects concerning the condition of equipment due to reduced operations and maintenance from cuts in the prior years, as well as significant damage to skillsets due to reductions in training, furloughs, and layoffs in the prior years.

DoD's 2014 budget proposal exceeded the defense spending caps by $52 billion. In DoD's original two 2015 budget proposals, one of the proposals was based on meeting the caps and the other proposal was built on the 2014 proposal that did not include sequestration.[7] As of the fall of 2013, under full sequestration, DoD would have needed to cut funds from its proposed FY 2014 budget, such that the defense budget would fall from $552 billion to $498 billion. Savings from entitlement programs or other forms of cuts, however, would have enabled the elimination of sequestration. Nevertheless, DoD expressed concern that it needed a line-by-line appropriation bill each year since, under the continuing resolution, DoD would have been kept at the same 2013 funding levels, which would prevent it from having the flexibility in terms of shifting money between different budget

5 US Department of Defense, 2014.
6 US Department of Defense, 2013a; US Department of Defense, 2013b.
7 Weisgerber, 2013b.

categories and prevent it from starting new programs. As a result, in the fall of 2013, DoD developed its budget for 2015–2019, one of which assumed that sequestration would continue.[8]

The bipartisan budget deal, which was passed in December, 2013, increased the cap on spending for defense and non-defense programs for FY 2014 and FY 2015. The cap for FY 2014 for defense and non-defense programs rose from $976 billion to $1.012 trillion, and the cap for FY 2015 rose from $995 billion to $1.014 trillion. As a result, spending would increase $62 billion over 10 years, but this would be offset by a variety of savings.[9] The bipartisan budget deal eliminated the potential that the government could shut down in the next two years and saved $28 billion over the next decade.[10] Some of the offset in the spending increase included: higher fees for airline passengers,[11] reductions in benefits for retirees for the military in the form of reduced annual cost of living adjustments for military retirees still of working age,[12] higher fees for firms to pay to the Pension Guarantee Corporation to guarantee pensions,[13] and larger pension contributions to be paid by federal employees hired after January 1, 2014.[14] Moreover, it lessened cuts linked to sequestration.[15]

In FY 2013, the budget for DoD was reduced by $37 billion due to sequestration, which was in addition to the cuts from the Budget Control Act, which instituted $487 billion in cuts over 10 years. The Bipartisan Budget Agreement from December 2013 lessened the impact of sequestration, although it imposed $75 billion in reductions for 2014 and 2015.[16] DoD spending, under the compromise budget which Congress passed and President Obama signed in December, was capped at $498 billion in 2014, which exceeds the original cap for sequestration by $21 billion, but is less than DoD's request by $29 billion. For 2015, DoD spending was capped at $9 billion more than the previous cap of $512 billion.[17] As a result, DoD had $30 billion more to spend during 2014 and 2015, due to the compromise budget, than would have been the case under full sequestration.[18]

If full sequestration occurs beginning in FY 2016, approximately another $50 billion would be cut annually through 2021.[19] Moreover, if full sequestration in 2016 returns, further reductions would include the retirement of 80 more aircraft by the Air Force, which would include elimination of the Global Hawk Block 40 fleet and the KC-10 tanker fleet, as well as reduced flying hours and a reduction in

8 Erwin, 2013a.
9 Weisman, 2013a.
10 Davis, 2013.
11 Hook, 2013.
12 McKinnon, 2013.
13 Hook, 2013.
14 McKinnon, 2013.
15 Weisman, 2013b.
16 US Department of Defense, 2014.
17 Weisgerber and Fryer-Biggs, 2014.
18 *Defense News* Staff, 2014.
19 Marshall, 2014; Simeone, 2014b.

F-35 purchases. The Army would reduce to 420,000. The Marines would further reduce to 175,000, while the Navy would need to eliminate one aircraft carrier starting in 2016.[20]

Comparison of FY 2013 and FY 2014 Enacted DoD Budgets and the FY 2015 Proposed Budget

The next several figures show the comparison between the enacted FY 2013 and FY 2014 budgets, as well as the proposed FY 2015 budget. These indicate some of the effects of both increasing budget cuts, as well as shifts in defense priorities. As of early November, 2013, the pressure of $1 trillion in defense cuts, including the $487 billion in cuts and the $500 billion in cuts from sequestration, suggested, according to Defense Secretary Chuck Hagel, that "the days of automatic, equitable allotment to the Army, Navy, and Air Force may be over," such that one of the services might be cut more than others.

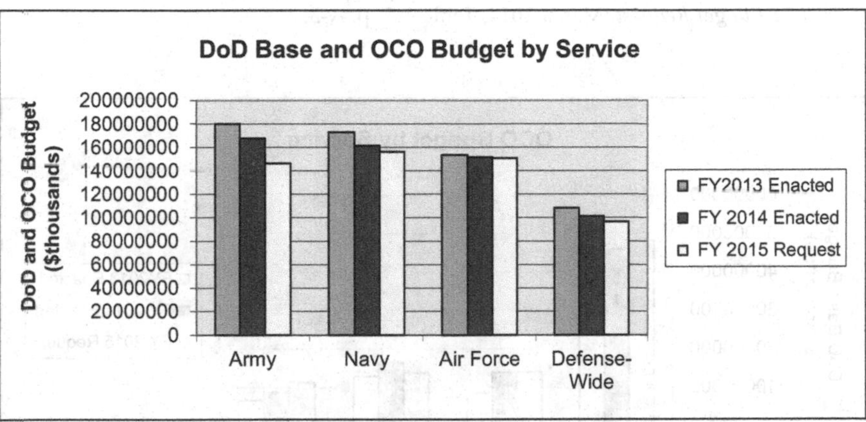

Figure 1.3 DoD Base and OCO Budget by Service

Source: US Department of Defense, Office of the Undersecretary of Defense (Comptroller), *Overview of Fiscal Year 2014 Budget Request,* April 2013, Table A-7, p. A-3: Department of Defense, Office of the Undersecretary of Defense (Comptroller), *Overview of Fiscal Year 2014 Budget Request: Addendum A, Overseas Contingency Operations*, May 2013, Table 2, p. 11; US Department of Defense, Office of the Undersecretary of Defense (Comptroller), *Overview of Fiscal Year 2015 Budget Request,* March 2014, Table A-7, p. A-3: Department of Defense, Office of the Undersecretary of Defense (Comptroller), *Overview of Fiscal Year 2015 Overseas Contingency Operations: FY 2015 Budget Amendment*, June 2014, Table 2, p. 12.

20 US Department of Defense, 2014.

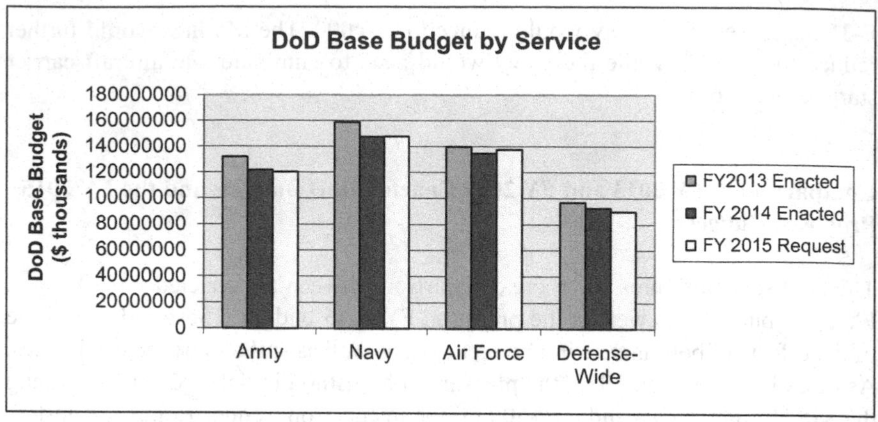

Figure 1.4 DoD Base Budget by Service

Source: US Department of Defense, Office of the Undersecretary of Defense (Comptroller), *Overview of Fiscal Year 2014 Budget Request,* April 2013, Table A-7, p. A-3; US Department of Defense, Office of the Undersecretary of Defense (Comptroller), *Overview of Fiscal Year 2015 Budget Request,* March 2014, Table A-7, p. A-3.

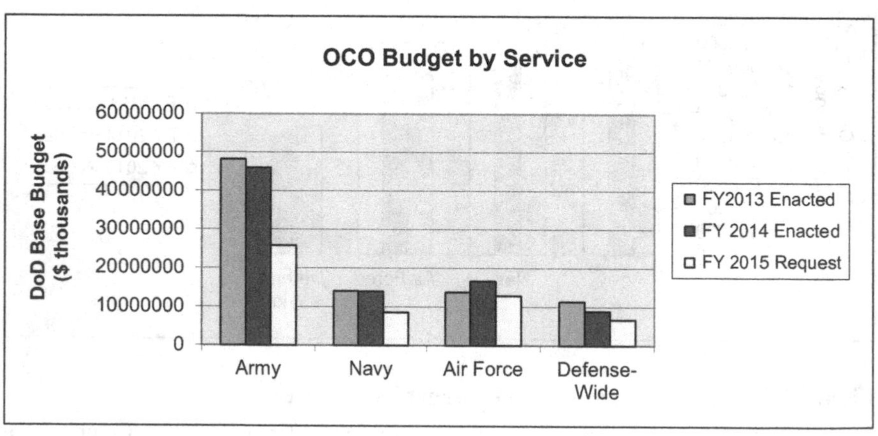

Figure 1.5 DoD OCO Budget by Service

Source: US Department of Defense, Office of the Undersecretary of Defense (Comptroller), *Overview of Fiscal Year 2014 Budget Request: Addendum A, Overseas Contingency Operations*, May 2013, Table 2, p. 11; Department of Defense, Office of the Undersecretary of Defense (Comptroller), *Overview of Fiscal Year 2015 Overseas Contingency Operations: FY 2015 Budget Amendment*, June 2014, Table 2, p. 12.

Figure 1.3 shows the combined DoD base budget and OCO budget by service – the Army, the Navy, the Air Force, and Defense-wide (the Joint Staff, etc.). All of the services had declining DoD and OCO budgets in recent years. The Army had the largest combined budget in the enacted FY 2013 and FY 2014 budgets, much of which was driven by their involvement in Iraq and Afghanistan. As a result, they experienced the most significant decline in the FY 2015 proposed budget. Under the FY 2015 proposed budget, the Navy and the Air Force had higher overall budgets and less of a decline from the FY 2014 enacted budget than the Army.

Figure 1.4 shows the DoD base budget by service, excluding the OCO budget. Unlike Figure 1.3, which focused on the combination of the base budget and the OCO budget, the Navy consistently had the largest base budget, followed by the Air Force, the Army, and the defense-wide forces. The reason why the Navy and the Air Force had higher base budgets than the Army is that despite the fact that the Army had the largest budget overall when combining the base and the OCO budget

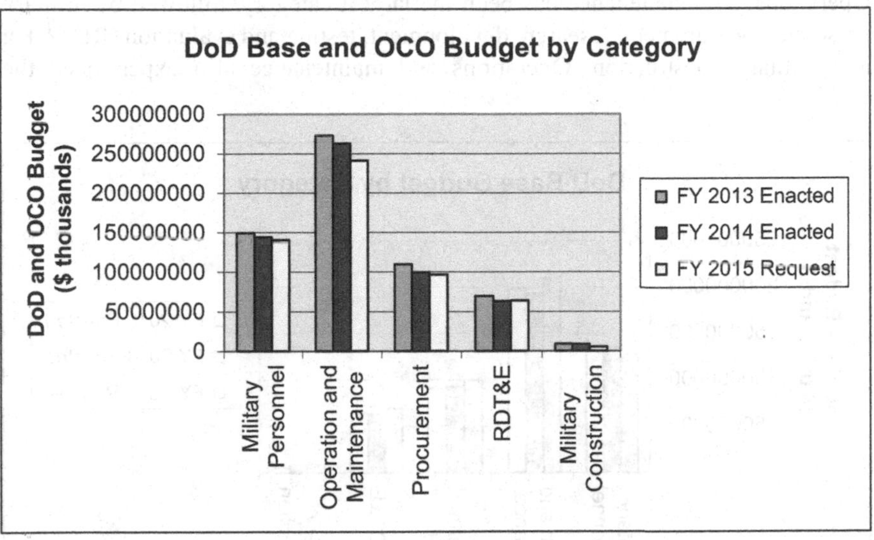

Figure 1.6 DoD Base and OCO Budget by Category

Source: US Department of Defense, Office of the Undersecretary of Defense (Comptroller), *Overview of Fiscal Year 2014 Budget Request,* April 2013, Table A-6, p. A-3; US Department of Defense, Office of the Undersecretary of Defense (Comptroller), *Overview of Fiscal Year 2014 Budget Request: Addendum A, Overseas Contingency Operations,* May 2013, Table 1, p. 11; US Department of Defense, Office of the Undersecretary of Defense (Comptroller), *Overview of Fiscal Year 2015 Budget Request,* March 2015, Table A-6, p. A-3; Department of Defense, Office of the Undersecretary of Defense (Comptroller), *Overview of Fiscal Year 2015 Overseas Contingency Operations: FY 2015 Budget Amendment,* June 2014, Table 1, p. 12.

(Figure 1.3), much of this was driven by its OCO budget, due to its involvement in Iraq and Afghanistan. Both the Army and the Navy had larger reductions than the Air Force in the FY 2013 enacted budget relative to the FY 2014 enacted budget. Under the FY 2015 proposed budget, the Army experienced the greatest reduction in the base budget of the services, while the Air Force experienced an increase.

Figure 1.5, which shows only the Department of Defense OCO budget by service, further reinforces the point that the Army has a substantial overseas contingency operation budget relative to the Navy and the Air Force, which rely on the base budget to a greater degree. The Army experienced the greatest decline in its OCO budget in the FY 2015 proposal, followed by the Navy and the Air Force. This is because the troops are completing their withdrawal from Iraq and Afghanistan, which will impact the Army to a significant degree; the Army is less likely to become a dominant force in the Asia-Pacific region than the Navy and the Air Force.

Figure 1.6 shows the DoD combined base budget and OCO budget by category. Operations and maintenance has been the largest category, followed by military personnel; procurement, research, development, testing and evaluation (RDT&E); and military construction. Operations and maintenance also experienced the

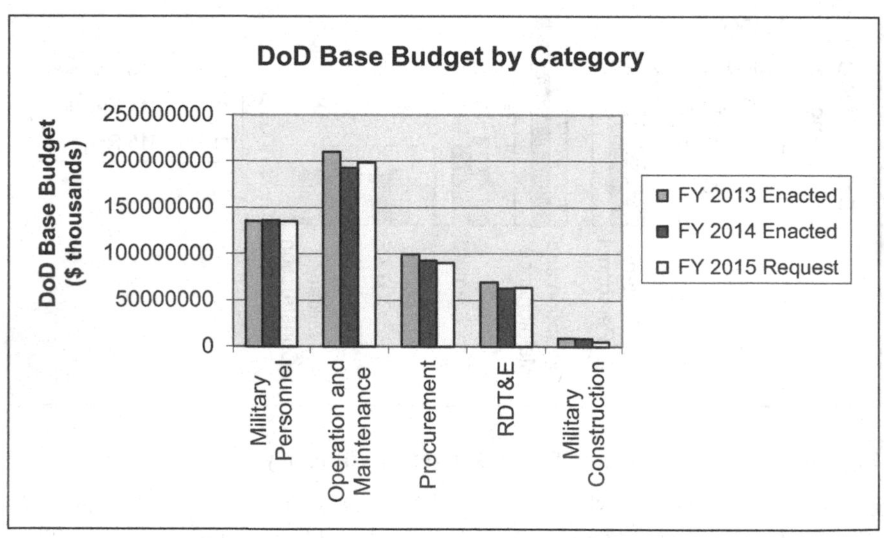

Figure 1.7 DoD Base Budget by Category

Source: US Department of Defense, Office of the Undersecretary of Defense (Comptroller), *Overview of Fiscal Year 2014 Budget Request,* April 2013, Table A-6, p. A-3; US Department of Defense, Office of the Undersecretary of Defense (Comptroller), *Overview of Fiscal Year 2015 Budget Request,* March 2015, Table A-6, p. A-3.

most significant decrease between the FY 2013 enacted budget and the FY 2015 proposed budget.

In comparing the enacted FY 2013 and FY 2014 base budgets with the FY 2015 proposed base budget, Figure 1.7 illustrates that the operations and maintenance category is the largest portion of the defense base budget, but that it increased between the enacted FY 2014 base budget and the proposed FY 2015 base budget, while procurement declined. This highlights the role of fiscal constraints in leading to greater sustainment of existing equipment with much less of an emphasis on RDT&E and procurement. This suggests that, with the reduction in the budget, it will be harder for DoD to emphasize innovation and new equipment, since much of its budget will be dedicated to operating existing equipment.

Similarly, operations and maintenance is by far the highest category in the DoD OCO budget in the enacted FY 2013 and FY 2014 budgets and in the FY 2015 budget request (Figure 1.8). It is distantly followed by military personnel and procurement. RDT&E and military construction are largely not included in the OCO budget. While operations and maintenance had the most substantive decline in the OCO budget, procurement and military personnel also experienced declines.

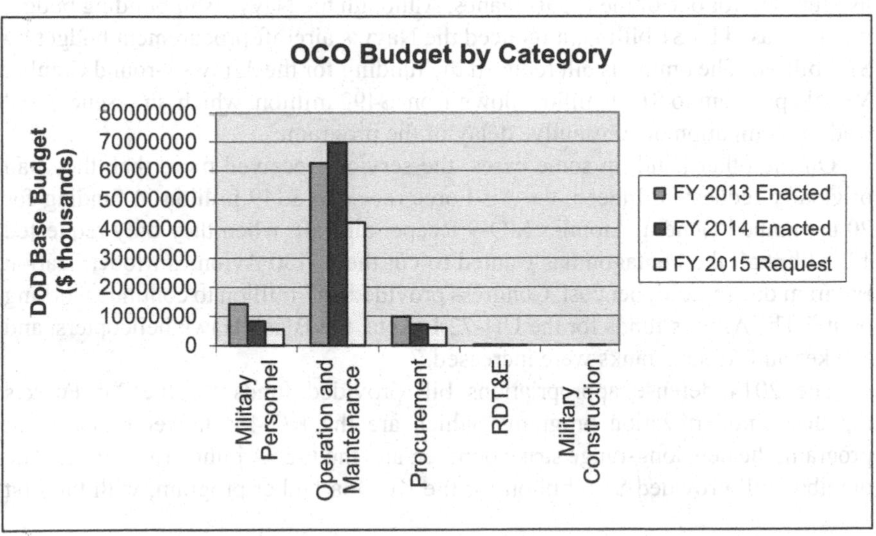

Figure 1.8 DoD OCO Budget by Category

Source: US Department of Defense, Office of the Undersecretary of Defense (Comptroller), *Overview of Fiscal Year 2014 Budget Request: Addendum A, Overseas Contingency Operations*, May 2013, Table 1, p. 11; Department of Defense, Office of the Undersecretary of Defense (Comptroller), *Overview of Fiscal Year 2015 Overseas Contingency Operations: FY 2015 Budget Amendment*, June 2014, Table 1, p. 12.

The FY 2014 Budget

The FY 2014 budget reflected the tradeoffs faced by DoD between fiscal constraints and its requests for funding for particular programs, with the shift in defense priorities. Moreover, the tradeoff between modernization and purchases of new equipment versus sustaining existing equipment was also reflected in the budget. The FY 2014 DoD budget, which was passed by the Senate on January 16, 2014 and signed by the President on January 17, provided DoD with $572 billion in the overall budget.[21] This included $486.2 billion in the base budget, which when the military construction bill of $9.8 billion was added to it, resulted in an overall enacted FY 2014 base budget of approximately $496 billion.[22] The enacted FY 2014 OCO budget was $85 billion.[23]

The proposed FY 2014 budget had requested $99.3 billion in procurement spending and $67.5 billion in R&D, but the enacted budget resulted in $93 billion in procurement and $63 billion in R&D.[24] In many cases, the specific equipment funding was less than the original request. For example, the Combat Rescue Helicopter program for the Air Force received $333.5 million, which is $60 million less than they requested. Moreover, the F-35A joint strike fighter program did not receive advance procurement funds to purchase two jets in fiscal 2015 and the Navy lost funding for one of the F-35C planes. Although the Navy's shipbuilding budget was increased by $1 billion, it reduced the Navy's aircraft procurement budget by $1.5 billion. The omnibus bill reduced the funding for the Army's Ground Combat Vehicle program to $100 million, down from $492 million, which, in essence, will lead to elimination or substantive delay of the program.[25]

On the other hand, in some cases, the services received more than they had originally requested. Indeed, the Air Force received $349 million in funding for 20 unmanned General Atomics MQ-9 Reaper aircraft, when they only requested 12. Although the Pentagon has wanted to cut the C-130 Avionics Modernization Program due to its higher cost, Congress provided $47 million to continue working on it.[26] The Army's funds for the UH-72 Lakota, the Black Hawk helicopters, and Stryker and Abrams tanks were increased.[27]

The 2014 defense appropriations bill provided funds for the Air Force's top three modernization programs, which are the KC-46a tanker replacement program, the new long-range strike bomber, and the F-35A joint strike fighter. The omnibus bill provided $1.6 billion for the KC-46a tanker program, with the first

21 Bennett, 2014a; Bennett, 2014b.

22 US Department of Defense, 2014a, Table A-6, p. A-3.

23 Department of Defense, 2014b, Table 1, p. 12.

24 Bennett, 2014b.

25 Weisgerber and Mehta, 2014; McLeary, 2014.

26 Weisgerber and Mehta, 2014.

27 Weisgerber and Mehta, 2014.

flight estimated to take place in 2015 and the fourth test aircraft under assembly beginning in January 2014. The long-range strike bomber received $359 million.

One of the most interesting differences between the FY 2014 enacted budget and the proposed budget was the reduction in base budget spending and an increase in OCO spending. The enacted FY 2014 OCO budget increased relative to the prior year for the first time since 2010, when it increased from $140 billion in 2009 to $160 billion in 2010.[28] This may be due to the fact that the OCO budget is not included in the rules for budget caps, although the base budget is.[29] For example, the OCO portion of the enacted FY 2014 budget provided significant support for the Army's industrial base because the $85 billion OCO portion of the budget included $6 billion in procurement of military equipment. The Army's portion of the enacted FY 2014 base budget for procurement, on the other hand, was reduced by 10% from the amount in the requested base budget. About $669 million for the Army's purchases of aircraft were in the OCO portion of the budget, which included funding for the Boeing CH-47 helicopters ($386 million), the Lockheed Martin Hellfire missiles ($54 million) and the Boeing AH-64 Apache helicopters ($142 million).[30] The Army also received $128.6 million to purchase missiles.[31] Consequently, using OCO for procurement and operations and maintenance funds, with the procurement and operations and maintenance funds in the base budget being cut to keep with the budget caps, led to an opportunity to maintain procurement and operations and maintenance spending for the Army because the OCO budget is not included in the rules for budget caps, although the base budget is.[32]

Since the OCO budget was not included in the rules for the budget caps, the procurement capabilities of the Navy, the Marines, and, to a lesser degree, the Air Force also received support. The OCO budget provides $211 million in support for the Navy's procurement account for aircraft and $125 million for the Marine Corps' total procurement. The OCO budget for the Navy included: purchases of work for the AV-8B Harrier jet ($57 million), purchases of Boeing F/A-18 fighters ($35.5 million), purchases of electronic warfare equipment for air-ground task force planes for the Marines ($20.7 million), procurement of MQ-8 Fire Scout drone helicopters from Northrop Grumman ($13 million), etc. The Air Force received fewer funds than the Navy or the Army in the OCO procurement budgets at $188.8 million. This included purchasing one CV-22 Osprey tilt-rotor aircraft ($73 million), as well as Hellfire missiles ($24.2 million), which were especially significant in their roles in pursuing al-Qaida leaders.[33]

28 Bennett, 2014c.
29 Bennett, 2014b.
30 Bennett, 2014b.
31 Bennett, 2014c.
32 Bennett, 2014b.
33 Bennett, 2014c.

The FY 2015 Budget

The FY 2015 DoD base budget request of $495.6 billion was approximately $0.4 billion less than the enacted FY 2014 budget, but met the current caps on the budget. Moreover, DoD also requested $26 billion to handle challenges with maintaining readiness, as well as modernizing equipment and the forces, through the larger, government-wide $56 billion Opportunity, Growth, and Security initiative. For FY 2016–FY 2019, DoD requested funding which was higher than the budget caps by a total of $115 billion, growing from $535 billion in FY 2016 to $559 billion in FY 2019.

In terms of minimizing the tradeoff between readiness and the size of the force, the DoD budget proposal suggested slower growth in compensation for the military, as well as another round of base realignment and closure. In the FY 2015 budget request, one-third of the budget ($159.3 billion) was for recapitalization of equipment and investments for future defense demands, while two-thirds of the requested budget ($336.3 billion) paid for compensation, healthcare, fuel, maintenance, training, logistics, etc. Almost 30% of the budget supported the Navy/Marine Corps, 27.8% supported the Air Force, and 24.2% supported the Army, while 18.1% supported defense-wide initiatives, such as Missile Defense Agency and Defense Advanced Research Projects Agency. The $26.4 billion request for the Opportunity, Growth, and Security Initiative would be used for modernization programs, such as the acquisition of additional P-8 aircraft and F-35's, as well as improved training, military construction, etc. In an effort to improve efficiency, the FY 2015 budget proposed reductions in civilian personnel, reduced funding for contractors, savings in healthcare for the military, a one-fifth reduction in the allocations for operating budgets for headquarters, and a Base Realignment and Closure program beginning in FY 2017. The proposal also included limiting basic pay for FY 2015 to 1% for most military personnel with no increase in General and Flag Officers basic pay and a slowing in growth in the Basic Allowance for Housing (BAH), as well as a reduction in subsidies for commissaries and changes to the TRICARE healthcare plans.[34]

Under the FY 2015 proposed budget, the Air Force emphasized modernizing its equipment, such that it included funds for purchasing 26 Joint Strike Fighters for $4.6 billion FY 2015, as well as $31.7 billion for 238 planes over the Future Years Defense Program (FYDP). Similarly, it provided funds for the Long Range Strike Bomber and the next-generation jet engine. Moreover, the budget request included funds for purchasing seven KC-46 tankers at $2.4 billion in FY 2015 and $16.5 billion for 69 aircraft over the FYDP. As a result of modernization, the proposal suggested that older equipment be retired: these included the elimination of the A-10 Warthog, the replacement of the 50 year old U-2 with the unmanned Global Hawk, and the reduction in purchases of Reapers and Predators.[35]

---

34 US Department of Defense, 2014.

35 US Department of Defense, 2014.

Similarly, the Navy faced the tradeoffs in the FY 2015 budget between modernization and retirement of older equipment. For example, the Navy plans to maintain the 11 carrier strike groups, although additional cuts in future years would result in the retirement of the USS George Washington. The modernization of the cruiser fleet would result in half of the fleet being placed in minimal operating status while they are being restored and improved.[36] Nevertheless, the Navy planned to support 288 ships in FY 2014 and 309 ships by FY 2019, as well as to continue investing in afloat forward staging bases, guided missile destroyers, and attack submarines. Indeed, the proposed budget requested:

a. $5.9 billion for two Virginia-class attack submarines in FY 2015 and $28 billion for two submarines annually through FY 2019;
b. $1.5 billion in FY 2015 to buy three Littoral Combat Ships and $8.1 billion over the FYDP to buy 14 littoral combat ships;
c. $2.8 billion in FY 2015 and $16 billion over the FYDP to acquire two DDG-51 guided-missile destroyers per year through FY 2019; and
d. $3.3 billion in FY 2015 for eight Joint Strike Fighters – two for the Navy and six for the Marine Corps – and $22.9 billion for 105 aircraft over the FYDP.

The 182,700 Marines would be supported by the request of $22.7 billion for FY 2015, which includes support for 900 Marines at US overseas embassies.[37]

The FY 2015 budget request accelerated the reduction of the Army forces from 520,000 troops to 440,000 to 450,000 – its smallest size since the period prior to World War II. These reductions assumed that lengthy operations overseas supporting stability on the scale of Iraq and Afghanistan would not take place.[38] This, when combined with the Marine Corps, was deemed to be adequate to defeat forces in one theater, support the other services in another theater, and protect the homeland. Moreover, the Army National Guard and Reserves would also be reduced by 5%.[39]As with the Navy and the Air Force, older equipment, such as the Kiowas, would be retired and the National Guard, which would obtain Blackhawks, would provide the Army with some of its Apaches and Lakotas to replace the Kiowas.[40] The Army National Guard's helicopter fleet would be reduced by 8%, while the active duty Army's helicopter fleet would be reduced by 25%, due to a reduction from seven helicopter models to four.[41] Finally, the Army proposed to cancel the Ground Combat Vehicle program.[42]

36 Simeone, 2014a.
37 US Department of Defense, 2014.
38 Simeone, 2014a.
39 US Department of Defense, 2014; Simeone, 2014a.
40 US Department of Defense, 2014.
41 Marshall, 2014.
42 Simeone, 2014a.

Finally, the proposed budget supported $5.1 billion to fund cyber operations and $7.5 billion for the Missile Defense Agency, as well as support the 69,700 Special Operations force with $7.7 billion (10% higher than FY 2014 funds).[43]

The FY 2015 defense budget approved by the House Appropriations Committee Defense subcommittee on May 30 approved a FY 2015 bill with $491 billion for the base budget – slightly lower than the proposed budget of $495.6 billion. The bill included $5.8 billion for 38 F-35's, which exceeded the 34 which were requested by the Pentagon. The 11 aircraft carrier fleet would be maintained such that $789 million would be shifted toward maintenance and refueling of the George Washington aircraft carrier. Although the Navy needed EA-18G Growler jets, they excluded them from their FY 2015 budget request; however, this bill provided $975 million for the 12 Growler electronic attack jets. It also provided funds of $1.6 billion for the KC-46A tanker, as well as $63.4 billion for researching and developing weapons, which was more than the Pentagon requested (by $171 million) and more than the 2014 enacted amount (by $370 million). The Pentagon requests for key items were funded. These included: the new long-range bomber program, the KC-46A, and the F-35 for the USAF; the unmanned carrier-based drone aircraft, the next-generation submarine, the P-8A multi-mission maritime aircraft, and the RQ-4 Triton UAV for the Navy; and the Joint Light Tactical Vehicle effort for the Army and the Marine Corps.[44]

The FY 2015 request for $58.6 billion in OCO funds was two-thirds of the amount in FY 2014, largely due to the drawing down of operations in Afghanistan.[45] Within the OCO budget, DoD had requested $6 billion for procurement. The 9800 troops in Afghanistan would cost $20 billion and the removal of equipment from Afghanistan and repair of the equipment would require funds for transportation costs for personnel, supplies, and equipment. Indeed, the cost of transporting equipment from Afghanistan could be between $5 and $7 billion.[46] Moreover, replacing and repairing older equipment will likely result in continued need for DoD OCO funding in subsequent years, especially due to the costs of forces in the Middle East, Southwest Asia, etc.[47] It is unclear how long OCO budgets will continue, although, as of early September, 2014 it was likely that they would end beginning in 2017.[48]

The requested OCO funding for FY 2015 also included $4 billion for President Obama's Counterterrorism Partnerships Fund, which would strengthen the capabilities of partner nations in operations combating terrorism and limit the expansion of terrorist threats, as well as deal with international operations combating terrorism in Iraq. It would support US training and training of foreign

43 US Department of Defense, 2014.
44 Bennet, 2014d.
45 Roulo, 2014.
46 Weisgerber, 2014d.
47 Roulo, 2014.
48 Weisgerber, 2014a.

troops, such as training forces in Yemen, supporting multinational forces in Somalia, training security forces in Libya with European allies, etc.[49] Finally, the European reassurance initiative of $1 billion would support rotational sea, land and area forces in Central and Eastern Europe, as well as support training with other allied NATO countries with concerns about the role of Russia in the Ukraine and the Crimean Peninsula.[50] Nevertheless, Secretary of Defense Hagel noted, in late August, 2014, that the FY 2015 budget proposal may need to be adjusted due to the actions taken in Iraq. Indeed, between August 8 and August 21, 2014, 89 airstrikes were conducted by the US against the self-styled "Islamic State," with the delivery of 636 bundles of supplies to the Iraqis and flights of unmanned and manned military aircraft for over 60 intelligence missions daily.[51]

Sequestration Concerns

One of the major concerns regarding the potential that the defense budget could have been delayed by Congress in January 2014 was that much of the production and delivery of military equipment would be delayed. The Chief of Naval Operations, Admiral Jonathan Greenert, noted that the potential continuing resolution, as well as the likelihood of sequestration continuing, could limit the production of newer equipment and lead to reductions in training and maintenance. Nevertheless, having greater flexibility in moving money from different areas of the budget into areas supporting maintenance and operations, as well as supporting modernization of the system would, from the perspective of the Chief of Naval Operations in the fall of 2013, be helpful. As of October 2013, he argued that, should sequestration continue, about 25 aircraft would be delayed and expressed significant concern about delays in submarine construction.[52] Fortunately, as discussed in the previous section, the FY 14 budget was passed in January, 2014.

The Chief of Naval Operations was also concerned about the costs of purchasing replacements for 14 Ohio class submarines with the SSBN-X, which reportedly has 16 missiles on it, each of which could carry five nuclear warheads. The Navy had been trying to reduce the cost of the first of 12 new submarines from $7.4 billion to between $4.9 and $5.4 billion. Other costs overruns included over $2 billion in overruns for the CVN-78, the USS Gerald Ford aircraft carrier, as well as the $1.2 billion in F-35 cost overruns.[53]

The FY 2014 full sequestration could have led to delays on the purchases by DoD of the F-35, which, in turn, would have had an impact on the entire supply chain associated with the production of F-35s. The Air Force would have delayed

49 Weisgerber, 2014b; Roulo, 2014.

50 Roulo, 2014.

51 Weisgerber, 2014c.

52 Parsons, 2013.

53 Pincus, 2013.

purchasing 4–5 F-35A joint strike fighters for FY 2014 (out of 19 for FY 2014), the Marine Corps would have delayed purchasing one of its F-35B aircraft, and the Navy would have postponed purchasing one of its F-35C aircraft and potentially cutting four EA-18G Growler electronic attack jets. Reductions in F-35 planned purchases could have hurt the contractors of the underlying components of the aircraft, such as Pratt & Whitney, which had just reached a $1.1 billion contract on its F-35 engines.[54] For FY 2013, the Air Force considered delaying purchases of 3–5 F-35A aircraft, but they were able to negotiate an improved cost structure with Lockheed Martin on the low-rate initial production lots six and seven. Moreover, due to the furloughs during the summer of 2013, the schedule for the F-35As lost 1–2 months.[55]

Nevertheless, sequestration has impacted the defense industry less than would be anticipated because "funds that are appropriated in one year are spent over many years"; for example, $3.2 billion in 2013 was authorized for a new submarine, but the expenditures of the authorized funds will occur over seven years, particularly between 2014 and 2016. As a result, while defense contractors reduced their overhead early, they had backlogs of projects from previous contract awards which were reflected in 2013 sales and earnings.[56] DoD continued to specialize its spending on computer network warfare, special operations, intelligence, surveillance, and reconnaissance, and space.[57] If the sequestration cuts had taken place in early January, 2013, DoD would have been forced to cut 108,000 civilian employees. It would have needed to reduce its 791,000 personnel by 13.7% in FY 2013 to achieve its $56.5 billion in sequestration reductions for FY 2013. Assuming proportionate cuts, the Army would have cut 38,000 jobs, the Air Force would have cut 23,000 jobs, and the Navy would have cut 24,000 jobs. Moreover, the FY 2013 budget would have been reduced to $491 billion from $546 billion.

Since sequestration in 2013 was delayed and did not take place until March 1, 2013, the growing uncertainty about when sequestration would occur led to reduced spending in the first few months of 2013. The Navy delayed sending the aircraft carrier – the USS Harry Truman – to the Middle East, as well as delayed the overhaul of a second carrier, which they later resumed. It evaluated deployments to Europe, the Mediterranean, South America, and the Caribbean which could affect the ability of the Navy to respond quickly to regional missions and planned to cut the number of flying hours by half on the aircraft carriers, as well as to reduce the naval presence in the Pacific by up to 35% and, due to the $8.6 billion shortfall, leave just one aircraft carrier in the Persian Gulf rather than two. The Army reduced its training, which could have ultimately impacted the readiness of three-quarters of its brigade combat teams and it planned to cut funding for surveillance aircraft

54 Mehta, 2013b.
55 Mehta, 2013c.
56 Erwin, 2013b .
57 Shanker, 2013.

and intelligence equipment. The Air Force planned to scale back about one-third of the total remaining flying hours between March and October (200,000 hours) by cutting $12.4 billion. It also planned to reduce aircraft maintenance by one third.[58]

Sequestration cuts exempted military personnel and focused on budget categories, which included: training, acquisition, research, and operations and maintenance. Nevertheless, as of 2013, the Pentagon had spent one third of its base budget – $181 billion per year – on military personnel ($107 billion in salaries and allowances; $50 billion for healthcare, and $24 billion for retirement pay) and estimates suggest that the other budget categories, such as research, acquisition, etc., could shrink as the costs of these benefits grow.[59] Moreover, although the military personnel category was the distant second largest category in the base budget and a very distant second largest category in the OCO budget, it was likely, as of the fall of 2013, to be cut further due to reductions in the Army and the Marine Corps. This was due to the withdrawal of forces from Iraq and Afghanistan and the greater focus on the Asia-Pacific region. While the Army had planned to cut the troops from 540,000 to 490,000 by 2017, this force reduction has been

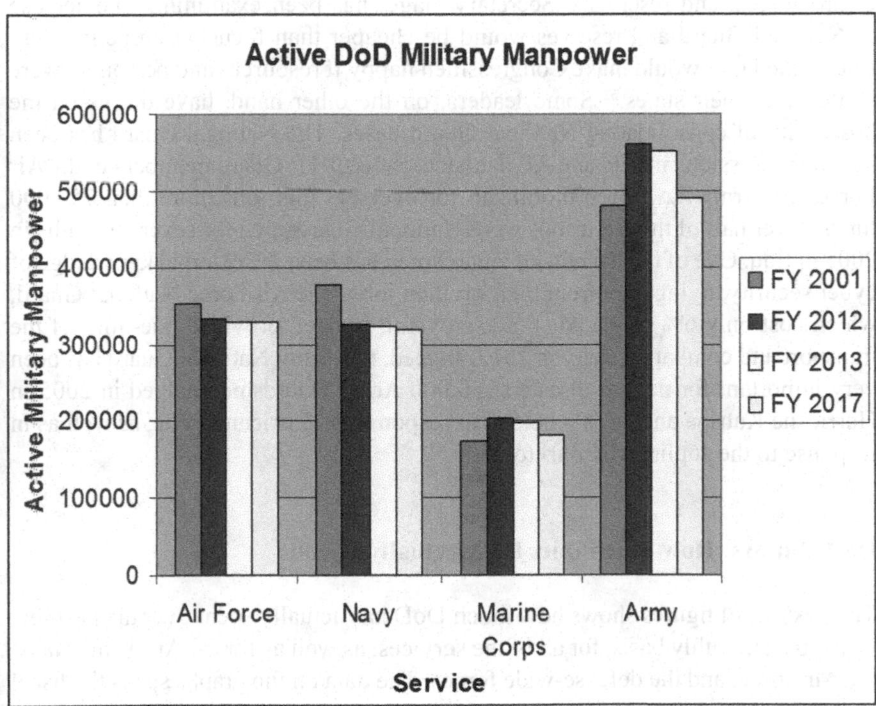

Figure 1.9 Active DoD Military Manpower

58 Nissenbaum, 2013; Vanden Brook and Davis, 2013; Londono and Rein, 2013.

59 Erwin, 2013b.

accelerated.[60] Two armored Army brigades – as many as 10,000 troops – would come back to the US from Europe over the following decade, leaving two brigades and support troops behind. A total of eight brigade combat teams were projected to be cut. Although the Pentagon asked Congress for another round of base closings for 2013 and 2015, these were rejected.

The Army has more active military that the Navy and the Air Force, which are similar in size. While the Army has increased in size since FY 2001, the Air Force and, to a greater degree, the Navy, have declined in size (Figure 1.9).

As of October 2013, General Odierno noted that the Army had only two combat-ready brigades and that the brigades which were being sent to Afghanistan were not qualified for combat, but rather for the trainer and adviser missions. Although General Odierno hoped to expand the number of equipped and trained brigades to seven by June 2014, he also observed that, "There is going to come a time when we simply don't have enough money to provide what I believe to be the right amount of ground forces to conduct contingency operations."[61]

One of the significant decisions for the military has been determining the appropriate and cost-effective balance between active duty forces, National Guard forces, and reservists. Secretary Hagel has been examining whether use of National Guard and reserves would be cheaper than focusing on active duty forces; the latter would make Congressmen happy if resources and personnel were shifted into their states.[62] Some leaders, on the other hand, have discussed the possibility of consolidating National Guard bases. The National Guard has been key in the missions in Iraq and Afghanistan. Since 9/11, Guard members in the Air Force and Army have been mobilized for overseas missions more than 750,000 times. Over half of the US troops were National Guardsmen for several months in 2005 in Iraq. One of the benefits of Guardsmen has been the external knowledge of cyber security, flying, etc. from their civilian jobs. The Air Force National Guard, which cost only 6% of the Air Force's overall budget, provided one-third of the transport and combat capacity in 2012. Indeed, the Army National Guard has been very important for natural disasters; 50,000 Army Guardsmen helped in 2005 in Hurricane Katrina and 46,000 helped in response to Hurricane Irene, as well as in response to the Joplin, Missouri tornado.[63]

DoD Outlays: How Much Does DoD Actually Spend?

The next set of figures shows how much DoD has actually spent over the past few years, on a monthly basis, for all of the services, as well as for the Army, the Navy, the Air Force, and the defense-wide forces. The data on the graphs spans the fiscal

60 *Defense News* Staff, 2014.

61 Bacon, 2013.

62 Shanker, 2013.

63 Chandrasekaran, 2013.

years for 2012 and 2013, as well as the first nine months of the fiscal year for 2014. FY 2013 was particularly difficult due to the actions regarding sequestration. FY 2014 began on a difficult note with the shutdown of the government.

During the 16-day US government shutdown in the fall of 2013, 400,000 DoD civilians were furloughed for the first four days. Estimates suggest $600 million in terms of lost productivity for DoD due to the government furlough program. As of mid-October 2013, Congress passed spending measures which raised the debt limit until February 7, 2014 and which funded the government through January 15. As a result, during the fall and winter of 2013, DoD operated under a continuing resolution which prevented the initiation for new programs.[64]

Unlike many of the other government agencies, DoD was not closed for the full period of the US government shutdown, beginning on October 1, 2013. Indeed, on October 5, the Pay Our Military Act, which had been passed by Congress on September 30, enabled DoD to recall over 90% of the furloughed 350,000 employees and bring them back to work. Despite this, several defense contractors announced that they would furlough employees; these included Lockheed Martin and BAE Systems, although the acquisition programs for DoD were not problematic since they were using FY 2013 funding rather than FY 2014 funding. Nevertheless, production slowed down.[65]

The shutdown of the federal government in the fall of 2013 hurt a number of urban areas. While a total of 800,000 federal employees were furloughed, about 350,000 civilian workers were called back to work under the Pay Our Military Act. One example of an area which is heavily dependent on the government is the Fayetteville area in North Carolina, where 38% of its GDP is dependent on Fort Bragg and where over $25 million is spent each day in the area by over 200 defense contractors and 57,000 soldiers.[66]

The government shutdown almost led to the Pentagon breaching its fixed price contract for the KC-46 tanker replacement. If the President had signed the continuing resolution 24 hours later, then the fixed price contract, which requires Boeing to cover any overruns beyond the $4.9 billion cap on the government's liability, would have been broken. Eighteen out of the 179 new tankers are scheduled to be delivered by 2017 and the production is scheduled for completion by 2027. The tankers would replace the aging KC-135 tanker fleet. The tanker program is one of the Air Force's three top modernization programs, with the other two being the F-35 Joint Strike Fighter and the new long-range bomber.[67]

The operations and maintenance outlays are the highest category in monthly DoD outlays from October 2011 to June 2014 (Figure 1.10). It was followed by military personnel outlays, procurement outlays, research and development outlays

64 Weisgerber, 2013a.

65 Weisgerber, 2013c.

66 "Government Shutdown Hits Military Towns Hard," 2013.

67 Mehta, 2013a.

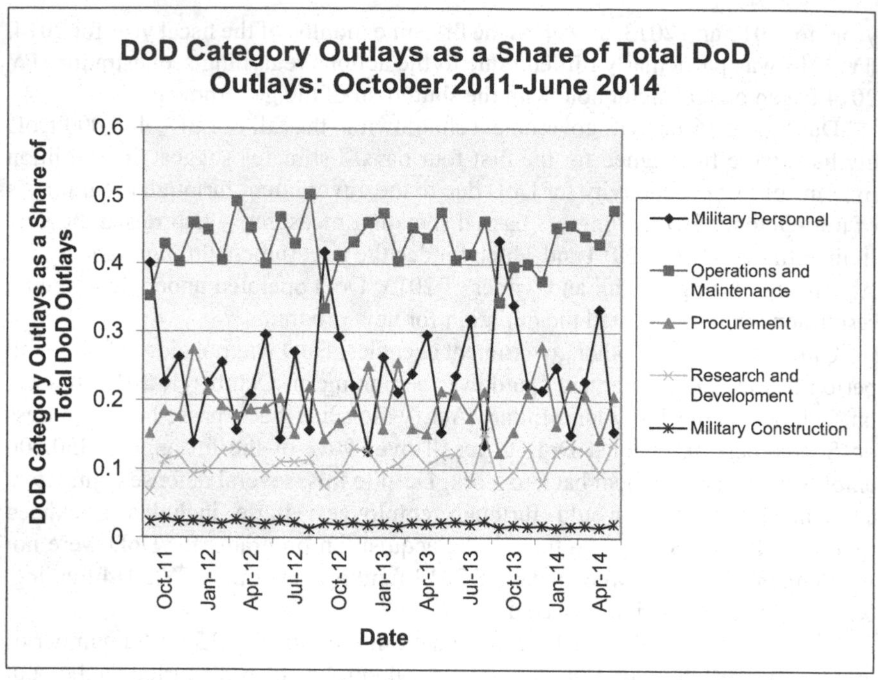

Figure 1.10 DoD Category Outlays as a Share of Total DoD Outlays: October 2011–June 2014

Source: Department of the Treasury, Bureau of the Fiscal Service, *Monthly Treasury Statement of Receipts and Outlays of the United States Government*, October, 2011–June, 2014, p. 8.

and, finally, military construction outlays. This is consistent with the pattern in base budget and OCO budget categories in Figures 1.7 and 1.8.

Operations and maintenance outlays as a share of total DoD outlays ranged, in FY 2012, from 35.5% of DoD outlays (October 2011) to 50.3% (September 2012). Throughout most of FY 2012, it was in the percentage range of the 40's. In FY 2013, it ranged from 33.5% (October 2012) to 46.2% (September 2013). The peaks and troughs of the shares were slightly lower in FY 2013 than in FY 2012. In the first nine months of FY 2014, it ranged between 34.3% (October 2013) and 47.9% (June 2014) and ranged between 40% and 46% in six of the nine months. These were often lower shares than in FY 2012 and FY 2013. Consistently, however, operations and maintenance outlays were at their highest shares of DoD outlays at the end of the fiscal year (September) and at their lowest shares of DoD outlays at the beginning of the fiscal year (October).

Military personnel outlays – the second highest share of DoD outlays – ranged in FY 2012 from 13.8% (January 2012) to 40% (October 2011). For most of FY

2012, the outlays tended to be between 21% and 29.6%. The outlays in FY 2013 were at their highest share of DoD outlays at 41.7% in October 2012; this was similar to FY 2012, when the highest share was in October 2011, at the beginning of the fiscal year. The lowest share of outlays in FY 2012 and FY 2013 were in January 2012 and January 2013. For the nine months of FY 2014, military personnel outlays as a share of DoD outlays ranged from 14% (March 2014) to 43.4% (October 2013). In six of the nine months, they ranged from 15% to 24%.

Procurement outlays were the third highest share of DoD outlays. In FY 2012, they ranged from 15.2% (in October 2011 and August 2012) to 27.4% (January 2012), although they usually varied between 18% and 21% of DoD outlays. In FY 2013, procurement outlays ranged from 14.4% (October 2012) to 25%–25.5% (in January 2013 and March 2013), although they usually varied between 15% and 21% of DoD outlays throughout FY 2013. For the first nine months of FY 2014, procurement outlays as a share of DoD outlays ranged from 12.3% (October 2013) to 27.9% (January 2014), although in six of the nine months, they ranged between 15% and 22%. Procurement outlays tended to have their lowest shares at the beginning of the fiscal year (October), which was similar to operations and maintenance. Procurement outlays had their highest shares in January. Military personnel outlays, on the other hand, often had their highest shares at the beginning of the fiscal year and their lowest shares in January.

Research and development outlays were the fourth highest share of DoD outlays, and were significantly lower than outlays for operations and maintenance, military personnel, and procurement. In FY 2012, they ranged from 6.7% (October 2011) to 13.3% (April 2012). For most of FY 2012, they ranged between 9% and 12%. In FY 2013, research and development outlays as a share of DoD outlays ranged from 7.5% (October 2012) to 15.4% (September 2013). Again, as in FY 2012, the outlays generally ranged between 9% and 13%. For the first nine months of 2014, research and development outlays varied between 7.7% (October 2013) and 13.9% (June 2014). They ranged between 9% and 13% in five of the nine months and their share of outlays tended to be slightly higher than in previous years. As was the case with procurement outlays and operations and maintenance outlays, research and development outlays were usually at their lowest share of DoD outlays at the beginning of the fiscal year (October).

Military construction outlays were the smallest share of DoD outlays. In FY 2012, they ranged from 1.25% (September 2012) to 2.8% (November 2011). For most of FY 2012, they were in the 2% range. In FY 2013, military construction outlays as a share of DoD outlays ranged from 1.6% (March 2013) to 2.4% (September 2013). For most of the year, they varied between 1.8% and 2.2%. For the first nine months of 2014, military construction outlays varied in a narrow range between 1.4% (February 2014) and 2% (June 2014). The outlays shares were lower than in FY 2012 and FY 2013 in that they were between 1.4% and 1.8% in eight of the nine months.

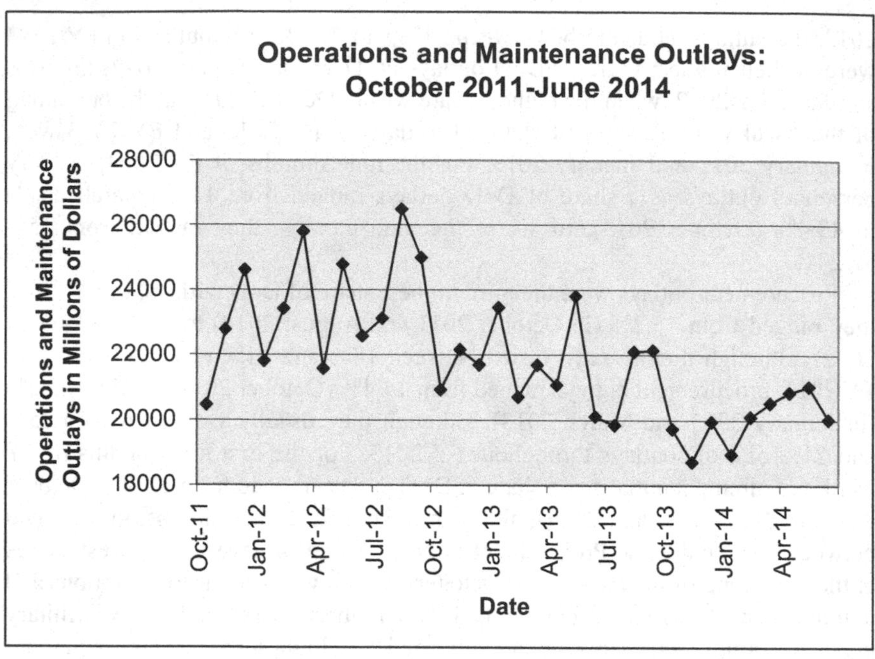

Figure 1.11 Operations and Maintenance Outlays: October 2011–June 2014

Source: Department of the Treasury, Bureau of the Fiscal Service, *Monthly Treasury Statement of Receipts and Outlays of the United States Government*, October, 2011–June, 2014, p. 8.

The next set of figures provides monthly detail of the trends in expenditures between FY 2012 and FY 2014 in each of the various budget categories. The trend in each budget category is examined across services, as well as by each service.

Operations and maintenance outlays, the largest of the broader DoD budget outlay categories, had less volatility than military personnel outlays, which were the most volatile. Figure 1.11 shows the operations and maintenance outlays aggregated across services in FY 2012, FY 2013, and the first nine months of FY 2014. In FY 2012, operations and maintenance outlays varied between 20,458 (October 2011; in millions of dollars) and 26,478 (August 2012; in millions of dollars), and usually remained in the 21,000–24,000 range. In FY 2013, they varied between 19,834 (July 2013) and 23,802 (May 2013), usually remaining in the 20,000–22,000 range. In the first nine months of FY 2014, they varied between 18,719 (November 2013) and 21,031 (May 2014). Consequently, the overall monthly outlays for DoD tended to be slightly lower in FY 2014 relative to FY 2013, as well as lower in FY 2013 relative to FY 2012.

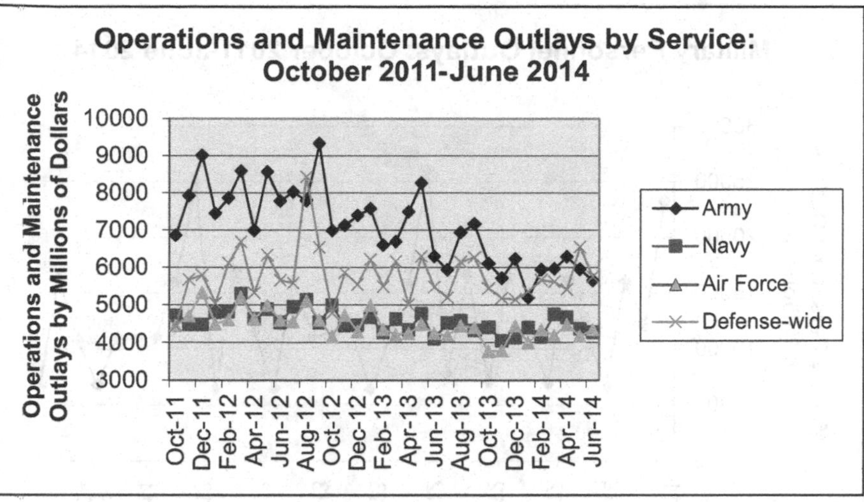

Figure 1.12 Operations and Maintenance Outlays by Service: October 2011–June 2014

Source: Department of the Treasury, Bureau of the Fiscal Service, *Monthly Treasury Statement of Receipts and Outlays of the United States Government*, October, 2011–June, 2014, p. 8.

Figure 1.12 shows the actual expenditures for the operations and maintenance category by service. Not surprisingly, the Army had the largest expenditures, due to its involvement in Afghanistan, much of which came from the OCO budget. With the withdrawal from Afghanistan, it is likely that the Army will have lower operations and maintenance outlays for MRAPS, JLTV's, etc. Similarly, the defense-wide forces have had the second highest monthly expenditures for operations and maintenance of equipment, largely due to the Joint Forces' involvement in Afghanistan and overseas. The Navy and the Air Force are very similar in their lower levels of operations and maintenance outlays, and have had less volatility than the Army and the defense-wide forces.

The military personnel outlays aggregated across services have exhibited significant volatility. As Figure 1.13 shows, in FY 2012, they varied from 23,231 (in millions of dollars) in October 2011 to 6,561 (in millions of dollars) in January 2013. The outlays were in the 16,000–17,000 range in four of the 12 months and were in the 6,000–7,000 range in four of the 12 months. In FY 2013, the outlays varied between 6,392 (January 2013) and 26,105 (October 2012). The outlays tended to be in lower ranges than in FY 2013 and were in the 11,000 range more than in the 16,000 range, unlike FY 2012. They were in the 11,000 range in five of the 12 months, in the 6,000 range in three of the 12 months, and in the 15,000–16,000 range in three of the 12 months. Consequently, military personnel outlays had slightly less volatility in FY 2013 than in FY 2012. In the first nine

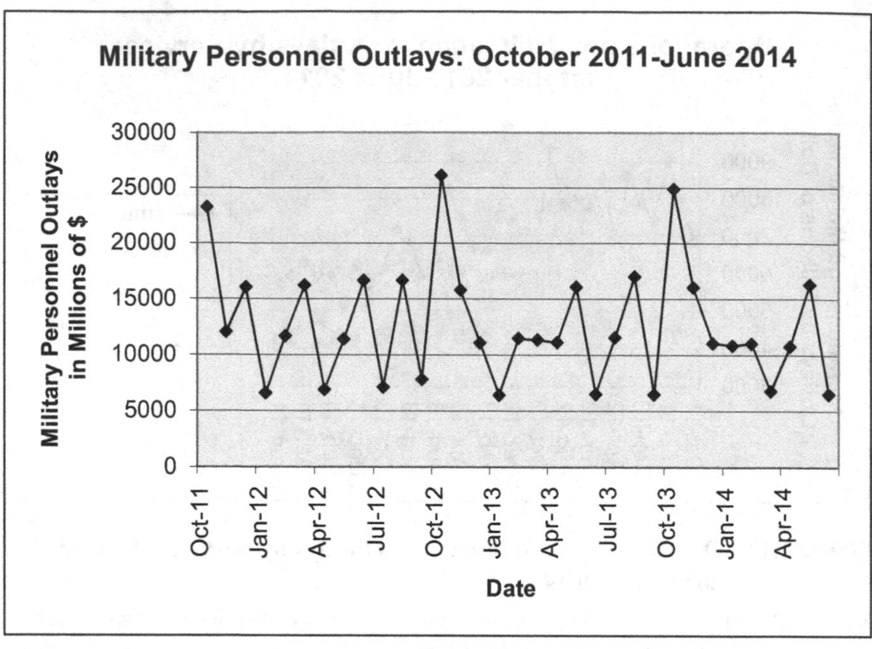

Figure 1.13 Military Personnel Outlays October 2011–June 2014

Source: Department of the Treasury, Bureau of the Fiscal Service, *Monthly Treasury Statement of Receipts and Outlays of the United States Government*, October, 2011–June 2014, p. 8.

months of FY 2014, the military personnel outlays varied between 6,493 (June 2014) and 24,882 (October 2013). They were in the 10,000–11,000 range in four of the nine months, in the 6,000 range in two of the nine months, and in the 16,000 range in two of the nine months.

Although military personnel outlays were highly volatile, Figure 1.14 shows that the Army, Navy, and Air Force were synchronized in their peaks and troughs. The Army consistently led in military personnel outlays, followed by the Navy and the Air Force.

Procurement outlays were the third highest outlay category for DoD. As is evident in Figure 1.15, procurement outlays varied between 7,849 (April 2012; in millions of dollars) and 12,998 (January 2012; in millions of dollars) in FY 2012. The expenditures remained in the 9,000–10,000 range in seven of the 12 months in FY 2012. During FY 2013, procurement outlays varied between 6,907 (February 2013) and 13,675 (March 2013) and tended to remain in the 8,000–10,000 range in eight of the 12 months. Indeed, outlays were more frequently in the 8,000–9,000 range in FY 2013 than in FY 2012. In the first nine months of 2014, procurement outlays varied between 7,030 (May 2014) and

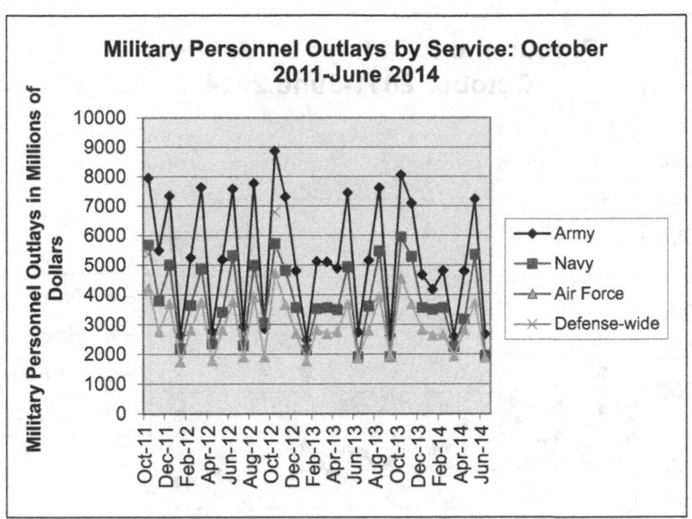

Figure 1.14 Military Personnel Outlays by Service: October 2011–June 2014

Source: Department of the Treasury, Bureau of the Fiscal Service, *Monthly Treasury Statement of Receipts and Outlays of the United States Government*, October, 2011–June 2014, p. 8.

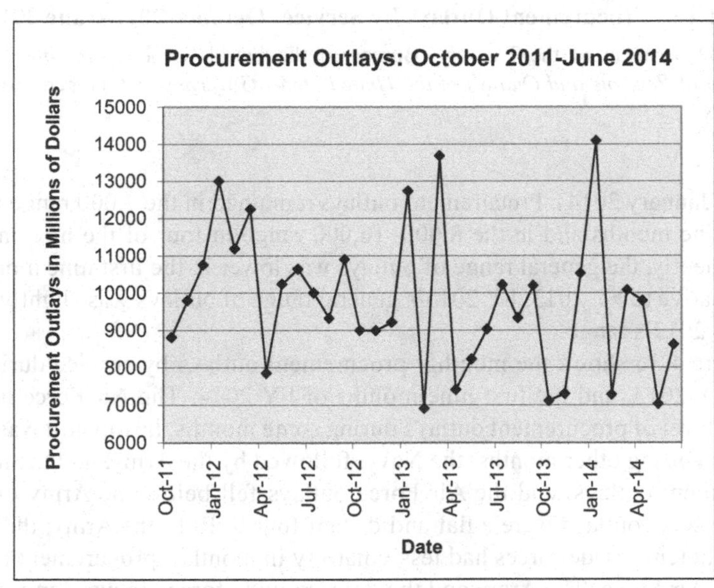

Figure 1.15 Procurement Outlays: October 2011–June 2014

Source: Department of the Treasury, Bureau of the Fiscal Service, *Monthly Treasury Statement of Receipts and Outlays of the United States Government*, October, 2011–June 2014, p. 8.

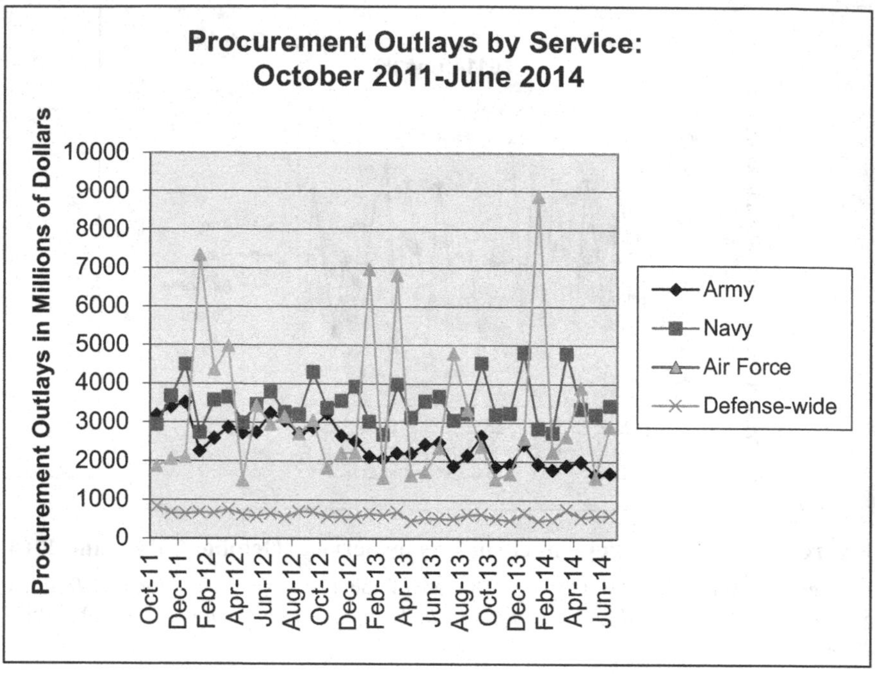

Figure 1.16 Procurement Outlays by Service: October 2011–June 2014

Source: Department of the Treasury, Bureau of the Fiscal Service, *Monthly Treasury Statement of Receipts and Outlays of the United States Government*, October, 2011–June 2014, p. 8.

14,086 (January 2014). Procurement outlays remained in the 7,000 range in four of the nine months and in the 8,000–10,000 range in four of the nine months. Consequently, the general range of outlays was lower in the first nine months of 2014 relative to FY 2013. FY 2013's general range of outlays was slightly lower than FY 2012's range.

Figure 1.16 shows the monthly procurement outlays by service during FY 2012, FY 2013, and the first nine months of FY 2014. The Air Force had the highest level of procurement outlays during some months, however it was fairly volatile. During other months, the Navy, followed by the Army, had the highest procurement outlays, and the Air Force outlays fell below the Army outlays. Defense-wide outlays were a flat and distant fourth. Both the Army, the Navy, and the defense-wide forces had less volatility in monthly procurement outlays than the Air Force. The Army and the defense-wide forces focused much more on operations and maintenance in places in Afghanistan, through the OCO budget, and placed much less emphasis on procurement of new equipment. In the first nine months of 2014, however, the Navy exceeded the Air Force in procurement outlays in seven of the nine months. The Navy exceeded the Air Force in eight out of the 12 months in FY 2013 and in nine out of the 12 months in FY 2012.

While R&D spending is very important to DoD, it has fallen over the years and may lead to the US, in the coming years, becoming less competitive and being less likely to develop cutting edge technology. R&D fell from $80 billion in 2009 to $68 billion in 2013, and then, due to sequestration, it dropped an additional 8% below the 2013 enacted budget. Indeed, although operations and maintenance outlays declined the most in the 2013 budget relative to prior years in terms of dollars, R&D experienced the greatest percentage decline of the other budget categories. Operations and maintenance outlays declined by $20 billion and R&D and procurement declined by $15.8 billion in 2013. By comparison, China, on the other hand, has doubled its budget for its defense department since 2006 and increased its defense spending for the past 22 years. Moreover, China has been

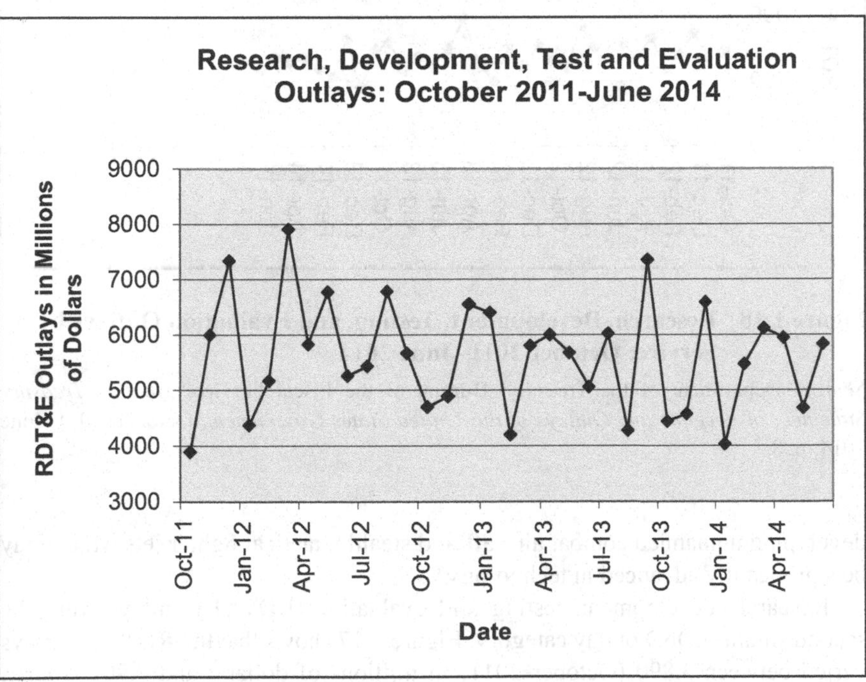

Figure 1.17 Research, Development, Testing, and Evaluation Outlays: October 2011–June 2014

Source: Department of the Treasury, Bureau of the Fiscal Service, *Monthly Treasury Statement of Receipts and Outlays of the United States Government*, October, 2011–June 2014, p. 8.

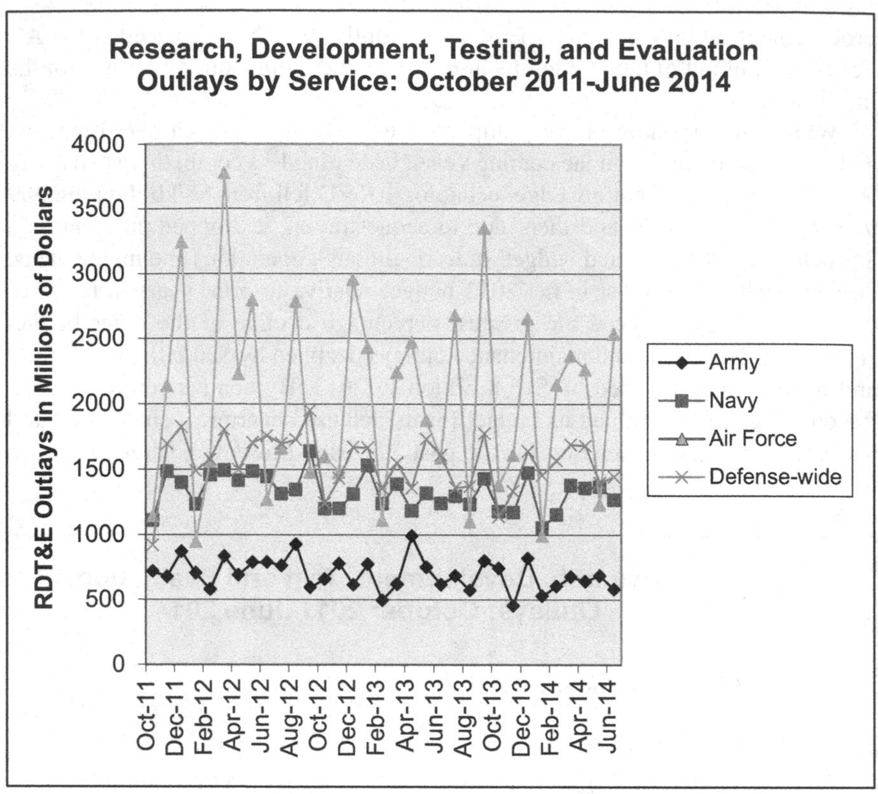

Figure 1.18 Research, Development, Testing, and Evaluation Outlays by Service: October 2011–June 2014

Source: Department of the Treasury, Bureau of the Fiscal Service, *Monthly Treasury Statement of Receipts and Outlays of the United States Government*, October, 2011–June 2014, p. 8.

developing unmanned combat aircraft and stealthy tactical fighter jets which may be significantly advanced in technology.[68]

Research, development, testing and evaluation (RDT&E) outlays were the second smallest DoD outlay category. Figure 1.17 shows that the RDT&E outlays varied between 3,890 (October 2011; in millions of dollars) and 7,904 (March 2012; in millions of dollars) in FY 2012, and remained in the 5,000–6,000 range in eight of the 12 months. In FY 2013, the RDT&E outlays varied between 4,196 (February 2013) and 7,352 (September 2013). In FY 2012, the RDT&E outlays were in the 4,000 range for only one month; in FY 2013, they were in the 4,000 range for three months. The outlays in FY 2013 were in the 5,000–6,000 range for

68 Weisgerber and Fryer-Biggs, 2014.

seven of the 12 months, which was similar to FY 2012. In the first nine months of FY 2014, RDT&E outlays varied between 4,030 (January 2014) and 6,587 (December 2013). The outlays were in the 4,000 range in four of the nine months and in the 5,000–6,000 range in five of the nine months. This suggests lower RDT&E outlays for FY 2014.

Figure 1.18 shows the monthly outlays of research, development, testing, and evaluation (RDT&E) over the past two years. The Air Force had the highest RDT&E outlays on a monthly basis, although it had significant volatility in the outlays, which is consistent with its same pattern in procurement (highest expenditure in procurement of the three services during many of the months, with the highest volatility). The defensewide forces and the Navy were the second and third in monthly expenditures of RDT&E. The Army's monthly RDT&E expenditures were last, which is consistent with their emphasis on operations and maintenance of equipment, especially in Iraq and Afghanistan.

Military construction outlays are the smallest of the main DoD expenditure categories. As Figure 1.19 shows, they ranged from 626 (September 2012;

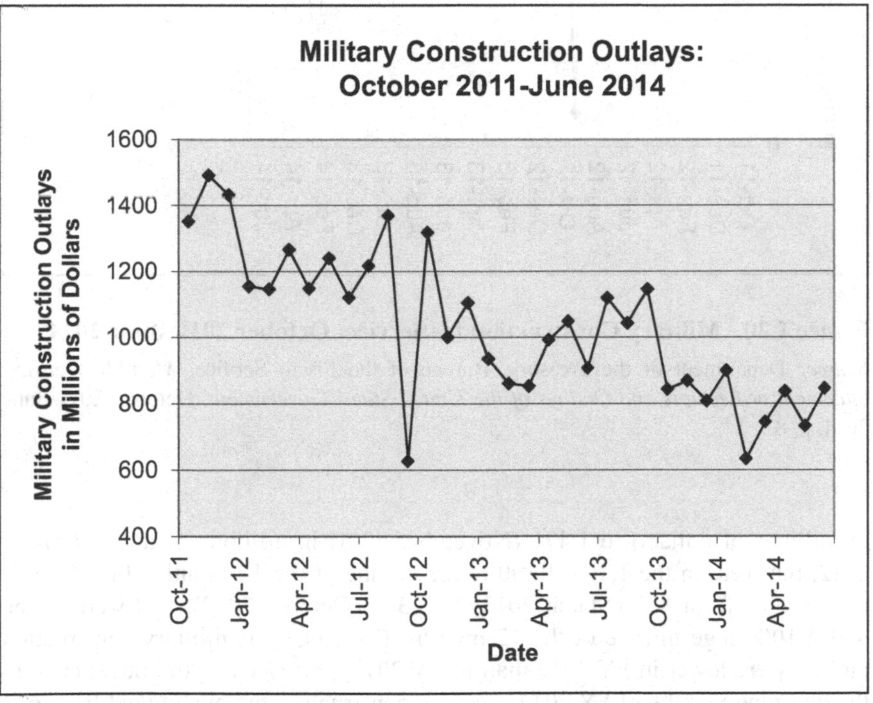

Figure 1.19 Military Construction Outlays: October 2011–June 2014

Source: Department of the Treasury, Bureau of the Fiscal Service, *Monthly Treasury Statement of Receipts and Outlays of the United States Government*, October, 2011–June 2014, p. 8.

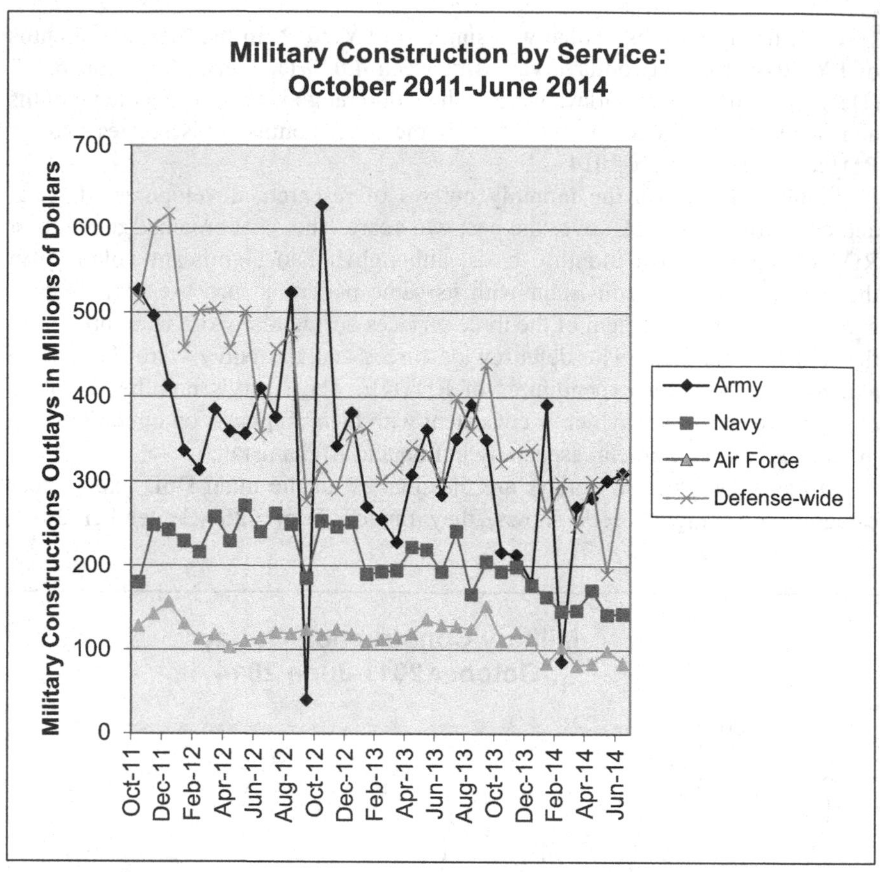

Figure 1.20 Military Construction by Service: October 2011–June 2014

Source: Department of the Treasury, Bureau of the Fiscal Service, *Monthly Treasury Statement of Receipts and Outlays of the United States Government*, October, 2011–June 2014, p. 8.

in millions of dollars) to 1,491 (November 2011; in millions of dollars) in FY 2012, but were in the 1,100–1,300 range in nine of the 12 months. In FY 2013, they ranged from 852 (March 2013) to 1,317 (October 2012), and were in the 900–1,100 range in nine of the 12 months. Consequently, military construction outlays were lower in FY 2013 than in FY 2012, partially due to budget cuts. In the first nine months of FY 2014, military construction outlays ranged from 633 (February 2014) to 900 (January 2014) and were in the 700–800 range in seven of the nine months. Therefore, military construction outlays in FY 2014 are likely to be lower than in FY 2012 and FY 2013.

Figure 1.20 shows military construction outlays by service. The defense-wide forces usually had the highest outlays for military construction. The Army, however, which had substantial volatility and often had the second highest outlays, had higher outlays than defense-wide forces for some months. The Navy was the service with the third highest outlays, followed by the Air Force.

Conclusion

With the concerns about sequestration and the defense budget, as well as shifting defense priorities, the Department of Defense unveiled the findings of their Strategic Choices and Management Review (SCMR) on July 31, 2013. They faced the difficult choice of cutting current capacity in order to protect and grow modernization programs or focusing on sustainment of current capabilities, while sacrificing future technologies. In other words, DoD has had to deal with the trade-off between various categories of the budget – operations and maintenance, procurement of new equipment, and RDT&E. There was less of an emphasis on Army platforms, such as the ground combat vehicle and the joint light tactical vehicle in the SCMR, although there was a greater emphasis on investments in countering anti-access and area-denial threats, such as submarine cruise missile upgrades, the JSF, and the long-range strike family of systems, as well as cyber capabilities and special operations.

The Air Force faces the challenges between cutting current capacity to modernize or sustaining current capabilities at the cost of developing new programs. On the one hand, the Air Force could cut or slow the innovative, new programs so that it can maintain its current force strength, thus sacrificing new procurement and RDT&E for existing operations and maintenance. On the other hand, the Air Force could create a smaller, more tactical force by cutting older bombers, such as the B-1B Lancer. The money saved by eliminating older equipment would be funneled into programs like the F-35 Joint Strike Fighter and the future long-range bomber, which, along with the new KC-46 tanker, has been identified as the service's top modernization program.

The tradeoffs between operations and maintenance, procurement, and RDT&E are likely to have a significant impact on the long-term viability of the defense industrial base, For example, in the case of the Army, when the Abrams and Bradley fighting vehicles end their new equipment production in 2015, there could be difficulties for General Dynamics Land Systems and BAE Systems in the demand for ground vehicles. If the budget cuts do not eliminate the Joint Light Tactical Vehicle (JLTV), the Ground Combat Vehicle (GCV), the Armored Multi-Purpose Vehicle (AMPV), and the new Marine Corps amphibious vehicles, then production would increase by 2019. But there would be a loss in skills between 2015 and 2019. There have already been layoffs among the major defense companies due to a reduction in wartime production needs, and sequestration could lead to further layoffs. Indeed, the Bradley manufacturing line in York, PA (BAE) will end its work in the middle of 2015 and it would remain dormant for over two years before any AMPV work comes along, if BAE wins the contract.[69]

The Navy is another example in which budget cuts and sequestration can impact its defense industrial base. Uncertainty about sequestration led the Navy to originally delay sending the aircraft carrier, USS Harry Truman to the Middle East and to delay the overhaul of a second carrier. Moreover, the Navy has evaluated deployments to Europe, the Mediterranean, South America, and the Caribbean which could affect the ability of the Navy to respond quickly to regional missions. Decreased procurement due to reduced defense spending can lead to closure of certain facilities, as discussed with the Army regarding the Bradley manufacturing line in York, PA. For example, Huntington Ingalls is closing its shipyard in Gulfport, MS due to the reduction in the DDG-1000 (Zumwalt class) ship construction, as well as the decision by the Navy to award the $212 million contract to the General Dynamics Bath Iron Works to build the deckhouse for the third (and final) Zumwalt class destroyer of steel.[70] Consequently, the reduction in orders for the DDG-1000's has made the competition between the few existing firms even more risky.

One of the legacies of reducing equipment due to the reduction in defense budgets is that a chain reaction can be triggered, which can lead to the elimination of other types of equipment. For example, on July 31, 2013, the possibility was introduced of reducing the number of carrier strike groups for the Navy from 11 to eight or nine. Such a reduction could lead to a chain reaction in that when carriers are cut, this can lead to the elimination of air wings with seven or so squadrons, cruisers, and three or four destroyers, as well as the crews. Moreover, it has implications for reduced operations costs. Furthermore, if the fleet dropped below 230–250 ships (it had 283 as of the fall of 2013), it would also need fewer bases or support facilities. Since 60% of the fleet will be Pacific-based, several facilities could close; examples include the Portsmouth Naval Shipyards in Kittery, Maine and Mayport Naval Station in Florida, as well as lesser facilities.[71]

Modernization of equipment illustrates the tradeoff between a reduction in older equipment with high operations and maintenance costs versus the rising costs for new technology and equipment. One example is the nuclear arsenal: there are five versions of the B-61 bomb, which, as General Robert Kehler, head of Strategic Command, noted, "is the only weapon in the stockpile that fulfills both tactical and strategic missions." The B-61, which the USAF has had since the 1970s, was designed as a short-range "tactical" weapon to deter Soviet forces from overrunning Western Europe. About 180 of the B-61 bombs are reportedly still in Europe on NATO bases in Turkey, the Netherlands, Italy, Belgium, and Germany since many of them were removed following the end

69 McLeary, 2013.
70 Cavas, 2013b.
71 Cavas, 2013a.

of the Cold War. The B61–12, which would have a smaller yield and more accuracy, would replace the older versions of the bomb under President Obama's planned upgrade, since the technology of the B-61 is over 25 years old and has been in need of continuous maintenance. As a result of the introduction of the B61–12, the more powerful atomic bomb, the B83, can be retired. Nevertheless, the estimated cost for modernizing the B61 bomb has grown from $4 billion to $8.1 billion and recent Pentagon projections suggest a cost of between $10 billion and $12 billion.[72]

Another example is the F-35. The US, which plans to purchase 2,443 F-35s, has experienced costs for the F-35 which have been just below the cost estimated by Congress. Nevertheless, although at the time of sequestration, the LRIP (low-rate initial production) 7, took around an 8%-10% reduction, the US was able to arrange a price where the aircraft would be bought within the cost cap. LRIP 6's costs were 4% less than LRIP 5's costs, which were below their estimated cost, and LRIP 7's costs were 4% less than the costs of LRIP 6.[73] In fact, the US potential fleet of 2,443 F-35's cost 70% more than the estimate in 2001, with the estimated costs now expected to be $395 billion.[74] DoD hopes that the F-35's cost will fall to $75 million per plane over the next ten years, which is much lower than the current cost of $115 million per plane. In the recent negotiations, the most recent set of planes sold to DoD were below $100 million per plane.[75]

In short, current defense budget cuts are likely to have a significant impact on the defense industrial base, particularly due to the rising costs of modernization. It is possible that defense cuts will lead to reduced procurement of key equipment and greater emphasis on life extension of existing programs, leading to more emphasis on operations and maintenance, rather than procurement. Nevertheless, although operations and maintenance will continue to be a key segment of the budget, it is possible that operations and maintenance can be reduced, as well. A key, related issue is whether defense budget cuts will impact innovation and new product development, especially since RDT&E is already low, and whether the next generation of equipment will be developed. Finally, a reduction in procurement can also lead to a negative impact on skillsets, which will impact innovation. Moreover, it can impact the ability to return to building equipment if defense priorities shift and place an emphasis on equipment for which funding had been cut back earlier.

In conclusion, this chapter evaluated the magnitude of DoD budgets by service and by budget category and explored some of the challenges due to budgetary fiscal constraints (including sequestration) and shifting defense priorities. The tradeoffs between military personnel, procurement, RDT&E, and operations and maintenance spending are numerous and the services have been evaluating

72 Rabechault, 2013.
73 Pittaway, 2013.
74 Cameron, 2013.
75 Kwaak and Cameron, 2013.

possible strategies. The tradeoffs between sustaining existing equipment at the cost of developing new equipment, versus cutting current capacity in order to protect and develop new equipment are evident across the services. In an environment of evolving defense priorities, this tradeoff will also have a significant impact on the strategies of defense contractors. The next chapter will explore the defense industrial base in the US and the opportunities and challenges facing defense contractors, as well as how their strategies are evolving toward greater diversification of their business base.

References

Bacon, Lance M., 2013. "Only Two US Army Brigades Now Combat Ready, Chief Says." *Defense News*, October 21.

Bennett, John T., 2014a. "House Passes Spending Bill with $572B for Pentagon." *Defense News*, January 15.

——. 2014b. "Senate Sends Omnibus, Pentagon-Funding Measure to Obama's Desk." *Defense News*, January 16.

——. 2014c. "War Funding Climbs in Omnibus Bill for First Time Since 2010." *Defense News*, January 14.

——. 2014d. "House Panel Deals US Defense Sector Another Winning Hand." *Defense News*, June 12.

Cameron, Doug, 2013. "Clipped by U.S., Lockheed CEO Aims Abroad." *Wall Street Journal*, December 7–8, pp. B1 and B3.

Cavas, Christopher P., 2013a. "A US Navy with Only 8 Carriers?" *Defense News*, August 4, 2013.

——. 2013b. "Gulfport Composites Shipyard to Close." *Defense News*, September 4.

Chandrasekaran, Rajiv, 2013. "Guarding the Guard." *Washington Post*, December 17, pp. A1, A14, A15.

Davis, Susan, 2013. "Budget Deal Wins Bipartisan Support." *USA Today*, December 13, p. 8A.

Defense News Staff, 2014. "2014 Defense Forecast: Many Decisions." *Defense News*, January 4.

Erwin, Sandra I., 2013a. "Defense Budget Headed for Dizzying Rollercoaster Ride." *National Defense*, October 20.

——. 2013b. "New Warnings on Sequester: Worst of the Damage Has Yet to Come." *National Defense*, October 14.

"Government Shutdown Hits Military Towns Hard," 2013. *Los Angeles Times*, October 7.

Hook, Janet, 2013. "Deal Brings Stability to U.S. Budget." *Wall Street Journal*, December 11, pp. A1 and A8.

Kwaak, Jeyup S. and Cameron, Doug, 2013. "Lockheed Zeroes in on Jet Contract." *Wall Street Journal*, November 23–24, p. B4.

Londono, Ernesto and Rein, Lisa, 2013. "Military Service Leaders Decry Cuts." *Washington Post*, February 22, p. A13.

Marshall, Tyrone C., 2014. "DoD Budget Request Adapts to Fiscal Realities, Hagel Says." *American Forces Press Service*, March 6.

Mehta, Aaron, 2013a. "KC-46 Came Within 24 Hours of Contract Breach During Shutdown." *Defense News*, November 6.

——. 2013b. "Pratt & Whitney, Pentagon Reach $1.1B Deal on F-35 Engines." *Defense News*, October 23.

——. 2013c. "Sequester Could Delay 'Four to Five' USAF F-35 Purchases." *Defense News*, October 23.

McGarry, Brendan and Taborek, Nick, 2013. "Pentagon Contracts Fall 67 Percent as Military Prepares for Budget Cuts." *Washington Post*, February 18, p. A-16.

McKinnon, John D., 2013. "Deal Includes $12 Billion in Pension Changes." *Wall Street Journal*, December 11, p. A8.

McLeary, Paul, 2013. "Army Sets Date to Release Much Anticipated Industrial Base Report." *Defense News*, September 8.

——. 2014. "US Army Chief Confirms: Ground Combat Vehicle is Dead (for Now)." *Defense News*, January 23.

Nissenbaum, Dion, 2013. "Pentagon Readies Budget Ax." *Wall Street Journal*, February 12, p. A6.

Parsons, Dan, 2013. "Chief of Naval Operations Warns About Program Cancellations." *National Defense*, October 30.

Pincus, Walter, 2013. "Time for a Fresh Look at Trimming Defense Costs." *Washington Post*, November 12, p. A13.

Pittaway, Nigel, 2013. "Australia's F-35 Buy Unaffected by US Sequestration." *Defense News*, October 31.

Rabechault, Mathieu, 2013. "US to Spend Billions 'Modernizing' Nuclear Arsenal." *Defense News*, November 6.

Roulo, Claudette, 2014. "Contingency Funds Support Operations, Recovery, New Missions." *DoD News, Defense Media Activity*, July 16.

Shanker, Thom, 2013. "Cuts Have Hagel Weighing Realigned Military Budget." *New York Times*, November 7, p. A 21.

Simeone, Nick, 2014a. "Hagel Outlines Budget Reducing Troop Strength, Force Structure." *American Forces Press Service*, February 24.

——. 2014b. "Hagel: Proposed Defense Budget Tailored to Meet Future Threats." *American Forces Press Service*, June 18.

US Department of Defense, 2013a. *Fiscal Year 2014 Budget Request: Summary of the DoD Fiscal Year 2014 Budget Proposal—Overview of FY 2014 Defense Budget*, April, pp. 1–1 to 2–6.

——. 2013b. "DoD Releases FY 2014 Budget Proposal." *US Department of Defense News Release, NR-223–13*, April 10.

——. 2014. "DoD Releases Fiscal 2015 Budget Proposal and 2014 QDR." *US Department of Defense News Release No: NR-111–14*, March 4.

US Department of Defense, Office of the Undersecretary of Defense (Comptroller), 2014a. *Overview of Fiscal Year 2015 Budget Request*, March.

——. 2014b. *Overview of Fiscal Year 2015 Overseas Contingency Operations: FY 2015 Budget Amendment*, June.

Vanden Brook, Tom and Davis, Susan, 2013. "Pentagon Warns of Huge Cuts." *USA Today*, February 7.

Weisgerber, Marcus, 2013a. "Comptroller: US Shutdown Cost DoD $600M in Productivity." *Defense News*, October 17.

——. 2013b. "Hagel Identifies Six Areas to Guide DoD Reform Effort." *Defense News*, November 5.

——. 2013c. "Shutdown Could Cost DoD Billions of Dollars." *Defense News*, October 13.

——. 2014a. "Magic Money': DoD Overseas Contingency Budget Might Dry Up." *Defense News*, June 30.

——. 2014b. "US Export Bank on $50B to $70B Overseas Contingency Ops Budget." *Defense News*, June 2.

——. 2014c. "Hagel: Iraq Crisis May Require DoD to Rethink 2015 Budget." *Defense News*, August 21.

Weisgerber, Marcus and Fryer-Biggs, Zachary, 2014. "Pentagon Seeks to Protect R&D Funding in '15 Budget." *Defense News*, January 11.

Weisgerber, Marcus and Mehta, Aaron Mehta, 2014. "DoD Wins Big in Omnibus Bill." *Defense News*, January 20.

Weisman, Jonathan, 2013a. "Budget Deal Heads for Senate Approval as More Republicans Give Support." *New York Times*, December 17, p. A 17.

——. 2013b. "Budget Vote Passes the Details to Two Panels." *New York Times*, December 19, p. A 28.

Whitlock, Craig, 2013. "Hagel Warns Cuts to Defense Budget Mean Trade-offs." *Washington Post*, November 6, p. A2.

Chapter 2
The US Defense Industrial Base

The US defense industrial base has evolved, in part, due to shifting defense priorities, away from seeing modern nation-states similarly equipped to the US as the immediate threat (as in the Cold War) and toward seeing insurgent forces in Iraq and Afghanistan as the immediate threat. As defense priorities shift to the Asia-Pacific region with potential conflicts between China and its neighbors due to China's expanding territorial activities, the defense industrial base is further evolving as the types of manpower and equipment differ from what was needed in Afghanistan and Iraq. The choice of defense priorities has had a significant impact on the types and volumes of equipment and, hence, has had a major role in developing the defense industrial base and building and reinforcing skillsets. The future evolution of defense priorities is likely to further impact the defense industrial base.

It is in this context that, over the past 20 years following the end of the Cold War, the defense industrial base in the US has witnessed many changes. First, reductions in defense budgets during the 1990s contributed to consolidation among US defense contractors and a contraction in the US defense industrial base. Many defense industry sub-sectors manifested a two-thirds reduction in the number of prime contractors and came to be dominated by larger defense giants formed from the consolidations, such as: Lockheed Martin, Boeing, Northrop Grumman, Raytheon, and General Dynamics. This will be discussed in the next chapter. Second, the post-9/11 period witnessed an increased emphasis on insurgent forces as the immediate threat to developed countries; however, defense priorities have shifted more to the Asia-Pacific region. Third, beginning with the onslaught of the global financial crisis in 2008, the trajectories of expansion in the global defense industrial base have shifted downward as the need for economic austerity and budget deficit/debt reduction has put greater pressure on various areas of the budget, including military spending. As a result, the challenges for large US defense contractors are growing, as they compete with smaller entrants in certain product areas, as they struggle to evolve their business base towards growing markets and growing product segments, and as they handle the dual role of foreign countries and foreign defense contractors as allies and customers on the one hand, and as competitors on the other hand.

Defense priorities have led to areas of growth and shrinkage in the defense industrial base, as well as a need to modernize certain aspects. Important equipment models for the Air Force require modernization: the average age of the F-15 fleet is 30 years,[1] strategic bombers average 38 years,[2] and KC-135 refueling aircraft average 53 years.[3] Similarly, the Navy has experienced a reduction in the size of its fleet; it had 316 ships in September, 2001 and had 291 ships by September, 2014.[4] Both the reduction in the number of ships, as well as the type of ships, are likely to have an impact on the defense industrial base and on the skillsets. Finally, the Army has Abrams and Bradley tanks, which date from 1980 and 1981, respectively, and which need replacement.

This chapter evaluates the evolution of the US defense industrial base in response to the shifting economic and strategic landscape. Specifically, it examines the impact of the shift in defense priorities on the defense industrial base and discusses areas of growth (UAVs, cybersecurity), as well as shrinkage (shipbuilding). It further discusses the impact of growth/shrinkage of the US defense industrial base on regions and employment. The chapter assesses the tradeoffs in demand for different types of equipment in the defense industrial base and examines the strategies for US defense contractors in combating defense industrial pressures – expanding into the civilian sector and expanding international sales. Finally, the chapter examines the US defense firms in the context of the global defense industry.

Areas of Transformation in the US Defense Industrial Base

The shift in defense priorities and hard budget constraints has led to the evolution of the defense industrial base and the attenuation of certain product areas. The elimination of the F-22 is one example of the impact of hard budget constraints and shifting defense priorities on significant defense equipment. The original concept for an advanced tactical fighter was conceived in 1981 during the Cold War, but the first F-22 actually entered USAF service at the end of 2005. Its original purpose was to field air-to-air combat in a traditional warfare situation so that troops on the ground would not be impacted by aerial attacks from the enemy and bombers could reach their target. The wars in Iraq and Afghanistan were different, however, from the Cold War in that the enemy was an insurgent force, not a modern nation-state, there were few strategic targets, and no opposing air force, such that the role of airpower emphasized protecting troops on the ground and airlifting supplies. By 2009, the Air Force already had 183 planes and four on the way. It wanted 20 more planes ($4 billion) for 2010 and hoped to have a total of 387 by the end of the decade. The Air Force studies justifying the 387 planes were based on the assumption that the US should be prepared to fight two major wars at the same time with foes that had similarly modern air forces. Nevertheless,

1 Everstine, 2012.
2 Neely, 2013.
3 Bolkcom, 2007.
4 Osborn, 2013; Status of the Navy website, 2014.

it was concluded that the F-35, which was smaller and cheaper at the time, could do a better job at destroying surface-to-air missiles, although not quite as good a job in air-to-air combat. On July 17, 2009, the Senate voted (58–40) to end the F-22 program, which had significant regional impacts because the F-22's contracts and subcontracts were in 46 states.[5]

Traditionally, the shift in defense priorities has led to the development of new areas in the defense industrial base. For example, the wars in Iraq and Afghanistan showed the potential for the market for UAVs. At the beginning of the war, the unmanned systems made up about 4% of the Army's flying hours; by 2009, they made up 40%. UAVs have lower purchase costs because they have less extensive electronics systems than manned aircraft. They also have lower operating costs because, in part, they require less fuel and don't need a lot of logistics support or big runways. Finally, UAVs require less pilot training and entail less personnel risk.[6]

The development of UAVs led to opportunities for smaller, innovative, younger firms, as well as for more established defense contractors, which expanded into the product space partially through acquisitions. Some examples of the smaller firms include: General Atomics, which makes the Predator and Reaper planes; Aerovironment, which makes the Raven and is also developing the Global Observer platform; the UK developer Qinetiq, which has been developing an ultra-long duration, high altitude UAS called Zephyr; and the Israeli UAS platform maker Elbit. Larger firms are also involved in this space, partially by enhancing their capabilities and reinforcing their skill sets through acquisitions of smaller firms. Northrop, which produces the Global Hawk, acquired some of its capabilities from its acquisition of Ryan Aeronautics, which had expertise in target drone production and design. It also bought Swift Engineering, which had expertise in designing blended wing UAVs. A second example is Rockwell Collins, which has become a significant supplier of avionics for unmanned and manned aircraft because it acquired Athena Technologies, a pioneer in flight control systems for UAVs, in 2004. A third example is Boeing, which acquired the Insitu Group in June, 2009, which originally designed the ScanEagle UAV for tuna fishing, but turned more to defense after 9–11. Boeing's UAV business includes the A160T autonomous helicopter and the Phantom Ray unmanned combat aircraft demonstrator. Boeing also acquired Frontier, an unmanned helicopter developer.[7]

UAVs have also stimulated other, related areas of the defense industrial base. For example, sensors are being designed around the constraints and advantages of UAVs. One example is the Forester foliage-penetration radar flown on the A160T, which takes advantage of the long endurance, high altitude and precise low speed control of unmanned helicopters. A second example is the Artemis 25-Ghz radar for Northrop's unmanned helicopter, the MQ-8B Fire Scout. Use of unmanned aircraft in Afghanistan and Iraq has led to increased demand for full motion video,

5 Kaplan, 2009.
6 Cole, 2009, p. B1.
7 Warwick, 2009.

such as the Gorgon Stare wide-area airborne surveillance pod on MQ-9 Reaper. Propulsion systems for UAVs are another area of development, since most of the engines have their origins in either non-aviation markets or manned aircraft. Solar cells and advanced batteries may be the engines for longer-endurance aircraft, such as Qinetiq's Zephyr. Liquid hydrogen is also considered as a source of power for high altitude UAVs like the Global Observer, made by Aerovironment.[8]

Another field of expansion has been cybersecurity and the development of the "cyber-industrial complex." Cybersecurity is an important area for a variety of government agencies, including the Energy Department, which has spent over $100 million in research on cybersecurity since 2010 in an effort to protect its critical energy infrastructure. Organizations in the energy arena in a variety of states, such as Washington, Tennessee, Virginia, North Carolina, Georgia, California, and New Jersey, are developing new services and frameworks with the support of the Department of Energy. Cooperation between various government agencies has been helpful in developing them. For example, the Department of Homeland Security worked with the Department of Energy in a White House initiative in 2012 to develop the Electricity Subsector Cybersecurity Capability Maturity Model (ES-C2M2).[9] A number of larger companies have acquired smaller companies in an effort to enter the cybersecurity industry, which has worldwide sales of over $67 billion, and which is estimated to reach sales of $94 billion in 2017. Sales of other types of cyber equipment, such as smartphone and tablet applications, have much smaller sales, exhibiting $25 billion in sales for 2013. Demand for jobs in the cybersecurity environment has increased 3.5 times faster than positions in the computer world, and 12 times faster than the broader environment for employment. Companies such as Cisco, Intel, and Google have acquired smaller companies in order to develop their role in the cybersecurity market. For example, Cisco purchased Sourcefire, Apple purchased AuthenTex, Google bought VirusTotal, and Twitter purchased Dasient.[10]

Cyber and information technology, including C4ISR and defense network equipment, are growing categories of military equipment, which have sustained significant growth in the recent spending bills. Indeed, in the FY 2014 spending bill for DoD signed by President Obama, there is significant emphasis on cyber and information technology. Cyber Command's budget has increased from $212 million in the FY 2013 budget to $447 million in the FY 2014 budget. Moreover, Cyber Mission Force teams will be stood up for Cyber Command by 2016, with 6000 personnel.[11]

Researchers have suggested that the global C4ISR market is valued at $83.1 billion and is likely to grow to $93 billion by 2019.[12] The C4ISR and defense

8 Warwick, 2009.

9 Department of Energy, 2013.

10 "Tech Titans See Opportunity in Cybersecurity," 2013.

11 Johnson, 2014.

12 "C4ISR Market to Reach $93B," 2014.

network equipment category includes several products that will do well under the approved FY 2014 budget. For example, as mentioned earlier, the Global Hawk Block 30's operational capability has been sustained based on funding in the budget, and $10 million is set aside to test whether U-2 sensors can be adapted to the Global Hawk Block 30 airframe. Other examples of C4ISR military equipment supported by the approved FY 2014 budget include the Air Force's MQ-9 Reaper unmanned aircraft (received $349 million, which is higher than the $272 million that was requested) and the High Performance Computing Modernization Program, which involves five supercomputing centers (received $221 million, when only $181 million was requested). Several programs, however, did not do as well in the approved FY 2014 budget. Some of these reductions in the C4ISR arena were likely partially based on the shifting role of the Army and the Marine Corps with the withdrawal of their forces from Afghanistan. One example is the 60% reduction in the amount requested relative to the amount which appeared in the approved budget for the Army's Distributed Common Ground System. The Army requested $267 million and received $111 million. Similarly, the Army and Marine Corps' main command and control system, the Joint Battle Command Platform, experienced a 32% reduction between the amount requested ($103 million) and the amount budgeted ($70 million).[13]

The military satellite market is another area of growth which has faced a number of difficulties in recent years. The significant cost overruns and delays in the DoD satellite purchases in 2009, as documented in a government audit report, combined with the elimination of TSAT, which was a significant satellite communications program, as well as the suspension of new plans for weather satellites, have led to challenges for the USAF, which oversees the bulk of military satellite programs. Payloads carried on other commercial satellites or on less expensive military spacecraft would cover traditional military satellite functions, such as navigation, weather forecasting, and communications, under the theory of "disaggregation." Two of the satellite constellations designed by Lockheed Martin – the Advanced Extremely High Frequency (AEHF) communications spacecraft and the Space Based Infrared System (SBIRS) missile warning satellites – may, as of the fall of 2013, be used to test the "disaggregation" theory. Indeed, three AEHF satellites are in orbit and the fourth is under development. Moreover, as of July 2013, the third of four SBIRS satellite payloads was delivered by Lockheed Martin. The development of the smaller, less expensive satellite industry is dependent on the commercial sector, since the ORS office, opened by the USAF in 2007, will be closed in 2014 due to budget cuts. Examples of efforts in the commercial world for satellites include Boeing's Phantom Phoenix, which is a "reconfigurable" satellite, and PlanetIQ, which hoped to launch a group of weather satellites.[14]

Some areas of the defense industrial base, however, have traditionally experienced less rapid growth. Figure 2.1 on the following page shows the

13 Rosenberg, 2014.
14 Erwin, 2013b.

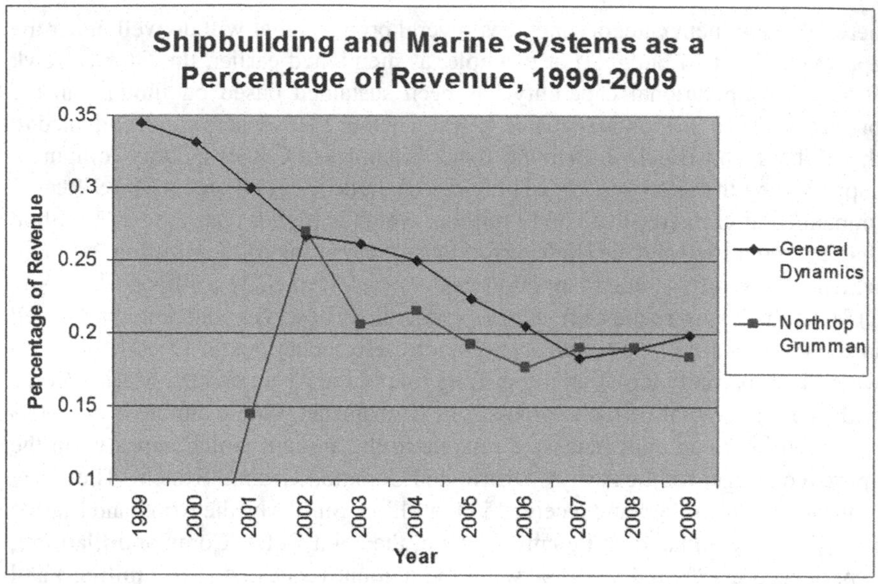

Figure 2.1 Shipbuilding and Marine Systems as a percentage of Revenue: 1999–2009

Source: Capital IQ database.

decline in the shipbuilding segment as a percentage of revenue for Northrop Grumman and General Dynamics, the two primary shipbuilders. Indeed, Northrop Grumman, which had entered the shipbuilding sector in 2001, spun off its shipbuilding unit as Huntington-Ingalls Shipbuilding in the spring of 2011 (which included Newport News, Ingalls, and Avondale). The Avondale shipyard in Louisiana, which has supported 5,000 direct jobs and 6,500 indirect jobs was scheduled to close in 2014 when LPD-25, Somerset, was to be completed, although some work would still be done on the LPD-27. Moreover, the Gulfport composite facility, with the current workforce of 450 to 500, was forecast to close in the second quarter of 2014.[15]

Huntington Ingalls has planned to consolidate Gulf Coast production at its Ingalls shipyard in Pascagoula in order to reduce overhead and to generate improved efficiencies and lower production costs. The remaining LPD's are to be produced by Ingalls. Newport News continues to have stable throughput since it is the sole manufacturer of nuclear-powered aircraft carriers and one of two shipyards which builds nuclear-powered submarines for the Navy.[16] Spinning off

15 Cavas, 2014b.

16 "Northrop CEO Bush Says Private Equity Firms May Be Interested in Ship Unit," 2010; Frost, 2010; Schmidt, 2010.

Northrop Grumman's shipbuilding unit as a separate company was more realistic than having General Dynamics purchase it. General Dynamics owns the other three major military shipyards – Bath Ironworks, Electric Boat, and NASSCO – and had run into antitrust difficulties when it attempted to purchase Newport News prior to Northrop's purchase of it in 2001.

Table 2.1 Major Shipyards and Equipment

	Electric Boat	Bath Iron Works	NASSCO	Newport News	Ingalls	Avondale
Aircraft Carriers				CVN-68		
Guided Missile Destroyers		DDG-51			DDG-51	
Submarines	SSN-774			SSN-775		
Amphibious Assault Ships and Landing Ships					LPD-19 LHA-6	LPD-17
Combat Logistic Ships			AKE-1			

Note: These examples are not exhaustive and date from 2008.

Source: Scott Arnold, Patricia F. Bronson, and Karen W. Tyson, *IDA Paper P-4393, "Infrastructure Rationalization in the US Naval Ship Industrial Base,"* November, 2008, p. 8.

Table 2.1 provides examples of US Navy shipyard product capabilities and highlights the specialization of the major yards. For example, Newport News is the sole manufacturer of aircraft carriers. Ingalls and Bath Iron Works make guided missile destroyers, Electric Boat and Newport News make submarines, Ingalls and Avondale make amphibious assault ships and landing ships, and NASSCO makes combat logistic ships. This further highlights the concern that if orders are reduced due to fiscal constraints, this could negatively impact skillsets since many of the major yards are so specialized and could ultimately lead to closure of some of the yards. If there is a greater demand for more of those particular ships in subsequent years, it could be difficult for the yards which had formerly had those skills to re-open due to the shortage of skilled workers, since the older ones would have retired and there would be few younger ones.

Historically, the closure of shipyards in the US has partially been due to cutbacks in orders, which impacts certain geographic areas. Of the 86 large shipyards, 23 are active (includes Avondale) and 63 are inactive.[17]

17 Colton, 2014.

Table 2.2 Active Major Shipbuilders

Shipbuilder	Location
General Dynamics: Bath Iron	Bath, ME
General Dynamics: Electric Boat	Groton, CT
General Dynamics: NASSCO	San Diego, CA
Huntington Ingalls: Newport News Shipbuilding	Newport News, VA
Huntington Ingalls: Ingalls Shipyard	Pascagoula, MS
Huntington Ingalls: Avondale Shipyard	New Orleans, LA

Source: See http://www.shipbuildinghistory.com

Table 2.2 shows the six major active shipyards, while Table 2.3 shows examples of some of the inactive major shipbuilders. The 63 inactive large shipyards are in various geographic areas: the Mid-Atlantic[18] (22), the Midwest[19] (17), the South/Southwest[20] (11), the West[21] (11), and New England[22] (two). Nevertheless, although many shipyards have closed, several shipyards have expanded. For example, with the significant demand for littoral combat ships for the Navy, the Austal shipyard (owned by Austal) in Mobile, AL and the Marinette Marine shipyard (owned by Fincantieri) in Marinette, WI, have grown.

Table 2.3 Inactive Major Shipbuilders

Shipbuilder	Location
Bethlehem Quincy (formerly Fore River SB, later GD Quincy)	Quincy and Squantum, MA
Bethlehem San Francisco (formerly Union Iron Works)	San Francisco and Alameda, CA
Bethlehem Sparrows Point (formerly Maryland Steel)	Sparrows Point, MD
Federal Shipbuilding & Dry Dock Company	Kearny and Newark, NJ
New York Shipbuilding Company	Camden, NJ
Sun Shipbuilding and Dry Dock Company	Chester, PA

Source: See http://www.shipbuildinghistory.com

18 The Mid-Atlantic includes: New York, New Jersey, Pennsylvania, Maryland, Delaware, West Virginia, and Virginia.

19 The Midwest includes: Illinois, Ohio, Michigan, Wisconsin, Minnesota, and Missouri.

20 The South/Southwest includes: Alabama, Texas, Florida, and Louisiana.

21 The West includes: Oregon, California, and Washington.

22 New England includes Massachusetts and Connecticut.

In short, the US defense industrial base has witnessed significant transformation over the years, partially due to the shift in defense priorities, as well as the development of new technologies, both in the government sector and the civilian sector. Areas of growth include: cybersecurity, C4ISR and defense network equipment, the military satellite market, and UAVs. On the other hand, the shipbuilding sector has witnessed shrinkage with the closure of shipyards and the consolidation of defense contractors. The risk of shrinkage in some of the defense industrial base sectors is that, if demand for the products in the sector increases in later years, both manufacturing facilities and skilled employees may be in significant shortage so that it would be difficult for the sector to expand to meet the burgeoning demand.

The Impact of Growth and Shrinkage in Areas of the Defense Industrial Base on Regions and Employment

The defense industry is very important to particular geographic areas and has a powerful impact on regional economic growth and job creation/job loss. Washington DC is an example of an area where defense spending has strong regional spillover effects and where reductions in defense spending can affect local labor market conditions. Federal spending in the DC area rose from 33% of the spending in the regional economy in 2000 to 37% in 2010. The volume of defense procurement contracts in the area rose from $12 billion to $35 billion between 2000 and 2010. Many defense contractors exhibited significant growth – for example, General Dynamics, headquartered in Falls Church, VA, expanded from 1,100 local employees in 2000 to nearly 6,000 by 2005. Similarly, Arlington-based CACI International had 2,600 area employees in 2001 and now has close to 6,200.[23]

Charleston, SC is a second example of an area where defense spending has strongly impacted the region. Many MRAPs are built by manufacturers based in Charleston (Force Protection) and some of the technology is developed by companies in the area (SAIC and SCRA). The MRAPS are outfitted with technology at Charleston's Space and Naval Warfare Systems Center. Between 2000 and 2007, IT employment in the area grew 52%, while IT employment only grew 9% nationally. The growth in engineers, architects, and scientist grew 52% in this area, but fell 3% nationally. As a result, partially due to defense spending, South Carolina is second only to Michigan in its concentration of industrial engineers.[24]

A third example is Newport News, VA. As of November, 2013, employment at the Newport News yard was at a post-Cold War high, with more than 23,000 employees working on carriers and submarines. Gerald R. Ford, the first ship of a new class of aircraft carriers, will be commissioned in 2016 and was christened on November 9, 2013. The next carrier, the John F. Kennedy, is scheduled to be

---

23 Censer, and Whoriskey, 2010.

24 Anonymous, 2009.

launched in 2019. Although Newport News has had little interest in taking ships apart and scrapping them, and more interest in building, the shipyard will perform the strip-and-defueling job on Nimitz starting in 2025 to deactivate it, as it is currently doing with the Enterprise. Newport News and General Dynamics Electric Boat in Groton, Conn. share in the construction of the Virginia-class submarines, and the yards alternate in final assembly.[25]

A fourth example is the impact that the Northrop/EADS proposal which won the 2008 aerial refueling tanker competition, would have had on Alabama. Airbus planned to assemble the A330 and the KC-45 in a $600 million plant in Mobile, bringing 1,500 jobs to the city and stimulating $360 million in economic activity.[26] This will be discussed in Chapter 5 and highlights the importance of the economic effect on a given region for Congressional representatives as they determine the types of equipment to fund.

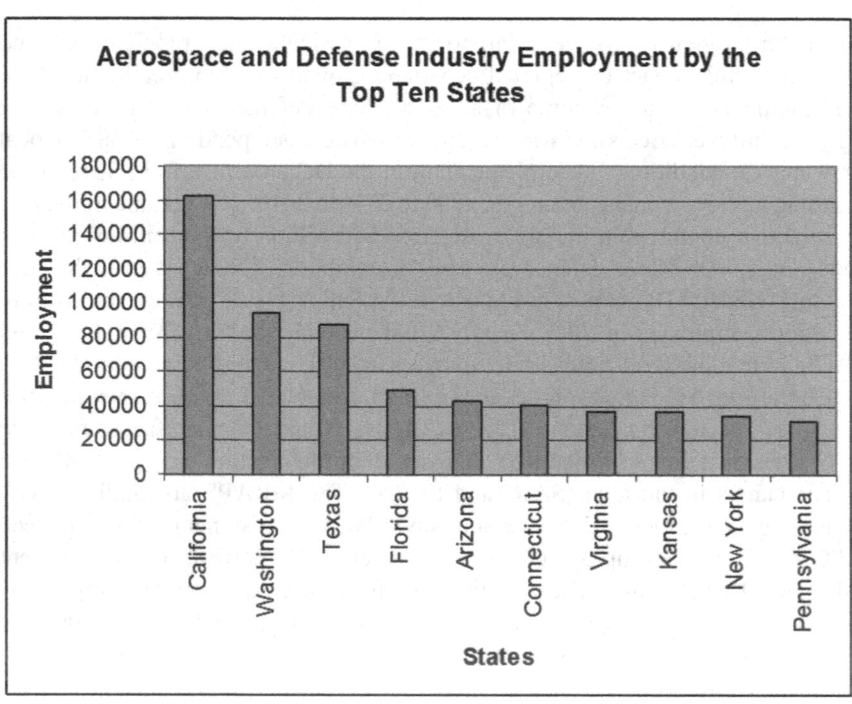

Figure 2.2 Aerospace and Defense Industry Employment by the Top 10 States

Source: Deloitte, *The Aerospace and Defense Industry in the US: A Financial and Economic Impact Study*, March, 2012, Figure 44, p. 72.

25 Cavas, 2013a.

26 Binns, 2008.

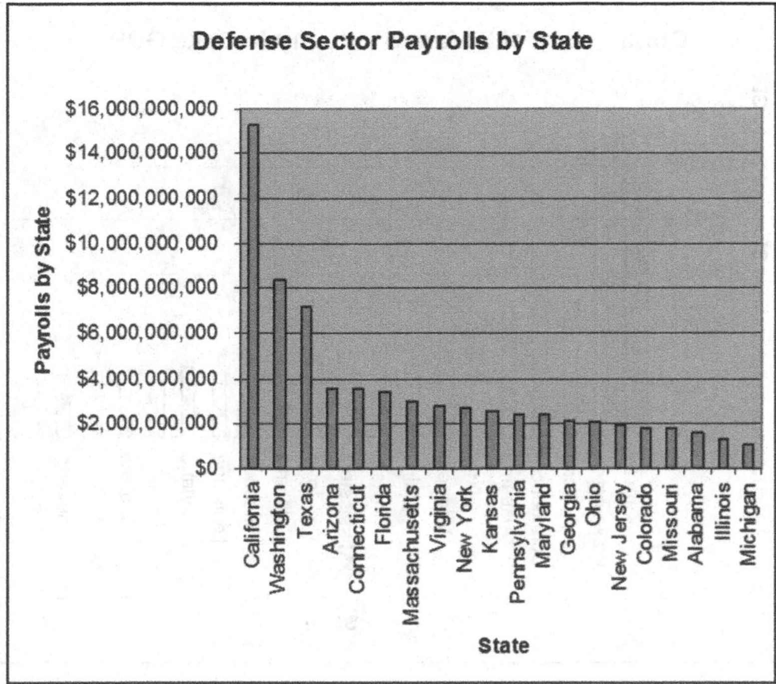

Figure 2.3 Defense Sector Payrolls by State

Source: Deloitte, *The Aerospace and Defense Industry in the US: A Financial and Economic Impact Study*, March, 2012, Figure 47, p. 74.

The defense sector has a significant impact in many of the states. Figure 2.2 shows the top states with substantive aerospace and defense industry employment and indicates the geographic areas which could experience significant damage if there is a reduction in contracts for the facilities in their areas.

The defense sector payrolls in California are significantly greater than the defense sector payrolls in the other states with significant defense employment, such as Washington and Texas (Figure 2.3). Some of the differential may be due to the quantity of equipment produced at the manufacturing facilities in different states. In addition, manufacturing facilities in some states have more skilled employees and management, which can result in higher payrolls than manufacturing facilities with many low-cost employees.

The importance of the defense sector to the GDP of the state is illustrated in Figure 2.4. Although Kansas is the eighth largest state in terms of defense employment, it has the largest defense industry as a share of its GDP (around 10%). Alabama is not one of the top 10 states in terms of employment, however it is the fifth largest in terms of the contribution of the defense industry to the GDP of the state (4.3%). California has the highest defense sector employment and payroll,

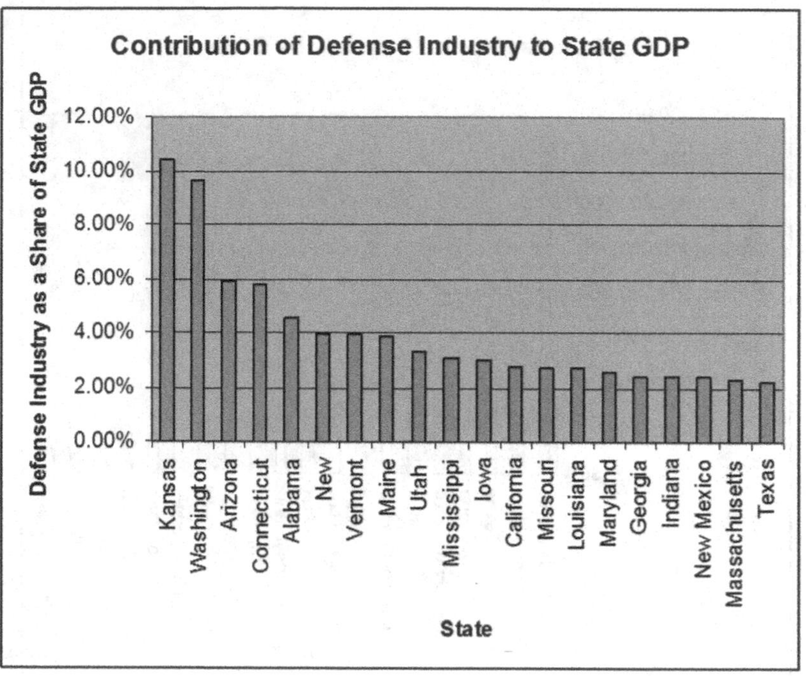

Figure 2.4 Contribution of Defense Industry to State GDP

Source: Deloitte, *The Aerospace and Defense Industry in the US: A Financial and Economic Impact Study*, March, 2012, Figure 46, p. 74.

but due to the size and diversity of the California economy, it is the twelfth largest state in terms of the contribution of the defense industry to the GDP of the state. Indeed, the defense sector comprises 2–3% of California's GDP, which is one-fifth to one-third of the size of the contribution of the defense industry to Kansas' state GDP of over 10%. This data provides a particularly valuable perspective on the impact that defense sector cutbacks can have on the economy of the state and how meaningful the defense sector is within the broader state economy.

As the contracts in the defense sector diminish due to reductions in defense budgets, it is likely that fewer younger workers will be hired, while older workers will retire. As a result, the opportunities for skillset transfer between generations could become less, which could make it more difficult for regeneration of the skillsets if defense priorities shift and different types of equipment are needed. Figure 2.5 shows the aerospace manufacturing employment by age group. Almost 35% of aerospace manufacturing workers are between the ages of 45 and 54 and are likely to retire in the coming years, while 20% are between the ages of 55 and 61 and are even closer to retirement. The share of younger workers is relatively

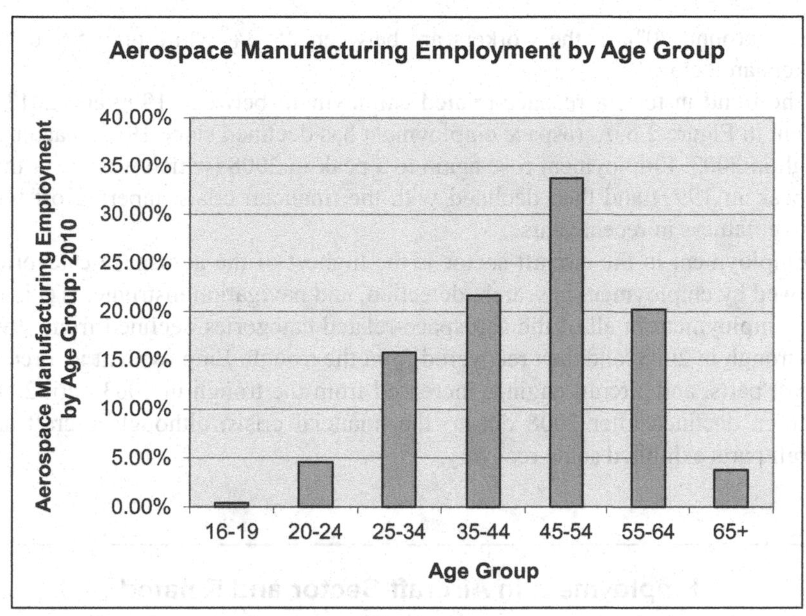

Figure 2.5 Aerospace Manufacturing Employment by Age Group

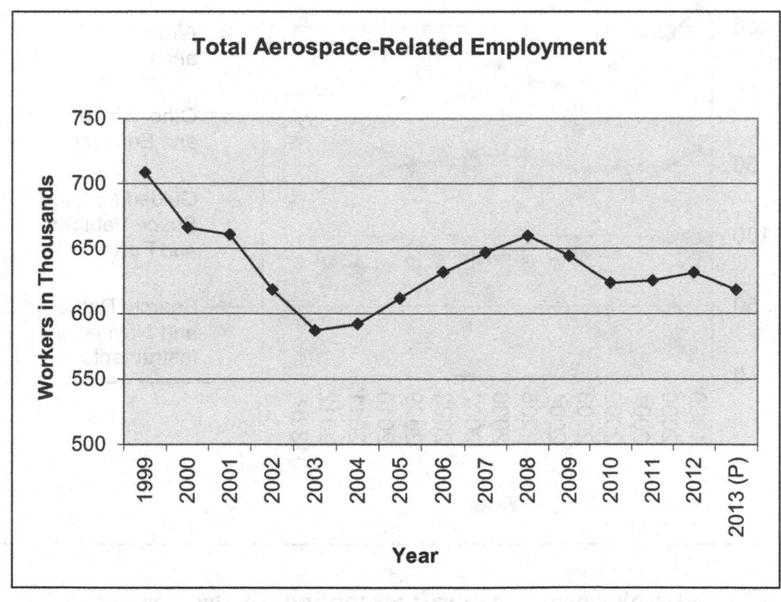

Figure 2.6 Total Aerospace-Related Employment

Source: AIA 2013 Year-End Review and Forecast; Table IX: Aerospace Related Employment, p. 18; workers are in thousands.

small – around 20% of the workers are between 25–34, while only 5% of the workers are below 25.

The trend in total aerospace-related employment between 1999 and 2013 is evident in Figure 2.6. Aerospace employment has declined since 1999, reaching a trough in 2003. Employment rose again to a peak in 2008 (which was lower than the peak in 1999) and then declined with the financial crisis, tapering off with relative flatness in recent years.

Employment in the aircraft sector is the highest of the aerospace categories, followed by employment in search, detection, and navigation instruments (Figure 2.7). Employment in all of the aerospace-related categories declined from 1999, hit a trough in 2003, and then recovered from the trough. Employment in aircraft, aircraft parts, and aircraft engines increased from the trough in 2003 until 2008, and then declined after 2008 due to the financial crisis, although aircraft and aircraft parts exhibited some recovery.

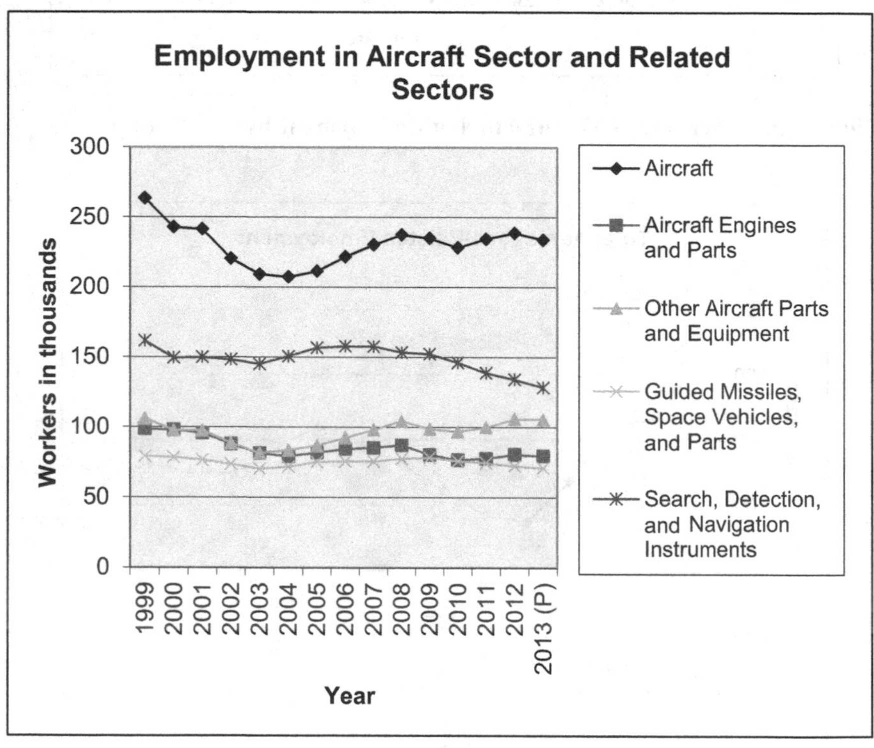

Figure 2.7 Employment in Aircraft Sector and Related Sectors

Source: AIA 2013 Year-End Review and Forecast; Table IX: Aerospace Related Employment, p. 18; workers are in thousands.

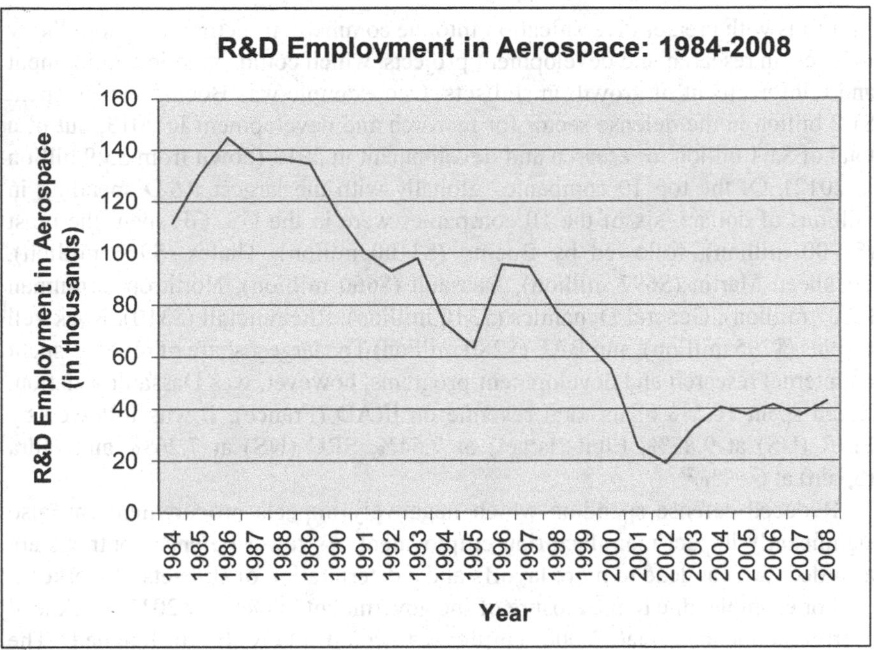

Figure 2.8 R&D Employment in Aerospace: 1984–2008

Source: Aerospace Industries Association. See http://www.aia-aerospace.org

Figure 2.8 shows the decline in R&D employment in aerospace from 1984 to 2008. R&D employment reached its trough from its downward slope in 2001–2002, and then increased, flattening out by 2004. R&D employment is very important for the next generation of defense equipment, especially since the defense budget funds for RDT&E are among the smallest of the DoD budget categories and R&D needs to be undertaken in the industry since the government expenditures are low and may become lower if budget cuts continue.

The role of R&D in defense firm budgets will impact employment for R&D. *Defense News* surveyed defense firms regarding expenditure between 2012 and 2013 on internal research and development (IRAD) projects. Eighteen of the 29 companies increased their expenditures on IRAD during this period, while 11 of them reduced their expenditures. The companies reducing IRAD over this period included Saab, Exelis, General Dynamics, Northrop Grumman, and Rockwell Collins. Firms like Lockheed Martin, however, increased their IRAD funding; indeed Lockheed spent $112 million more on IRAD spending in 2013 relative to 2012, although, despite its three consecutive years of expanding its spending on research and development, Lockheed still spends less than its peak of IRAD spending, which was $822 million in 1999.[27]

27 Weisgerber, 2014

Firms with greater diversification into the commercial sector were more likely to invest in research and development projects, which could assist in employment and reinforcement of growth in skillsets. One example was Boeing, which spent $1.2 billion in the defense sector for research and development in 2013, out of a total of $3.1 billion in research and development in 2013 (down from $3.9 billion in 2012). Of the top 10 companies globally with the largest R&D spending in millions of dollars, six of the 10 companies were in the US. GE spent the most ($4700 million), followed by Boeing ($3100 millon), Thales ($966 million), Lockheed Martin ($697 million), Dassault ($660 million), Northrop Grumman ($507 million), General Dynamics ($310 million), Rheinmetall ($310), Rockwell Collins ($295 million), and BAE ($280 million) The largest share of revenue spent on internal research and development programs, however, was Dassault Aviation, which spent 10.5% of its total revenue on IRAD (France). It was followed by FLIR (US) at 9.87%, Elbit (Israel) at 7.54%, SRC (US) at 7.26%, and Indra (Spain) at 6.69%.[28]

Reduced defense spending which negatively impacts procurement can also significantly impact manufacturing employment. As fewer defense contracts are awarded, this can lead to more layoffs and a deterioration of skillsets of workers.

For example, due to the closure of the government in October 2013, Lockheed Martin announced that 3,000 employees would likely be furloughed. The subsidiary of United Technologies Corp (UTC), Sikorsky, announced that 2,000 employees would be furloughed by the second week of October, while Pratt and Whitney and UTC Aerospace System also announced that 2,000 employees would be furloughed and that, if it continued for a week after that, another 1,000 employees would be furloughed. Due to a reduction in contracts since 400,000 civilians were furloughed within the Defense Department, furloughs for workers in the defense sector increased. At the end of the first week of federal government furloughs, UTC announced that over 5,000 employees could be furloughed if the shutdown lasted into November. BAE Systems Inc. indicated 10% to 15% of its 35,000-person workforce could be impacted by the shutdown.[29]

Reductions in defense spending have resulted in a loss of $1.5 billion in contracts for defense contractors in St. Louis between 2010 and 2012. Indeed, Boeing has not increased its employment levels of 15,000 workers over the past two years. At Boeing's Defense, Space and Security division, $2 billion in savings had been achieved since 2010 by undertaking a variety of strategies to reduce costs, including integrating facilities. The Aerospace Industries Association conducted a survey of small and medium-sized contractors in 2013; the results indicated that half of these contractors had laid off workers and 88% of these contractors had experienced some downswing from reductions in defense spending, partially due to difficulty in accessing capital. Examples of defense contractors which closed due to reductions in contracts include Labarge Products Inc. in St. Louis, while

28 Weisgerber, 2014.

29 Herb, 2013.

Sabreliner laid off over 230 positions in Perryville and Ste. Genevieve. Companies like Boeing are emphasizing international sales; indeed, the possibility of a $7.4 billion contract to South Korea for 60 F-15s will enable production in St. Louis to continue through 2020, which will also support suppliers in the area. [30]

In mid-November, 2013, Lockheed Martin announced that it would eliminate 4,000 jobs – 2,000 of them would be linked to office closings, which would occur in 2015 and 2,000 of them would be lost through downsizing of missions systems, space businesses, and information systems by the end of 2014. Plant and office closures include: Lockheed's facilities in Goodyear, AZ; Akron, OH; Newtown, PA; and Horizon City, TX by mid-2015.[31] Similarly, due to concerns of declining military spending, Lockheed Martin has eliminated 30,000 employees since 2008 (including the upcoming 4,000 job cuts), reducing the number of employees to 116,000.[32] Lockheed has also eliminated 1.5 million square feet of space since 2008 and, in November 2013, announced factory and plant closures which would result in the elimination of a total of 4 million square feet. As the information systems and space systems divisions are re-organized, some Lockheed personnel will move to facilities in Valley Forge, PA and Denver, CO.[33] The layoffs have largely been due to the delay in awarding of contracts by DoD. The space division of Lockheed had the best margins in the third quarter of 2013 relative to the other four business units. Lower commercial orders resulted in a 5% decline in sales during the first nine months of 2013. The space business, located in Newtowne, PA, which Lockheed expanded three years ago, will transfer 800 of its jobs to Denver over the next 12–18 months, and an additional 350 employees will be hired in Colorado.[34]

The assembly line for the Boeing C-17 in Long Beach, CA would close towards the end of 2014 because Boeing delivered its final 223rd C-17 plane to the USAF in September, 2013 and has foreign orders of only five planes. In 1981, the Air Force provided Boeing with the contract for this large, four engine jet and the first C-17 flew in 1991. Douglas Aircraft built the Long Beach factory, which had built MD-80s, DC-8s, and DC-3s, and which Boeing acquired when it purchased McDonnell Douglas in 1997.[35] Boeing has noted that C-17 line closure would be the end of the "last large military-jet aircraft production line in the US."[36]

The cargo-lifting capabilities of the C-17 were very important during the Cold War, but the last C-17 order from the Air Force was placed in 2006.[37] In the current environment of sustained and high unemployment, Congressional representatives

30 "Defense Industry Belts Getting Tighter," 2013.
31 Cameron and Rubin, 2013.
32 Halzack, 2013, p. A9; Cameron and Rubin, 2013.
33 Drew, 2013b, B2.
34 Cameron and Rubin, 2013, p. B3.
35 "Boeing to Deliver Final C-17 Cargo Jet to Air Force," 2013.
36 Butler and Norris, 2009.
37 "Boeing to Deliver Final C-17 Cargo Jet to Air Force," 2013.

have become increasingly concerned about the labor market impacts of reducing certain defense programs. The development of an industrial base across many states by a defense contractor for a particular system can be a strategy for survival because, especially in the current labor markets, it can mobilize Congressional representatives to encourage preservation of the weapons system. Senator Barbara Boxer had encouraged funding for the plane; the Long Beach C-17 assembly plant employed 5000 workers. About 650 suppliers were supported, and the impact of closing the line has been estimated to be $5.8 billion per year.[38] Consequently, appropriations were subsequently added by Congress since the C-17 supported 30,000 jobs in 44 states.[39]

This occurred until 2010, when Congressional lawmakers put aside funding for 10 planes, and the Secretary of Defense informed Congress of his "strong recommendation that the president veto legislation that contains money to pay for any additional C-17s." The C-17 program did not receive any new US orders after this and, in 2012, Boeing received $500 million for post-production transition from the US Air Force. With the reduction in orders, jobs were reduced and Boeing, in an effort to extend the assembly line for a longer period of time so that it could attempt to increase overseas sales, reduced production rates from 15 aircraft annually to 10 aircraft. Examples of overseas C-17 sales included sales to: Qatar, the UK, Canada, India and Australia. The C-17s, each of which cost $200 million, face lower cost competition from EADS' A400M and Lockheed Martin's C-130J.[40]

Boeing has played a dominant role in the industrial base of Washington State. The company was founded in 1916 in Seattle and developed the Puget Sound area beginning in the 1940s. About half of Boeing's 171,000 employees are located in Washington.[41] In February, 2014, Boeing finally announced that its new commercial airplane, the 777-X would be built in the Everett, Washington plant.[42] Boeing had spent significant time in evaluating where it should build the 777X because Boeing and the International Association of Machinists and Aerospace Workers had disagreements regarding the extension of an $8.7 billion 777X contract. This contract would freeze the pensions (and limit pay raises) to increasing 1% every other year.[43] As a result, Boeing considered other locations, such as: Alabama, Utah, Kansas, and North Carolina. As of mid-December, 2013, Boeing, which pledged to deliver its first 777-X in 2020, had $95 billion in orders. The plan would be that the production line of wide body 777s would be gradually eliminated when the 777-X production begins; currently, 20,000 employees build the wide body 777s.[44] Boeing has a second line for assembly of 787s in North

38 Elgin and Epstein, 2009, p. 46.
39 "Boeing to Deliver Final C-17 Cargo Jet to Air Force," 2013.
40 "Boeing to Deliver Final C-17 Cargo Jet to Air Force," 2013.
41 Greenhouse, 2013, p. B6.
42 Gates, 2014.
43 Drew, 2013a, p. B3.
44 Johnson, 2013, pp. A14 and A19.

Charleston, SC and owns significant land in North Charleston.[45] The construction of the 777-X in Washington will likely create 10,000 jobs.[46]

Tradeoffs in Demand for Different Types of Equipment in the Defense Industrial Base

The defense industrial base faces a number of challenges due to the tradeoffs in demand from the Department of Defense and the various services. On the one hand, DoD needs certain types of equipment for various defense priorities; on the other hand, there are tightening budgetary limits regarding how much of the equipment can be bought and what the equipment should be. Moreover, DoD is faced with the tradeoff between keeping older equipment and handling the operations and maintenance costs, or buying new equipment and dealing with the high procurement costs. Furthermore, the challenge of funding the next generation of equipment faces significant budgetary constraints. As the defense priorities shift away from insurgent warfare in Iraq and Afghanistan, and more toward the Asia-Pacific region, the Army will have less demand for certain types of equipment from the defense industrial base which were needed in Iraq and Afghanistan. The Navy and the Air Force, however, are likely to play an even greater role in the Asia-Pacific region and will have greater demand for modernizing and replacing certain types of equipment. Understanding the shifting demands for equipment by the services helps defense contractors to develop strategies to sustain the defense industrial base, manage skillsets, and continue developing innovative products.

The Army

The Army faces significant changes in 2014 and 2015, since, by 2015, only at most a few thousand soldiers will remain in Afghanistan. Since 2001, the Army has dealt with counterinsurgency and combat missions, but they are less likely to be involved with the recent shift in defense priorities toward the Asia-Pacific region. As a result, the defense industrial base has different demands for particular types of equipment. Since salaries, benefits, and health care are the largest portion of the Army budget, which is flattening, the Army is facing slow modernization of helicopters and ground vehicles. As the Marine Corps moves away from land operations in Afghanistan and more into operations from ships, it faces the same challenges as the Army.[47]

With the strategic re-positioning of the role of the Army, a key issue for the Army has been the conversion of MRAP vehicles into trucks and finding uses for surplus MRAPS. This has implications for the defense industrial base due

45 Drew, 2013b, p. B3.

46 Greenhouse, 2013, p. B6.

47 *Defense News* Staff, 2014

to the reduced demand for MRAPS since defense contractors are faced with the challenges of whether they should wind down production lines for the MRAPS and, if so, how quickly it should be done. There are a variety of other uses for the MRAPS – they could be left in brigades as troop transports or used for training. Indeed, 60% of them could be put in storage. The surplus MRAPS could even be sold to foreign military establishments.[48]

The need to acquire new technologies is also very important and the budget constraints can present tradeoffs, which heighten the challenge regarding innovation for the defense industrial base. For example, the Army has a strong need to acquire new vertical takeoff and landing aircraft to replace its fleet of aging helicopters since older platforms are reaching the end of their service life. Should the Army continue rejuvenating existing platforms or acquire new weapons? Indeed, budget issues are leading to uncertainly regarding the need for the Armed Aerial Scout helicopters which may lead to a delay in the decision from Pentagon procurement officials until the summer of 2015 at the earliest. This heightens the uncertainties for contractors in the defense industrial base.

The Army is formulating strategies to deal with developing and procuring key equipment, with the shift in defense priorities as it exits Afghanistan, as well as in response to fiscal constraints. For example, the passage of the omnibus budget bill in January, 2014, suggested that the Ground Combat Vehicle for the Army would likely be terminated,[49] and, indeed, it ended in March, 2014.[50] This highlights the issue for defense contractors concerning whether they should retain the skillsets and plan for further development in the future. While the Army requested $492 million for the Ground Combat Vehicle, it only received $100 million in 2014. These funds, however, could be used to slowly develop technology so that in three to four years the Army can return to building a vehicle to replace the older Bradley vehicles. New technologies for the Ground Combat Vehicle program have been funded by the Army and developed by General Dynamics and BAE Systems. Mobile and light ground vehicles which can withstand rockets and bombs along the road are key due to the expeditionary nature of the troops potentially moving in small groups in remote regions, as noted by General Odierno, the Chief of Staff of the Army. Other ground vehicle programs are likely to continue, such as the Armored Multi-Purpose Vehicle and the Joint Light Tactical Vehicle, which can replace the Humvee and the M113 infantry carrier.[51] Consequently, the defense industrial base will not only have to adjust its schedule for developing the Ground Combat Vehicle, but it also needs to focus on the development of mobile and light ground vehicles.

Nevertheless, if the service stops the $10 billion program which would have replaced the M113 infantry carriers, then there would be no procurement funding

48 Magnuson, 2013a, p. 30.
49 McLeary, 2014a.
50 McLeary, 2014b.
51 McLeary, 2014a.

for modernizing the Bradley, Abrams, and Stryker vehicles. Moreover, the Armored Multi-Purpose Vehicle (AMPV) program may be slowed; the AMPV is the most important procurement from the perspective of ground vehicle manufacturers due to the ending of the Ground Combat Vehicle. As of June 27, 2014, the Army suggested that, in order to develop new requirements for the vehicles, funds would have to be reduced from modernizing the GDLS-made Abrams, the Stryker, and the Bradley.[52]

Although the withdrawal of US forces from Afghanistan has reduced the demand for ground vehicles, US and foreign defense contractors continue to manufacture ground vehicles and provide alternative uses for them. For example, Nexter (France) developed TITUS, a Tactical Infantry Transport & Utility System, which is a personnel carrier that can be re-adjusted to serve a variety of functions, such as cargo transport or medical evaluation. Patria (Finland) developed a new combat vehicle that can "be outfitted with modular ballistic, mine and improvised explosive device protection systems depending on customer requirements."[53]

Despite the upcoming budget cuts, as of the fall of 2013, the Army wanted to replace its light tanks and the development of solutions to replace them has been important for the defense industrial base. Production of Sheridan tanks ended in 1970, by which time the Army had 1700 Sheridan tanks, and the last division to operate them, the 82nd Airborne Division, ceased operating them in 1997 when the Division was deactivated. Although the Army bought six Armored Gun Systems later, the program was terminated. The benefit of acquiring the light tanks would be that portions of the Army, such as the XVIII Airborne Corps, could quickly assist in crises in combat zones since the light tanks could be carried on C-130 planes and parachuted into a warzone. As of the fall of 2013, about 140 candidates to replace the Sheridan tank were being considered. In the fall of 1999, then-Army Chief of Staff Gen. Eric Shinseki discussed acquisition of lighter vehicles to enable the Army to transition into a lighter force. This plan led to the $200 billion Future Combat Systems, however, due to difficulties with costs and with technologies, Future Combat Systems was terminated in 2009.[54]

Moreover, as of the fall of 2013, the Army had significant demand for acquiring new vertical lift aircraft – the Joint Multirole Demonstrator – which would ultimately replace the Army's current helicopters, especially the Apache attack helicopter and the Blackhawk utility chopper by the mid-2030s. As a result, the Army has awarded technology investment agreements to Karem Aircraft, AVX Aircraft Co., Sikorsky Aircraft, and Bell Aircraft. The funding agreements match investments from the companies over six years with funding from AMRDEC (the Army's Aviation and Missile Research, Development, and Engineering Center). Bell Helicopter has teamed with Lockheed Martin to produce the V-280 Valor, which will be the successor to the V-22 Osprey, while Sikorsky has teamed with

52 McLeary, 2014b.
53 Insinna, 2013.
54 Erwin, 2013d.

Boeing to provide a helicopter closely modeled on Sikorsky's X-2 demonstrator. Neither AVX Aircraft nor Karem Aircraft have built operational aircraft.[55] Moreover, investment for helicopters, such as the Apache Block III attack helicopters, the CH-47F Chinooks, and the Black Hawks, is likely to continue.[56]

The Navy

With the greater emphasis on the Asia-Pacific region, the Navy faces the need to handle the tradeoff between funding modernization and new equipment versus maintaining older equipment. The demand for modernizing ships and aircraft continues to stimulate innovation in the defense industrial base, although defense contractors have to continue to focus on maintaining skillsets in production and maintenance for existing equipment. For example, updating aircraft in the fleet is a key process. Indeed, the EA-18G Growler is replacing the older EA-6B Prowler, the MH-60R and the MH-60S are replacing older helicopters, the P-8A Poseidon is replacing the aging P-3C Orion aircraft, and the FA-18 E/F Super Horner is replacing the strike fighters.

In an effort to reduce costs and to take advantage of new technology from the defense industrial base, the Navy plans that by 2023, a carrier strike group would deploy with five models of aircraft with five types of engines. This is in contrast to 2005, when 10 models of aircraft and eight different types of engines were flown by the carrier strike group, which increased maintenance costs. Moreover, the first ship in the new Ford class of aircraft carriers, the USS Gerald R. Ford (CVN 78), will cost $4 billion less to maintain than the earlier per ship costs in the Nimitz class, although the cost of purchasing the ship is high.[57] The USS Gerald R. Ford is the first entirely new carrier design since the 1960s, and is also the most expensive ship currently under construction. Indeed, the cost for the carrier has increased over 22% since construction was authorized in 2008. The rising costs are partially due to incorporating new technologies, such as a new dual-band radar system and new launch and recovery systems. As of the fall of 2013, these costs were projected to reach $12.8 billion by 2016, when it is scheduled to be delivered. Indeed, a GAO report from 2013 suggested that until the cost growth of the CVN 78 is understood, the production of the CVN 79 (the John F. Kennedy), should be slowed down.[58]

In July, 2013, Secretary of Defense Chuck Hagel noted that the tradeoff between high-end capability and size could lead to a reduction to 8–9 carrier strike groups, rather than 11. Although this reduction in demand would reduce the production of this new type of equipment for the defense industrial base, operations and maintenance functions by defense contractors would be sustained. If the carrier

55 Parsons and Insinna, 2013.
56 McLeary, 2014a.
57 Erwin, 2013c.
58 Cavas, 2013b.

strike groups were reduced, this would be consistent with the historical trend. In 1962, the carrier force had 26 flattops, which included smaller World War II ships. By 1991, the carriers increased from 13 ships – developed during the 1980s – to 15 ships. Then the ships reduced to 14 (1992), 13 (1993), 12 (1994) and 11 ships (2007). Temporarily, due to the retirement of the Enterprise prior to the completion of the Gerald Ford, there are currently 10 carriers. Carriers have a lifespan of 50 years, and most major systems in the ship are replaced or rebuilt during the one refueling overhaul included in the lifespan.[59]

The tradeoff between reducing the carrier fleet by decommissioning carriers versus the need for refueling/operations and maintenance for carriers highlights the difficulties of military modernization/defense priorities and fiscal tightness. It also highlights the tradeoff for the defense industrial base in terms of building new equipment versus sustaining older equipment, and the implications for the skillsets which are needed by their workforce. One of the current budgetary issues is the cost of over $3 billion for the three year refueling overhaul for the carrier George Washington (commissioned in 1992) at Newport News Shipbuilding, which is scheduled to begin in 2016. Since the carrier Ronald Reagan will replace the George Washington in Japan, the choice of decommissioning the carrier at the cost of $2 billion or undertaking the three year refueling overhaul for $3 billion would not impact the deployment of a forward-based carrier in Japan. The possibility of carrier reduction by the Navy could be blocked by Congress and could lead to some modification of the reductions in budgets in subsequent years. Indeed, when, in 2008, the Navy asked Congress to allow the carrier force to temporarily decrease to 10 due to the gap between the retirement of the Enterprise and the completion of the Gerald Ford, Congress did not support it, however, in 2009, it was quietly approved.[60] The fleet currently has 10 carriers and is waiting for the delivery of the Gerald R. Ford in early 2016, which will be available for deployment in 2017. Although the Lincoln aircraft carrier began its overhaul in March 2013, the George Washington will be next up for the maintenance/refueling and will come out in 2019.[61] As of the end of May, 2014, the FY 2015 defense budget approved by the House Appropriations Committee Defense subcommittee would maintain the 11 aircraft carrier fleet such that $789 million would be shifted toward maintenance and refueling of the George Washington aircraft carrier.[62]

The demand for littoral combat ships has resulted in significant development in the defense industrial base for ships. There are two variants of the littoral combat ship: the Independence class built by General Dynamics and Austal in the Austal shipyard in Mobile, AL, and the Freedom class, which is being built in Wisconsin by Lockheed Martin. The littoral combat ships are different from other ships in that they are fast and maneuverable, they don't need rudders due to their

59 Cavas, 2014a.
60 Cavas, 2014a.
61 Larter, 2014.
62 Bennett, 2014b.

propulsion system, and they are modular, so parts can be interchanged for different mission purposes, ranging from antisurface warfare against small surface ships to detecting submarines. By 2028, the current plan is that the littoral combat ship will make up about one-third of the Navy's surface combat fleet.[63]

The Littoral Combat ship has experienced some difficulties, however, in terms of costs and, in the case of the USS Freedom, in terms of propulsion difficulties, burst pipes, etc. which delayed the entry of the USS Freedom in international naval exercises in the fall of 2013. DoD has internally suggested that the Navy may need to purchase 32 ships, rather than the 52 that were originally planned at $40 billion. The original cost estimates were much lower than the estimated $440 million for each of the basic models, and much cheaper than the cost of a carrier at around $13 billion or the cost of a destroyer at $1.9 billion. Nevertheless, the Navy is still determining what the appropriate uses of the littoral combat ship would be.[64]

Another area of demand for skillsets in the defense industrial base involves the maintenance, equipment conversion, and service life extension for Navy destroyers and cruisers. In 2012, 94 destroyers and cruisers were needed by the Navy over the next three decades, but the recent Navy report to Congress projects the need to have 88 destroyers and cruisers for the next three decades. This may lead to the Navy being chronically short of destroyers over the next 10 years, since it will not have enough ships to carry out all of its missions until 2023–2028 and the Flight III's will not be commissioned until 2016. Due to the cost of shipbuilding programs, for example, the service lives of some of the earlier Flight I and Flight II destroyers may be extended to 40–45 years, relative to 35–40 years.[65]

The Navy plans to shorten the long cruises for ships to reduce the operations and maintenance that would be needed for resetting hulls, etc. Under the Optimized Fleet Response Plan, the carrier strike group's deployment cycle would be expanded to 36 months with an eight month cruise, which, although higher than the original six month deployments, is lower than the current 9–10 month deployments for destroyers and carriers.[66]

The tradeoffs concerning acquisition of new technologies due to budget challenges are also reflected in other decisions faced by the Navy. This has implications for the focus of the shipping construction sector of the defense industrial base in terms of building smaller vessels relative to larger vessels. From the perspective of the Navy, although larger vessels face budget challenges, smaller vessels, which are less expensive, provide significant benefits. Smaller, more affordable ships for the Navy can provide larger ships with capabilities to patrol shallower waters and rivers, feed information to the fleet, and operate in littorals within reach of missiles that could threaten large vessels, and assist with overseas installations where the Navy has large concentrated assets. It is

63 Hoffman, 2013.

64 Nissenbaum, 2014, pp. A1 and A14.

65 Magnuson, 2013b.

66 Fellman, 2014.

likely that small boat purchases will continue at a brisk pace as larger vessels face significant reductions in procurement funding. One example is the Navy's small patrol craft, the Defiant, which are built by Metal Shark. Metal Shark provided the Army with 54 boats during the Iraq War and builds small security vessels for the Air Force.[67]

The reduction in budgets may impact significant Navy programs, such as Virginia-class submarines, destroyers, and littoral combat ships. Defense contractors will be faced with maintaining skillsets for developing new equipment, as well as for maintaining older equipment. Several new ships will be launched in the upcoming years. A significant ship delivery in 2014 will be that of the Zumwalt destroyer, which is the lead ship of the DDG-1000 class. The testing of the mission modules for the LCS will be continuing in 2014, with Fort Worth – the lead ship of the LCS Independence class – deploying in the Pacific. The deployment of the first JHSV, the Spearhead, which is the first of the 10 ships, is scheduled to occur in 2014. Nevertheless, longer cruises, and, as a result, a shift in maintenance, deployment and training cycles, have led to some imbalances for ships, combined with deferrals in availability of maintenance to save funds.[68]

The Air Force

The Air Force, like the other services, faces the challenges of modernization and new technology relative to maintaining older equipment in an environment of fiscal constraints, which results in the demand for a variety of different types of skillsets in the defense industrial base. The most important modernization programs for the Air Force are the long-range strike bomber, the F-35 Joint Strike Fighter, and the KC-46 tanker program. Indeed, on July 9, 2014, the USAF issued their proposal for the long-range strike bomber, with the plan that 80–100 of these next generation bombers would be operational in the mid-2020s and replace the B-2 Spirit and the B-52 Stratofortress.[69] Nevertheless, to maintain and develop these programs, other programs, such as the A-10, which has been supported by Congress, may, from the perspective of the Air Force, need to be cut to maintain the important modernization programs. Retirement of the KC-10 tanker may also be necessary to afford modernization.[70] General Welsh has noted, "We don't exist to fight a counterinsurgency [such as the wars in Afghanistan and Iraq] … We can help in that, but air forces – major air forces – exist to fight a full-spectrum conflict against a well-armed, well-trained, determined foe."[71]

One of the key issues in the discussions associated with the FY 2015 budget has been the way that the Department of Defense will handle sequestration.

67 Parsons, 2013.
68 *Defense News* Staff, 2014.
69 Everstine, 2014.
70 *Defense News* Staff, 2014.
71 Whitlock, 2013, p. A8.

Should sequestration occur, the KC-10 tanker fleet would be retired by the USAF, as would the adaptive engine program which would have cost $1 billion over five years. The USAF would reduce the purchases of the KC-46a tanker by three in FY 2017 and by two in FY 2018. Moreover, the purchase of 39 MQ-9 Reapers would be cut in FY 2018 and FY 2019, while the purchases of the F-35A would be reduced in FY 2016 by 16 and in FY 2017 by one.[72]

The Air Force has emphasized the need for manned spy planes, satellites, and advanced stealth drones which have newer surveillance sensors.[73] The continued demand by the Air Force in these equipment categories involves an emphasis within the defense industrial base on innovation, as well as skills in computer systems and electronics. As older systems are retired, defense contractors face the shift in demand to newer types of equipment. As of the fall of 2013, the Air Force was planning to replace F-16s and A-10s – older equipment – from its aircraft inventory, many of which are used by the Air National Guard, and transition them more to UAVs, such as Global Hawks.[74] As of May, 2014, the USAF plans to cut the 65 UAV combat patrols to 55, and to further cut them if full sequestration continues in 2016.[75]

One of the key issues of retiring UAVs, such as the Predator, is the determination of what their future function would be if they were retired. For example, the USAF began its Predator purchases with one in 1996, seven by 1998, reaching 11 in 2002, 2003, and 2005, before spiking to 39 in 2007, 24 in 2008, and 17 in 2009. After 2009, the Air Force only purchased Reapers. The USAF plans to retire the current fleet of 154 Predators between FY 2015 and FY 2017 since the MQ-9 Reapers offer greater range and payload capacities than the Predator. Nevertheless, the USAF is unclear where to place its functional, but outdated, Predators. One source is the Davis-Monthan AFB in the "boneyard," while another possibility is transferring them to either the Customs and Border Patrol of Homeland Security or providing them to the Coast Guard. Nations, such as France, Italy and the UK are purchasing Reapers (MQ-9s) rather than Predators (MQ-1s).[76]

The Air Force suggested that retiring 286 aircraft over the next five years would save nearly $9 billion, of which 227 aircraft were to be eliminated in fiscal year 2013. Moreover, the 2013 National Defense Authorization Act directed that the Air National Guard reduce 65 aircraft, the Air Force Reserve cut 57 aircraft, and the regular Air Force cut 122 aircraft. As of the fall of 2013, the opportunities for newer companies to design new jets at reasonable prices had expanded. For example, Textron AirLand unveiled the new Scorpion aircraft, about which the National Guard indicated strong interest, such that the Scorpion, rather than unmanned aircraft, could replace the A-10s and the F-16s.

72 Mehta, 2014c.
73 Whitlock, 2013, p. A8.
74 Erwin, 2013a.
75 Mehta, 2014d.
76 Mehta, 2014d.

AirLand Enterprises, whose funding comes from private investors, is the partner to Textron's Cessna Aircraft. One of the National Guard's strongest arguments regarding manned aircraft relative to unmanned aircraft is that, when involved with emergencies, it is much easier for manned aircraft to quickly deploy to help in natural disasters, etc.[77]

Fiscal constraints have led to a tradeoff between modernizing equipment and maintaining the current equipment, with the risk of not being prepared for defense priorities in the coming years due to older technology. As the workers in the defense industry retire, there will be an attenuation in certain skillsets if there is not sufficient demand in those areas to hire newer workers. This tradeoff becomes more complex since the perspectives of Congressional representatives often differ from the perspectives of the various services regarding these equipment procurement issues, which leads to uncertainty for defense contractors about continuation of production. One example is the Air Force, which is considering "vertical cuts" or the removal of entire platforms in an effort to save money. In some cases, the Air Force has had a strong interest in preserving a particular type of equipment, but has had to balance the interests of Congress. For example, the Air Force has felt that the Global Hawk, a drone, is expensive and supported having U-2 spy planes which are manned, however many Congressional representatives have supported the Global Hawk. Hence, the Global Hawk was protected from a reduction in funding in the FY 2013 and FY 2014 budgets because Northrop Grumman, its manufacturer, galvanized its support among Congressional representatives. The Air Force felt that optimally, both platforms could be kept due to their complementarity, however, they have agreed to place funding for the U-2 in the FY 2015 budget proposal at the cost of maintaining the Global Hawk.[78] Indeed, the cost per flying hour of the Global Hawks – their sustainment costs – fell to $24,000 in FY 2013, while the U-2's cost per flying hour was $32,000.[79] Moreover, the FY 2014 omnibus appropriations bill, signed on January 17, also contained $10 million to examine whether the U-2 sensors could be used on the Global Hawk, which would further support the argument that the U-2 was unnecessary. Nevertheless, although the Global Hawk Block 30 is likely to be funded in FY 2015, it is less clear whether the more advanced version of the Global Hawk – the Global Hawk Block 40 – will likely receive funding.[80]

A second example of the conflict between the perspectives of Congress and those of the Air Force, which impacts the demand for equipment in the defense industrial base, is the A-10. From the perspective of the Air Force, the A-10 has been among the most likely candidates for elimination, however, members of Congress have protested in many cases. Retirement of the A-10 would save $4

77 Erwin, 2013a.
78 Mehta, 2014a.
79 Mehta, Aaron and Brian Everstine, 2014.
80 Mehta, 2014a.

billion, and 80% of the missions of close-air support in Iraq and Afghanistan had been accomplished by other types of aircraft. Nevertheless, the House Armed Services Committee in May 2014 inserted language which suggested that the A-10 would be continuing for another year.[81]

One of the significant concerns regarding the potential continuation of the A-10 program in the FY 2015 budget is that the Air Force may have to terminate the Combat Rescue Helicopter (CRH) program as an offset. The FY 2014 enacted budget provided funding for the Combat Rescue Helicopter, which was designated a "congressional special interest item" and which, although the Air Force found it to be helpful, was not one of the Air Force's main priorities.[82]

The CRH program has $345 million in the FY 2015 budget, although the program would be worth about $7 billion overall. About 112 CRH helicopters were to replace the aging H-60 Pave Hawk combat search-and-rescue helicopters. Another possible offset to inclusion of the A-10 would be the B-1B Lancer fleet.[83] Other examples of equipment that the Air Force has considered eliminating to sustain the A-10 are the MC-12, made by Beechcraft, and the KC-10, which is an older transport plane.[84]

The case of the C-27J also illustrates the tradeoffs between costs, defense priorities, and Congressional perspectives, as well as the uncertainties facing the defense industrial base over whether it would continue to produce particular equipment. Since 2007, the Air Force has spent $567 million on a total of 27 C-27J aircraft, of which 16 had been delivered by the end of September, 2013.[85] In February, 2012, the Pentagon's proposed FY 2013 budget terminated the C-27J in an effort to save $400 million. Despite DoD's protests regarding its lack of need for C-27J aircraft, members of Congress, such as Ohio's Senate delegation, defended purchases of the C-27J. This was partially because the production of the C-27Js impacted 800 positions and the Mansfield Air National Guard Base was heavily influenced by the manufacture of C-27Js. L-3, the primary contractor, and Alenia North America, which is part of Finnmeccanica, are the defense contractors involved in the production of the C-27Js. Pentagon officials argued that the C-130, which had a 25 year lifespan and which cost $213 million, could perform all the functions of a C-27J, which, at $308 million per aircraft, was more expensive.[86] By March 2013, the USAF indicated that it planned to eliminate its fleet of the 21 C-27J aircraft by the end of 2013. Some of the C-27Js had not been delivered at that point. Also in March, 2013, a spending bill was passed in Congress to require that the Air Force continue to buy C-27Js. Some C-27Js were delivered to the USAF in May 2013, and, per the congressional directive, the Air Force solicited

81 Mehta, 2014e.
82 Bennett, 2014a.
83 Mehta, 2014f.
84 Mehta, 2014b.
85 "New Air Force Planes Parked in Arizona 'Boneyard,'" 2013.
86 "New Air Force Planes Parked in Arizona 'Boneyard,'" 2013.

that the company build more C-27Js. Nevertheless, by July, 2013, the first C-27Js entered storage[87] at the Davis-Monthan Air Force Base in Tucson, AZ. The Air Force has the plans to do the same for the five additional planes, which will be completed by August 2014.[88]

As of the fall of 2013, the Senate was likely to eliminate the cost of C-27Js from the 2014 defense budget. The planes were to be kept in workable condition by the 309th Aerospace Maintenance and Regeneration Group, which, although located in Tuscon, is overseen by Materiel Command at Wright-Patterson AFB. The Tuscon complex has 13 aerospace vehicles and 4,400 unused aircraft from NASA and the branches of the military. Other equipment which DoD is willing to eliminate, but which is supported by Congressional representatives from Ohio, are the Global Hawk drone and the M-1 Abrams tank.[89] Therefore, the C-27J example highlights the uncertainty facing defense contractors over continued production of certain types of equipment, which has implications for their investments on production lines and in hiring workers with particular skillsets.

Consequently, the Air Force, like the Navy, will be involved in the Asia-Pacific region to a greater degree than the Army. The shifting defense priorities, as well as the desire to modernize and replace existing equipment, puts significant demand on the defense industrial base in terms of modernizing equipment, continuing production of existing models, and sustaining current equipment. The uncertainty for defense contractors regarding fiscal constraints, as well as differences between perspectives of the Air Force and Congress, can lead to greater risk for them in continuing to innovate.

The Coast Guard

The Coast Guard, like the Army, is another example of a service facing the challenges of modernizing equipment with fiscal constraints, while simultaneously dealing with new developments in defense priorities. Defense contractors are facing increasing demand from the Coast Guard to replace older equipment, as well as to heighten the Coast Guard's capabilities in operations, including the Arctic region. In late November, 2013, DoD issued its first "Arctic Strategy" for several reasons. First, rising temperatures have led to shrinkage of the icecaps, and fishing, shipping, etc. could take advantage of the new and developing sea routes. Second, the Arctic region has important natural resources; it has almost one-third of the world's undiscovered natural gas, $2 million in minerals buried under the ice, and 13% of the undiscovered oil in the world.[90] As the new routes open, countries, such as China and Russia,

87 "New Air Force Planes Parked in Arizona 'Boneyard,'" 2013.

88 "New Air Force Planes Parked in Arizona 'Boneyard,'" 2013.

89 "New Air Force Planes Parked in Arizona 'Boneyard,'" 2013.

90 Myers, 2014.

are likely to compete for the routes. The DoD Arctic strategy focuses on the capabilities to "detect, defer, prevent, and defeat threats to the United States," while simultaneously exercising "US sovereignty in and around Alaska."[91] This strategy involves the use of portions of the Coast Guard and the Navy. As of the end of 2013, the US had nuclear submarines and C-130 planes in the area, as well as 27,000 personnel from the reserves, the National Guard, and the active duty forces.[92]

The new Arctic strategy will impact the Coast Guard's deployment of equipment and affect their demand for certain types of equipment, including the replacement of older equipment, such as the cutters in the polar icebreaker fleet, of which the youngest is 15 years old.[93] The demand to replace Coast Guard cutters is key for the defense industrial base to maintain the relevant skillsets. The new national security cutter program will replace the older endurance cutters, which have been in service since the 1960s. The Coast Guard requested funding for eight cutters, although the eighth one, as of January 2014, was still unfunded. The fourth cutter, the Hamilton, was commissioned; the fifth cutter, the James, is projected to enter service in 2017; the sixth cutter, the Munro, will have fabrication beginning in the fall of 2014; the seventh cutter, the Kimball, will have its funding in the pipeline for 2014.[94]

In short, the US defense industrial base faces challenges regarding the shift in defense priorities and fiscal constraints. The tradeoffs between modernization and maintenance of existing equipment in an environment of tight budgets lead to costs and benefits. There can be a significant impact from the reduction in orders in shipbuilding and aircraft on skillsets. When the older employees with the skillsets retire, due to the reduction in orders for equipment, there will be fewer younger workers to whom the employees can transfer their skillsets. Moreover, the uncertainty in demand for new products can lead to greater challenges in the future for defense contractors, since investment in innovation can be expensive and could be difficult to recoup if, due to fiscal constraints or Congressional and DoD perspectives, the orders for the equipment are reduced or eliminated. On the other hand, the US defense industrial base does face opportunities for increasing the possibilities of extending product life for existing products. As the next section discusses, defense contractors are developing new strategies for expanding into the civilian sector to diversify the risks of fiscal constraints, which could provide strong opportunities for continued growth and innovation in UAVs and cybersecurity. Defense contractors are also developing new strategies for extending existing product lines in the US and overseas.

91 Shanker, 2013, p. A7.
92 Shanker, 2013, p. A7.
93 Myers, 2014.
94 Myers, 2014.

Strategies for Defense Contractors in Combating Defense Industrial Pressures: Expanding into the Civilian Sector and into Overseas Markets

Defense contractors have developed a variety of strategies for handling fiscal constraints, as well as for meeting the demand for different products due to shifting defense priorities. Increasing sales to the commercial sector is one potential solution. Diversification of defense firms away from the defense sector and toward the civilian sector will help to mitigate the risk of the reduction in the defense budget. For example, designs for airframes or electronic equipment which have been used for military purposes can be modified and used in the civilian markets. Cybersecurity, UAVs, and airframes which are similar to airframes used among the military have many uses in non-military markets. Moreover, operations and maintenance parts and equipment have similar uses in such cases between civilian and military markets, enabling economies of scale and lower costs in maintaining the equipment. Furthermore, expanding into overseas markets can be profitable with the budgetary pressures in the US, although many of the other countries have similar fiscal challenges, as will be discussed in subsequent chapters.

The risk to defense contractors of investing in new technologies and products would be that DoD may not buy them, which suggests that developing products with both government and civilian uses helps to diversify risk.[95] For example, Textron is developing a light-attack surveillance aircraft, the Scorpion for the military, through a partnership of 22 vendors, which is based on the design of a commercial Cessna business jet and which can carry spy sensors. This reduces the cost and enables greater diversification of government risk.[96]

Figure 2.9 shows the trends in the share of sales to the US government for US defense contractors between 2010 and 2013. General Dynamics, Boeing, Raytheon, and Lockheed Martin all experienced declines in the share of US government sales to their total firms sales over this period, and an increase in commercial sales. Lockheed Martin's share of US government sales as a share of total sales declined from 85% in 2010 to 82% in 2013. Raytheon's share declined from 76% in 2010 to 72% in 2013. General Dynamics' share declined from 72% in 2010 to 62% in 2013, while its share of US commercial sales increased from 10% in 2010 to 18% in 2013. Boeing, which had an extensive presence in the commercial sector, experienced a decline in the US government sales as a share of total sales from 49.7% in 2010 to 38.3% in 2013.

The aerospace sector is one example of profitable expansion into the civilian markets. Figure 2.10 shows total aircraft sales, as well as sales by product, from 2003 onwards and indicates the significant growth in sales in civilian aircraft, and the flat to declining sales in military aircraft. Military aircraft sales overall were higher than sales of civilian aircraft between 2003 and 2011, with civilian aircraft sales exceeding military sales in 2012, but civilian aircraft sales grew much faster

95 Erwin, 2013e.
96 Erwin, 2013e.

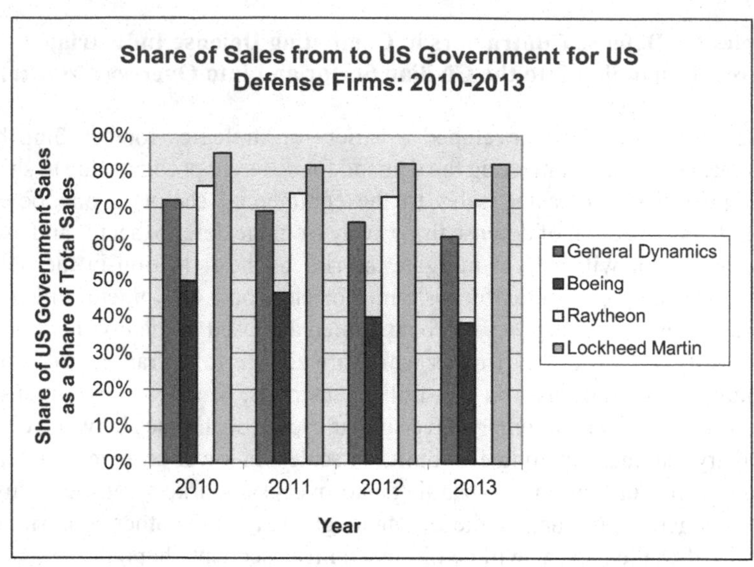

Figure 2.9 Share of Sales to the US Government for US Defense Contractors: 2010–2013

Source: Annual Reports for Lockheed Martin, Boeing, General Dynamics, and Raytheon from 2010–2013.

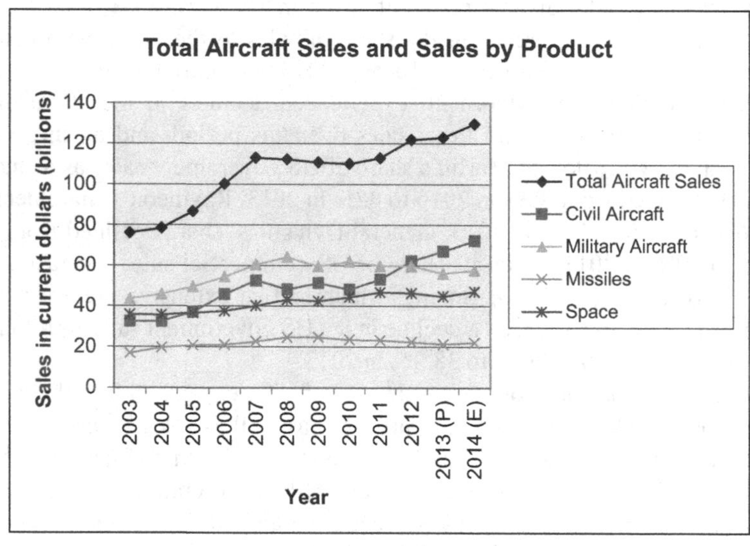

Figure 2.10 Total Aircraft Sales and Sales by Product

Source: AIA 2013 Year-End Review and Forecast; Table I: Aerospace Industry Sales by Product Group, p. 10; current dollars in billions.

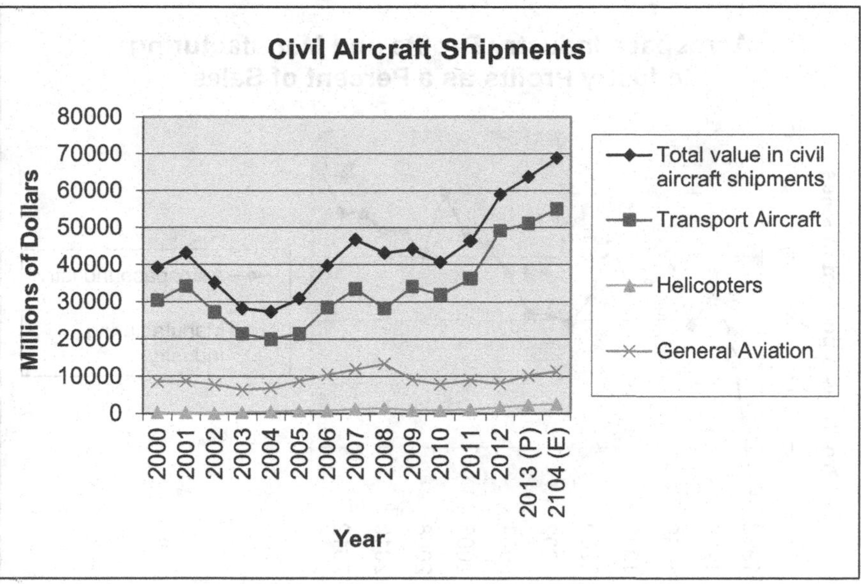

Figure 2.11 Civil Aircraft Shipments

Source: AIA 2013 Year-End Review and Forecast; Table V: Civil Aircraft Shipments, p. 14.

during this time period than the sales of the other categories. The largest gaps between civil aircraft and military aircraft sales were in 2008 and in 2010. This may have been due to slowing economic growth and the reduced demand for sales of civil aircraft, while military aircraft sales were driven by different priorities. By 2012, however, civil aircraft sales exceeded military aircraft sales. This was largely due to the fiscal constraints on DoD budgets, as well as increasing demand for civil aircraft due to improving profitability of airlines. Sales of space products had the third highest sales of the four categories, and were similar in value to sales of civil aircraft during 2003–2005, but then the gap expanded, with sales of civilian aircraft growing faster than sales of space products.

Figure 2.11 shows the total value of civil aircraft shipments, as well as the shipments of categories of civil aircraft – transport aircraft, helicopters, and general aviation. Not surprisingly, shipments of transport aircraft were the largest component of civilian aircraft shipments. Transport aircraft peaked in 2001, declined, reaching a trough in 2004, then recovered, peaking in 2007. Nevertheless, with the beginning of the financial crisis, civilian aircraft shipments declined, and then began to recover, with significant growth in 2011 and 2012.

Expansion into the civilian aircraft market is profitable and is a significant driver of profitability in the aerospace industry. Figure 2.12 compares aerospace industry profits as a percentage of sales relative to broader manufacturing industry profits as a percentage of sales. About two-thirds of the time (14 out of

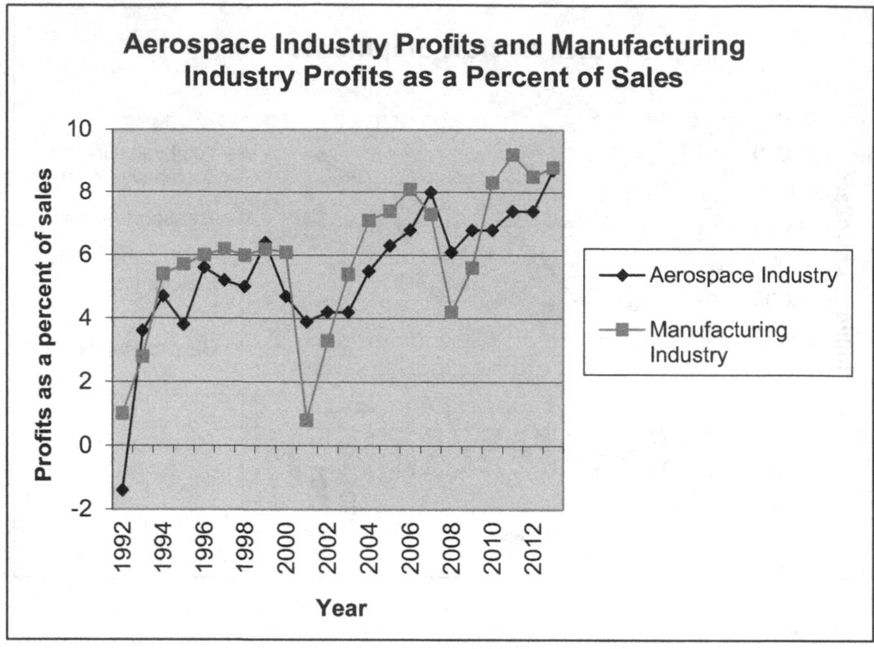

Figure 2.12 Aerospace Industry Profits and Manufacturing Industry Profits as a Percentage of Sales

Source: AIA 2011, 2012, and 2013 Year-End Review and Forecast Table X: Net Profits After Taxes, p. 19.

the 21 years), the broader manufacturing industry had higher industry profits as a percentage of sales. Nevertheless, the spread between manufacturing industry profits as a percentage of sales and aerospace industry profits as a percentage of sales was not substantive, and there were some years (2001 and 2008) when the manufacturing industry's profit as a percentage of sales was substantially below that of the aerospace industry. Consequently, the average for the manufacturing industry's profits as a percentage of sales (5.74%) was only slightly higher than the average for the aerospace industry's profits as a percentage of sales (5.3%)

As the slowdown in government spending continues, it reinforces the need for defense contractors to diversify into the civilian markets. The financial results of Boeing are a good case study. Boeing, one of the most diversified of the major defense contractors, announced a 4% increase in revenues between the first half of 2013 and the first half of 2014. Much of this increase in revenues was driven by the commercial airplanes segment, which had an 11% increase in revenues between the first half of 2013 and the first half of 2014 and a 12% increase in commercial airplane deliveries over the same period. Indeed, the first 787–9 Dreamliner was delivered in the second quarter of 2014. Moreover, Boeing had more deliveries of

737s (239 vs. 218) in the first half of 2014 relative to the first half of 2013, as well as more deliveries of 787s (48 vs. 17).[97]

As discussed earlier, one of Boeing's most innovative new jets is the 777-X, which seats 350–400 passengers and is the longest jet that Boeing has built and is slightly larger than the Boeing jumbo 747. Like the 787 Dreamliner, the 777-X is built with new carbon fibre composite, and the GE9X engines, made by General Electric, will be 10% more fuel efficient than the existing 777s. The plane will be larger than the 787, which carries less than 300. The 777–300ER is the most expensive thus far of the 777s at $320 million, so it is unclear how much the 777-X will cost. As will be discussed in the next section, Boeing sold a number of 777-Xs to overseas airlines, especially in the Middle East, which is currently the market with the most significant growth. Indeed, in July 2014, Qatar Airways and Emirates Airlines arranged to buy 200 777-X airplanes.[98]

On the military side, on the other hand, Boeing's Defense, Space & Security division fell 6% in revenues between the first half of 2013 and the first half of 2014, with Boeing Military Aircraft's revenues declining by 8% and its Network and Space Systems declining by 5%. Moreover, Boeing also had a $272 million after-tax charge to complete additional installation and energy work for the KC-46A tanker. Although there were fewer deliveries of P-8s in the second half of 2014 relative to the second half of 2013, there were more deliveries of F-15s and CH-47s.[99]

Lockheed Martin, General Dynamics, and Raytheon, which had a much greater share of their market in the government sector than Boeing and a smaller share in the commercial sector, experienced declines in sales in the second half of 2014 relative to the second half of 2013. Lockheed Martin's net sales declined by 2.3% in the first half of 2014 relative to the first half of 2013. Its aeronautics segment, however, increased by 9.8% in the first half of 2014, which was helped by higher net sales for F-35 production contracts. Its other segments, however, declined: Information Systems and Global Solutions declined by 8.5%, Space Systems declined by 8.4%, Missiles and Fire Control decreased by 6.8%, and Missions Systems and Training declined by 5.6%.[100] Similarly, General Dynamics' revenues declined by 2.7% for the first half of 2014 relative to the first half of 2013. Like Lockheed, its Aerospace sector's revenues increased by 7.5%. Its Marine Systems increased by 2%. Nevertheless, its Information Systems sector declined by 11.1% and its Combat Systems declined by 7.2%. General Dynamics' orders were helped by a contract for $645 million to support the Canadian Helicopter Project and $17.8 billion to construct 10 additional Virginia class submarines.[101] Finally, Raytheon's net sales declined 6.5% between the first half of 2013 and the first half of 2014.

---

97 PRNewswire, 2014a.
98 PRNewswire, 2014a.
99 PRNewswire, 2014a.
100 Lockheed Martin News Release, 2014.
101 General Dynamics News Release, 2014.

The net sales for its Integrated Defense Systems decreased by 9%, its Intelligence, Information, and Services segment declined by 4%, its Missile Systems declined by 6%, and its Space and Airborne Systems declined by 9%. Nevertheless, its bookings increased by 23.9% with the $764 million booking for TOW missiles for the Marines, the Army, and international customers; $289 million for the SM-6 missile for the Navy; $259 million for AIM-9X missiles for the Navy, etc.[102]

Diversification away from defense business and towards the civilian sector will not necessarily hedge risk for firms because the decisions to reduce the defense budget and, by extension, the federal budget can lead to broader closures of the federal government. Indeed, the closures of the other federal agencies besides DoD can impact the civilian business of a given defense company. For example, during the closure of the federal government in October, 2013, Boeing faced

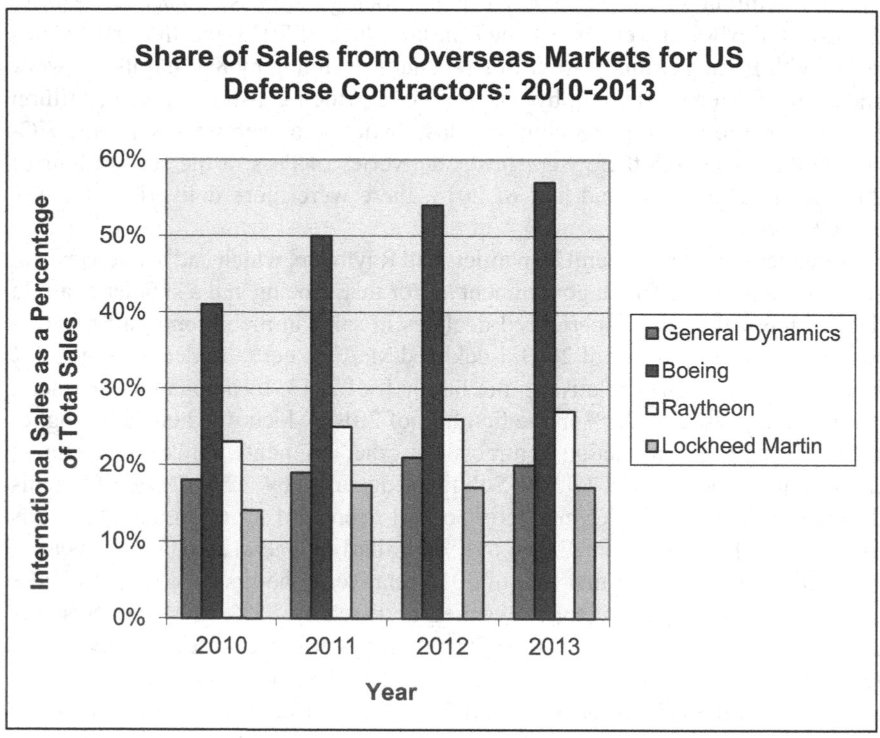

Figure 2.13 Share of Sales from Overseas Markets for US Defense Contractors: 2010–2013

Source: Annual Reports for Lockheed Martin, Boeing, General Dynamics, and Raytheon from 2010–2013.

102 PRNewswire, 2014b.

the risk of delays of shipments of its 787 planes from its South Carolina factory because the FAA personnel who would sign off on the completed aircraft were furloughed. Similarly, JetBlue Airways Corp noted that a delivery of its first new Airbus 321 jet was delayed in Germany after Airbus could not transfer the jet to the US because the FAA office in Oklahoma City was closed.[103]

Figure 2.13 suggests that many defense contractors have moved toward strategies involving growth in overseas sales, since the domestic markets may have limitations in procurement due to fiscal constraints. This was evident in the increase in sales to overseas markets for the top defense contractors in the US between 2010 and 2013. Boeing had the largest share of sales from overseas markets, growing from 41% in 2010 to 57% in 2013. Furthermore, Lockheed Martin, General Dynamics, and Raytheon all experienced an increase in sales to overseas markets between 2010 and 2013. Indeed, Lockheed Martin's international sales as a share of total sales increased from 14% in 2010 to 17% in 2013 and, as of the winter of 2013, it estimated that it could increase international sales to 20% of total sales over the next two to three years.[104] Similarly, Raytheon's international sales increased from 23% in 2010 to 27% in 2013 and General Dynamics' international sales increased from 18% in 2010 to 20% in 2013. Raytheon's overseas sales increased the most in the Middle East and North Africa, where they grew from 7.4% of its total sales in 2010 to 10.11% in 2013. General Dynamics also experienced growth in Africa and the Middle East, expanding from 1.75% of its total sales in 2010 to 2.4% by 2013. General Dynamics also expanded in Europe (increasing from 8.7% to 9.1% of total sales) and in Asia-Pacific (increasing from 3.4% to 4.8% of total sales).

US Firms in the Context of the Global Defense Industry

The US firms play a valuable role in the global defense industry and an examination of their placement in this context is important in understanding the challenges and opportunities faced by US defense contractors. Moreover, it provides perspectives on the degree to which key defense suppliers on the global market can be impacted by slower economic growth and fiscal constraints in their respective countries.

Figure 2.14 shows the top 100 global defense firms as ranked by their 2013 defense revenues. Almost half of the top 100 defense firms were in the United States; the UK, on the other hand, which had the second largest number of the top 100 defense firms, only had 10 firms. The UK was followed by Russia (eight), France (five) and Japan (five).

Figure 2.15 shows the top 20 global defense firms as ranked by their 2013 defense revenues. Significantly, 13 of the top 20 firms were from the US, two were from the UK, two were from France, one was from the Netherlands, one was from Italy, and one was from Russia. Consequently, the cuts in the US budget are likely

103 Ostrower and Stynes, 2013, p. B3.

104 Cameron, 2013, p. B1 and B3.

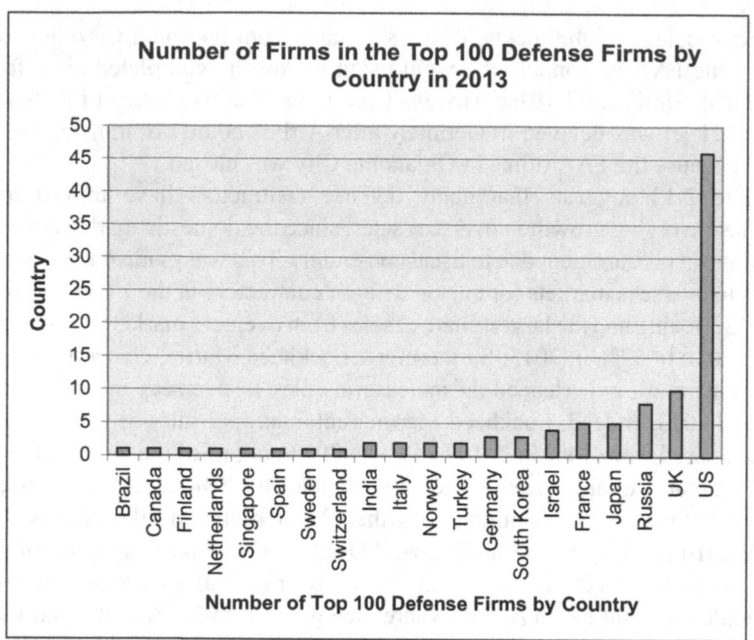

Figure 2.14 Number of Top Global 100 Defense Firms by Country in 2013

Source: Defense News, Top 100. Available at: http://special.defensenews.com/top-100/

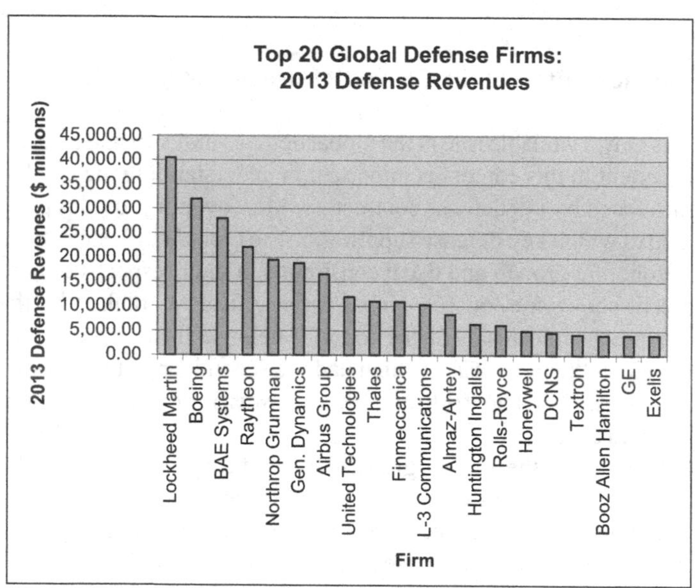

Figure 2.15 Top 20 Global Defense Firms: 2013 Defense Revenues

Source: Defense News, Top 100. Available at: http://special.defensenews.com/top-100/

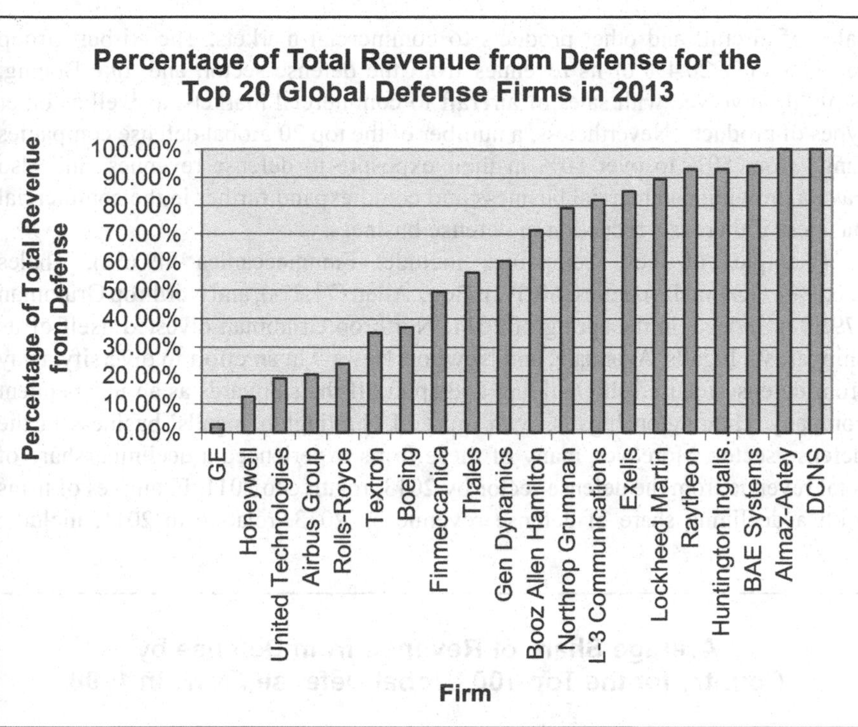

Figure 2.16 Percentage of Total Revenue from Defense for the Top 20 Global Defense Firms in 2013

Source: Defense News, Top 100. Available at: http://special.defensenews.com/top-100/

to have a significant impact on the global defense industrial base, as well as the US defense industrial base. Similarly, the reduction in defense spending by European nations will also lead to cutbacks in the European defense industrial base.

Figure 2.16 shows the percentage of total revenue from defense for each of these top 20 global defense firms. This is important in that it shows the degree to which many, but not all, of these top 20 global defense firms, could be exposed to potential cuts in defense spending in their respective countries. For example, firms whose total defense revenues were around 90% or more of total revenue included: Lockheed Martin (89.3%), Raytheon (93%), Huntington Ingalls (93%), BAE (94 %), Almaz-Antey (100%), and DCNS (100%); these firms could experience substantive reductions in skillsets due to the disproportionate risk of being exposed to defense spending cuts.

On the other hand, other firms have much greater diversity in their defense revenues relative to their commercial revenues and are less likely to experience the risk from defense budget reductions. For example, about 37% of Boeing's total revenue is from defense sales, while over 60% of its revenue comes from sales of aircraft and other products to commercial markets. The Airbus Group receives only 20.4% of its revenues from the defense sector, and, like Boeing, is highly involved with sales of aircraft to commercial markets, as well as other types of products. Nevertheless, a number of the top 20 global defense companies range from 50% to over 80% in their exposure to defense revenues, and also have a growing commercial business and could expand further in the commercial business if there is a reduction in defense business.

Examples of these companies include: Finnmeccanica (49.6%), Thales (56.3%), General Dynamics (60.3%), Booz Allen (71.2%), and Northrop Grumman (79.1%). Indeed, in the spring of 2011, Northrop Grumman divested itself of its shipyards – Ingalls, Avondale, and Newport News – in an effort to diversify away from defense-focused shipbuilding and spun off the shipyards as an independent company, Huntington Ingalls, with most of Huntington Ingalls' business in the defense sector. Moreover, many of these firms experienced a declining share of total revenue from the defense sector by 2013, relative to 2011. Examples of firms with a declining share of defense revenue by 2013, relative to 2011, include:

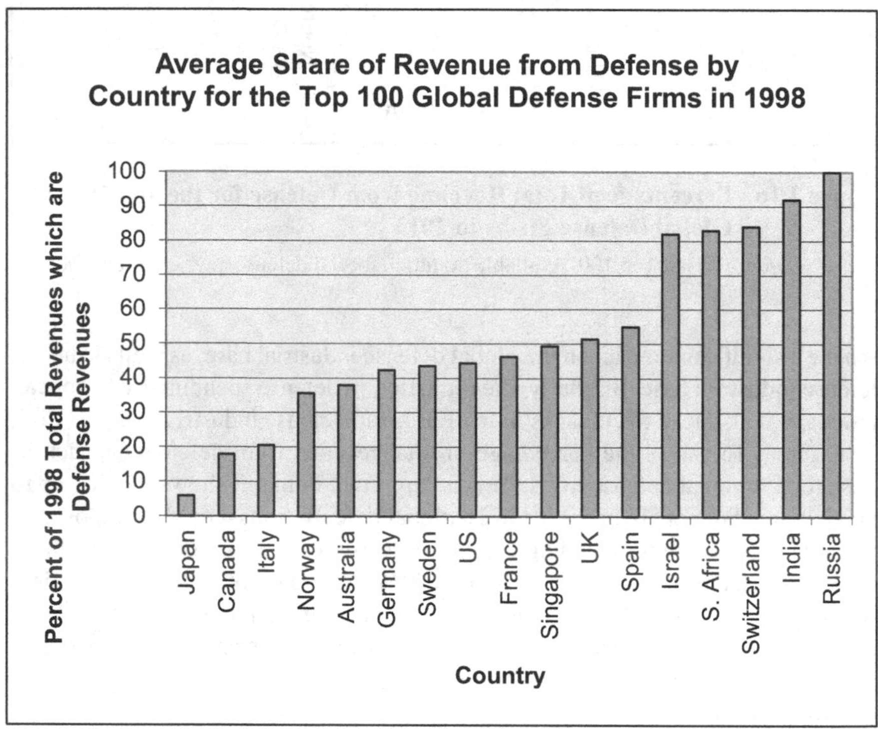

Figure 2.17 Average Share of Revenue from Defense by Country for the Top 100 Global Defense Firms in 1998

Source: Defense News, Top 100. Available at: http://special.defensenews.com/top-100/

Finmeccanica (declined from 60.5% to 49.6%), General Dynamics (declined from 78% to 60.3%), and Booz Allen (declined from 82.4% to 71.2%).

Nevertheless, other companies had a comparatively small focus on the defense sector, but a substantive exposure to the commercial markets, which significantly lessens their risk as defense budgets decline. Examples of these companies included: GE (2.8%), Honeywell (12.5%), United Technologies (19%), the Airbus Group (20.4%), and Rolls-Royce (24.1%).

Figure 2.17 shows the share of revenue from defense in 1998 for the top 100 global defense firms based on the country in which these defense firms were located. In 1998, those firms which were among the top global defense firms in countries such as Russia, India, Israel, South Africa, and Switzerland achieved, on average, between 80% and 100% of their revenues from the defense sector. This would suggest that defense firms in these countries would be exposed to greater risk from defense budget cuts from their country. Other defense firms among the top 100 global defense firms in countries, such as Canada, Italy, and Japan, achieved, on average, between 6% and 18% of their revenues from defense and were less exposed to the risk of budgetary cuts. About 44% of the defense revenue

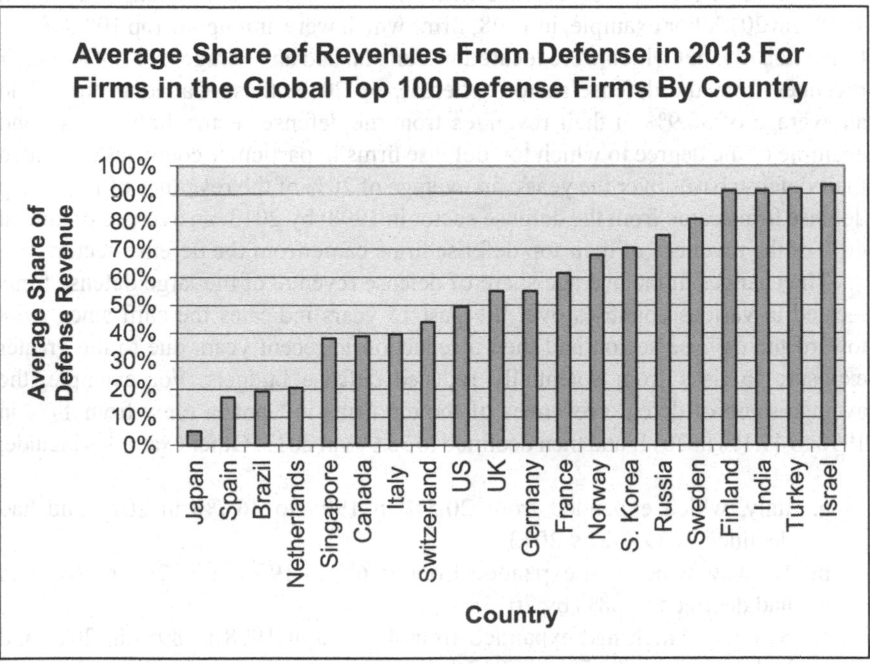

Figure 2.18 Average Share of Revenue from Defense by Country for the Top 100 Global Defense Firms in 2013

Source: Defense News, Top 100. Available at: http://special.defensenews.com/top-100/

of those companies in the top 100 global defense firms which were located in the US came from the defense sector

Figure 2.18, on the other hand, shows the average share of revenue from defense in 2013 for the top 100 global defense firms, based on the country in which these defense firms are located. Unlike Figure 2.17, which depicted the firms in 1998, by 2013, a number of top firms in the defense market were located in countries which did not have significant defense firms in 1998. This indicates the broadening of the global defense market and the development of the defense industry in countries which did not have a top defense firm in their country in the late 1990s. Examples of countries with larger, more profitable defense firms in 2013 relative to 1998 include: Brazil, Finland, the Netherlands, and Turkey.

In 2013, the countries whose firms were in the top 100 and that had the highest average share of total revenues in the defense sector were: Israel (93.2%), Turkey (91.8%), Finland (91%), Russia (75%), South Korea (70.5%), Norway (68%), and France (61%). The countries whose firms in the top 100 were the most diversified with the lowest share of defense revenues were: Japan (5%), Spain (17%), Brazil (19.1%), the Netherlands (20.4%), Singapore (38%), and Italy (39.6%).

The average of the share of defense revenues relative to total revenues for firms which were in the top 100 defense firms in various countries changed between 1998 and 2013. For example, in 1998, firms which were among the top 100 defense firms globally, but which were located in Canada, had an average of 17.9% of their revenues from the defense sector, whereas, by 2013, these Canadian firms had an average of 38.9% of their revenues from the defense sector. Italy is a second example of the degree to which top defense firms in particular countries expanded their defense bases over the years; an average of 20% of the revenues of Italy's top defense firms came from the defense sector in 1998; by 2013, an average of almost 40% of the revenues of their top defense firms came from the defense sector.

The changes in the average share of defense revenue of the large defense firms located in various countries over the past 15 years indicates the shift since 1998 toward the defense sector, and then a reduction in recent years due to the greater exposure to risks from potentially reduced defense budgets. For example, the average share of defense revenues of the top firms in Canada grew from 17% in 1998 to 49.1% in 2011, and then declined to 38.9% in 2013. Other examples include:

a. Italy, which expanded from 20.5% in 1998 to 46.7% in 2011 and had declined to 39.6% by 2013;
b. Norway, which had expanded from 35.6% in 1998 to 76.7% in 2011 and had declined to 68% by 2013;
c. Sweden, which had expanded from 43.53% in 1998 to 89% in 2011 and had declined to 80.7% in 2013;
d. the UK, which had expanded from 51.42% in 1998 to 61.4% in 2011 and had declined to 54.9% in 2013; and
e. the US, which had expanded from 44.3% in 1998 to 52.8% in 2011 and had declined to 49.2% by 2013.

The leading firms in some countries, however, have expanded their average share of defense revenues over the past 15 years. Examples include:

a. France, which had expanded from 46.2% in 1998 to 59.8% in 2011 and 61.4% in 2013; and
b. Israel, which had expanded from 81.9% in 1998 to 92% in 2011 and 93% in 2013.

Firms in other countries, however, experienced a decline in average share of defense revenues. Examples include:

a. Singapore declined from 49.3% in 1998 to 41% in 2011 and 38% by 2013;
b. Spain declined from 55% in 1998 to 19% in 2011 and 17% in 2013;
c. Switzerland declined from 84.1% in 1998 to 42% in 2011 and 44% in 2013; and
d. Russia declined from 100% in 1998 to 72.9% in 2011 and 75% in 2013.

Other countries, such as Japan and India, remained fairly stable over the past 15 years. Figure 2.19 shows the share of defense revenue in 2013 for the 46 US firms in the top global 100 defense firms. About 16 of these firms had shares of defense revenues out of total revenues of over 70%, which could expose them to significant risks as defense budgets in the US and overseas decline. About 15 of the firms, however, had defense revenues which were 30% or less of their total revenues, indicating greater diversification.

Figure 2.20 shows the average share of defense revenue of the top 20 global defense firms in 2013 by country. The 13 US firms had an average share of defense revenues of 58.4%. Russia's leading firm had 100% of its revenue in defense, while the top firms in France and the UK had 78.2% and 59.1%, respectively, in defense revenues as a share of total revenues.

Figure 2.21 shows the average percentage growth or contraction in defense revenues for these top global defense firms in various countries. In some cases, the failure of these firms to grow between 2012 and 2013 was largely due to the increase in defense budgetary cuts and the resulting decrease in orders for equipment and services. Countries whose defense firms in the top 100 global defense firms experienced a decrease in defense revenues between 2012 and 2013 (negative growth) included: Switzerland (-22.2%), Japan (-11.9%), India (-8.5%), Canada (-7.4%), the US (-5.2%), and the UK (-5%). On the other hand, countries whose firms in the top 100 defense firms globally experienced more than a 10% increase in defense revenues between 2012 and 2013 included: Russia (22.8%), Finland (20.8%), France (20.1%), South Korea (17.9%), Turkey (15%), Spain (14%), Italy (10.6%), and Israel (9.9%).

Figure 2.22 shows the percentage change in defense revenue between 2012 and 2013 for US firms in the top 100 global defense firms. Almost 70% of the firms experienced declining revenue between 2012 and 2013, of which 37% had

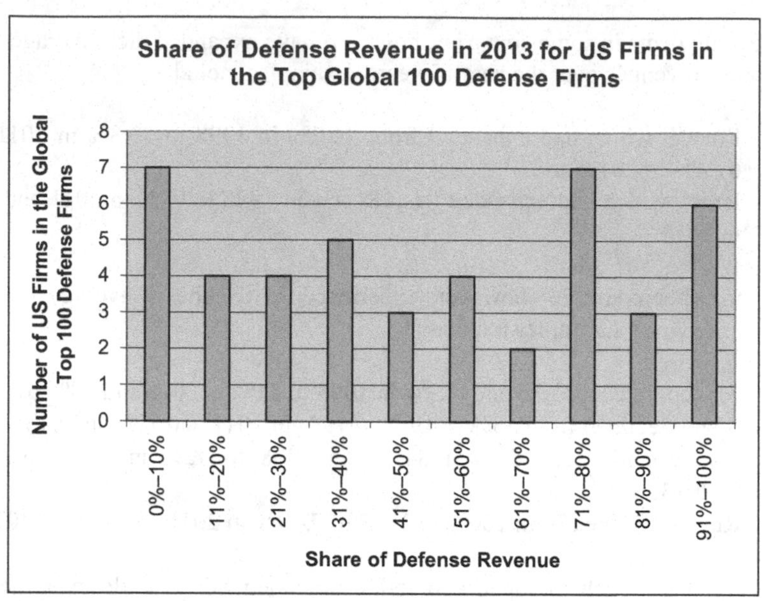

Figure 2.19 Share of Defense Revenue in 2013 for US Firms in the Top Global 100 Defense Firms

Source: Defense News, Top 100. Available at: http://special.defensenews.com/top-100/

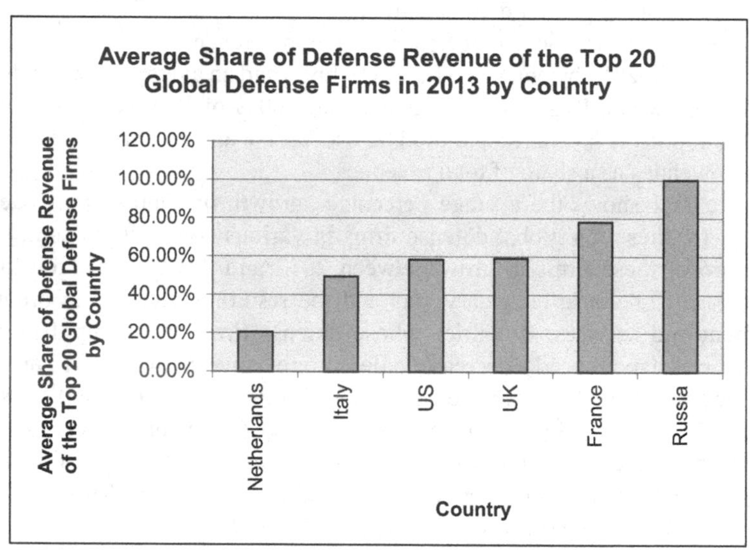

Figure 2.20 Average Share of Defense Revenue of the Top 20 Global Defense Firms in 2013 by Country

Source: Defense News, Top 100. Available at: http://special.defensenews.com/top-100/

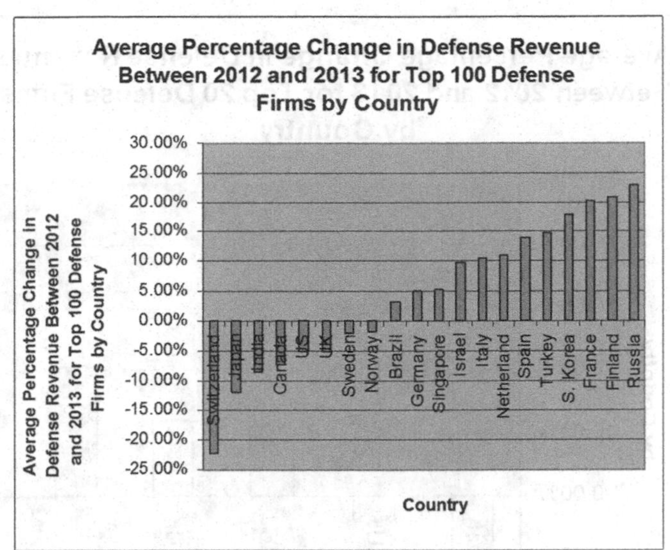

Figure 2.21 Average percentage Growth in Defense Revenues by Country of Top 100 Defense Firms from 2012 to 2013

Source: Defense News, Top 100. Available at: http://special.defensenews.com/top-100/

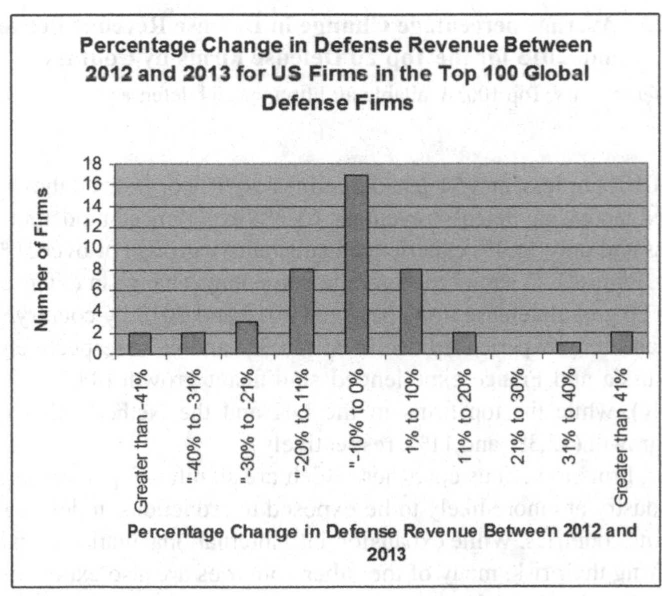

Figure 2.22 Percentage Change in Defense Revenue between 2012 and 2013 for US Firms in the Top 100 Global Defense Firms

Source: Defense News, Top 100. Available at: http://special.defensenews.com/top-100/

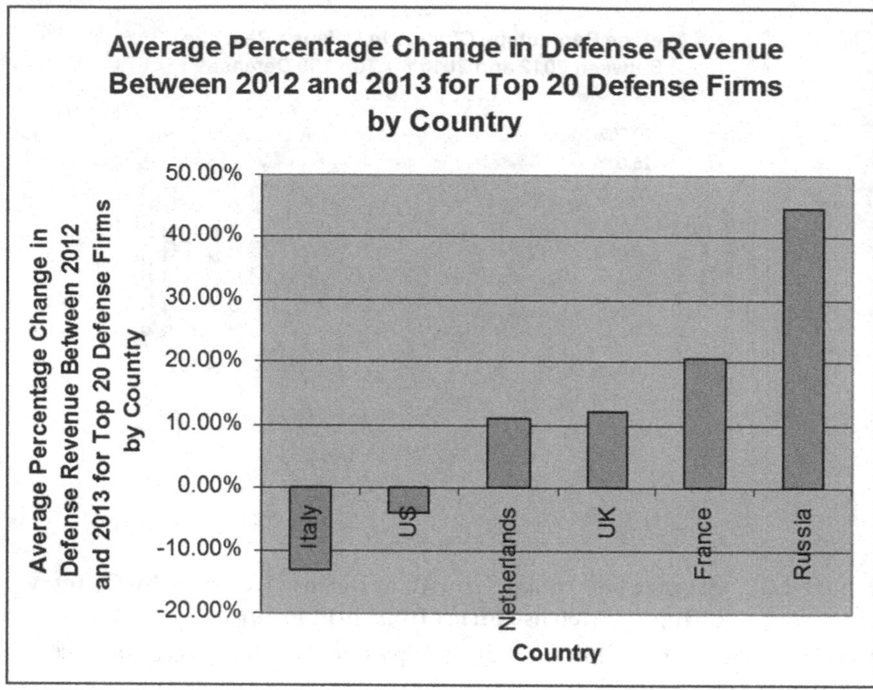

Figure 2.23 Average percentage Change in Defense Revenue between 2012 and 2013 for the Top 20 Defense Firms by Country

Source: Defense News, Top 100. Available at: http://special.defensenews.com/top-100/

declines of 10% or less, and 54% had declines of 20% or less. Of the firms which experienced increasing defense revenues, 61.5% experienced moderate growth of 10% or less and only 15.4% experienced substantive growth of over 50%.

Finally, Figure 2.23 shows the average percentage change in defense revenues for the top 20 global defense firms between 2012 and 2013 by country. The firms in the US and Italy experienced declines: -4.03% and -1%, respectively. The top firms in Russia and France experienced significant growth (44.7% and 20.6%, respectively), while the top firms in the UK and the Netherlands experienced moderate growth (12.3% and 11%, respectively).

In short, firms in various countries which are significant players in the global defense industry are more likely to be exposed to reductions in defense spending of their home countries. While expansion into international markets can be helpful in diversifying their risk, many of the other countries are also experiencing fiscal constraints. As discussed in the previous section, expansion of the firms into commercial markets can mitigate the impact of fiscal constraints. Indeed, about half of the average share of revenue for the firms which are in the top 100 global defense firms in the US is from defense, which suggests the success of expansion into markets which are broader than the US defense markets.

Conclusion

In conclusion, fiscal constraints and evolving defense priorities have impacted the defense industrial base over the years. Demand for the development of new types of equipment has been vital to the continued growth of the defense industrial base. Moreover, the maintenance and sustainment of existing equipment is also necessary for the strength of the defense industrial base. Due to increasing fiscal constraints, however, fewer domestic orders for new equipment are likely. This, in turn, can lead to an attenuation of skillsets as employees retire, which will lead to challenges in the coming years if there is demand for certain types of equipment. Diversification of defense contractors into the commercial sector will help to maintain skillsets, as well as will promote the growth of equipment categories with defense and civilian uses, such as UAVs and cybersecurity. Moreover, expansion of defense contractors into overseas markets will also help to promote the continued production of key equipment. The US defense contractors play a valuable role in the global defense market, and their competitors face challenges similar to those faced by US firms. This chapter evaluated the areas of growth and shrinkage in the US defense industrial base, the challenges faced by the defense industrial base due to budget constraints and shifting defense priorities, the resulting impact of shifts in the defense industrial base on regions and employment, and the tradeoffs between rising costs and new products in response to demand for different types of equipment as defense priorities evolve. The chapter also evaluated the strategies for defense contractors in combating defense industrial pressures by expanding commercial sales and international sales in a global context. The flexibility of the defense industrial base in adjusting to changing constraints is key. The balance between developing new products and maintaining production of more traditional products is vital in promoting the health of the US defense industrial base and in preparing it to meet the challenges of the future.

References

Anonymous, 2009. "United States: A Turn in the South: Charleston." *The Economist*, January 3, Vol. 390, Issue 8612, p. 24.

Barnes, Julian E., 2013. "Warning over Pilot Training." *Wall Street Journal*, December 20, p. A4.

Bennett, John T., 2014a. "Senate Sends Omnibus, Pentagon-Funding Measure to Obama's Desk." *Defense News*, January 16.

——. 2014b. "House Panel Deals US Defense Sector Another Winning Hand." *Defense News*, June 12, pp. 1, 7.

Binns, Jessica, 2008. "A330 Gets Return Ticket." *Traffic World*, October 13.

"Boeing to Deliver Final C-17 Cargo Jet to Air Force." 2013. *Los Angeles Times*, September 12.

Bolkcom, Christopher, 2007. "Air Force Aerial Refueling, *CRS Report for Congress*, March 20. Washington, DC: Library of Congress Congressional Research Service.

Butler, Amy and Norris, Guy, 2009. "A Neglected Base." *Aviation Week and Space Technology*, July 20, 2009, Vol. 171, Issue 3.

"C4ISR Market to Reach $93B." 2014. *Defense News*, January 22.

Cameron, Doug., 2013. "Clipped by U.S., Lockheed CEO Aims Abroad." *Wall Street Journal*, December 7–8, pp. B1 and B3.

Cameron, Doug and Rubin, Ben Fox, 2013. "Lockheed Plans to Cut 4000 Jobs." *Wall Street Journal*, November 15, p. B3.

Cavas, Christopher P., 2013a. "Birth, Rebirth, and Death of a Flattop." *Defense News*, November 3, 2013.

——. 2013b. "Shipyard President: We Are Driving Cost Out of New Carrier." *Defense News*, October 27.

——. 2014a. "Carrier Cut Could Be Back on Table." *Defense News*, January 26.

——. 2014b. "Ingalls' Transition Sees Katrina's Effects Waning." *Defense News*, January 13.

Censer, Marjorie and Peter Whoriskey, 2010. "Defense Cuts Could Slow DC Economy for Years." *Washington Post*, September 11.

Cole, August, 2009. "Makers of Military Drones Take Off." *Wall Street Journal*, August 24, 2009, p. B1.

Colton, Tim, 2014. "Large Shipbuilders." Available at: http://www.shipbuildinghistory.com/history/shipyards [accessed March 10, 2014].

"Defense Industry Belts Getting Tighter." 2013. *St. Louis Post-Dispatch*, September 9.

Defense News Staff, 2014. "2014 Defense Forecast: Many Decisions." *Defense News*, January 4.

Department of Energy, 2013. "Energy Department Announces New Investments of over $30 Million to Better Protect the Nation's Critical Infrastructure from Cyber Attack." September 19.

Drew, Christopher, 2013a. "Boeing Warns Washington State Could Lose 777X if Labor Deal is Rejected." *New York Times*, November 9, p. B3.

——. 2013b. "Lockheed Will Cut 4000 Jobs by 2015." *New York Times*, November 15, p. B2.

Elgin, Ben and Epstein, Keith, 2009. "It's a Bird, it's a Plane, it's Pork." *Business Week*, November 9, 2009, Issue 4154, p. 46.

Erwin, Sandra I., 2013a. "As Fighter Inventory Shrinks, National Guard Eyes Commercial Alternatives." *National Defense*, September 30.

——. 2013b. "Military Space Business Headed in New Direction." *National Defense*, October 13.

——. 2013c. "Naval Aviators' Worst Enemy Is the Budget." *National Defense*, September 12.

——. 2013d. "U.S. Army in the Market for Light Tanks." *National Defense*, October 7.

——. 2013e. "Companies See Bright Spots in Bleak Market." *National Defense*, November.

Everstine, Brian, 2012. "F-15 Upgrades Aim to Double Service Life." *Air Force Times*, October 29.

——. 2014. "USAF Sends Next-Gen Bomber Requirements to Defense Companies." *Defense News*, July 14, p. 4.

Fellman, Sam, 2014. "US Navy's New Plan Aims to Lock in 8-Month Carrier Deployments." *Defense News*, January 15.

Frost, Peter, 2010. "Navy Cautious on Foreign Ownership of Northrop's Shipyards." *Daily Press*, July 19.

Gates, Dominic, 2014. "Boeing, Politicians, Vow to Heal Rift with Machinists over 777X." *Seattle Times*, February 20.

General Dynamics News Release, 2014. "General Dynamics Reports Second Quarter 2014 Results." July.

Greenhouse, Steven, 2013. "Rejection Puts Future Home of Boeing Jet in Doubt." *New York Times*, November 15, p. B6.

Halzack, Sarah, 2013. "Lockheed to Cut 4000 Jobs, Close Facilities Amid Reduced Federal Spending." *Washington Post*, November 15, p. A9.

Herb, Jeremy, 2013. "Defense Firms to Begin Furloughs Monday." *The Hill*, October 4.

Hoffman, Michael, 2013. "Navy Leaders Try To Rescue LCS from its Own Report." *DoD Buzz*, May 9.

Insinna, Valerie, 2013. "Combat Vehicle Manufacturers Adapt to New Market Realities." *National Defense*, September 11.

Johnson, Kirk, 2013. "Boeing Looks Around, and a State Worries." *New York Times*, December 11, pp. A14, A19.

Johnson, Nicole Blake, 2014. "5 Key Facts About DoD's IT and Cyber Budget." *Defense News*, February 3.

Kaplan, Fred, 2009. "Attack of the Drones: Now that Congress Has killed the F-22, the Air Force is Facing Another Shock to the System: Planes without Pilots." *Newsweek*, Vol. 154, Issue 13, September 28.

Larter, David, 2014. "GW Decision Has Ripple Effects." *Defense News*, April 28, p. 13.

Lockheed Martin News Release, 2014. "Lockheed Martin Reports Second Quarter 2014 Results." July 22.

Magnuson, Stew, 2013a. "The MRAP: Reuse, Recycle, Reduce." *National Defense*, February, p. 30.

——. 2013b. "When it Comes to the Navy's Destroyers, it's a Numbers Game." *National Defense*, April.

McLeary, Paul, 2014a. "US Army Chief Confirms: Ground Combat Vehicle is Dead (for Now)." *Defense News*, January 23.

——. 2014b. "US Armored Vehicle Battle Intensifies." *Defense News*, June 30, pp. 2, 7.

Mehta, Aaron, 2014a. "Global Hawk Wins in 2015 Request, Sources Say." *Defense News*, January 26.

——. 2014b. "USAF Leaders Hint at Platforms, Personnel Cuts." *Defense News*, February 3.

——. 2014c. "USAF Faces More Tough Choices in 2016." *Defense News*, May 2, p. 20.

——. 2014d. "Ready for Retirement, Can Predator Find a New Home?" *Defense News*, May 12, p. 18.

——. 2014e. "A-10 Fleet May Be Reprieved, for Now." *Defense News*, May 12, p. 20.

——. 2014f. "Senate Appropriators Likely to Protect A-10 over CRH." *Defense News*, June 16, p. 23.

Mehta, Aaron and Brian Everstine, 2014. "USAF's Hostage: Manned, Unmanned Mix Needed in Iraq." *Defense News*, August 4, p. 32.

Myers, Meghann, 2014. "New Cutters, Arctic Presence Top USCG Priority List." *Defense News*, January 16.

Neely, Meggaen, 2013. "But What About the Bombers? A Reminder of Flexibility in the Triad." CSIS, January 17.

"New Air Force Planes Parked in Arizona 'Boneyard.'" 2013. *Dayton Daily News*, October 6.

Nissenbaum, Dion, 2014. "Navy's Ship of the Future Faces Rough Budgetary Seas." *Wall Street Journal*, November 13, pp. A1, A14.

"Northrop CEO Bush Says Private Equity Firms May Be Interested in Ship Unit." 2010. Bloomberg, July 19.

Osborn, Kris, 2013. "Navy's Plan for 306 Ship Fleet Fading Away." *Military.com News*, July 31.

Ostrower, Jon and Tess Stynes, Tess, 2013. "Commercial Deliveries Rise at Boeing." *Wall Street Journal*, October 4, p. B3.

Parsons, Dan, 2013. "Small Boats Mean Big Business for Shipbuilders." *National Defense*, April.

Parsons, Dan and Insinna, Valerie, 2013. "Army Selects New Helicopter Designs." *National Defense*, October 2.

PRNewswire, 2014a. "Boeing Reports Second-Quarter Results and Raises 2014 EPS Guidance." July 23, 2014. Available at: http://boeing.mediaroom.com.

——. 2014b. "Raytheon Reports Solid Second Quarter 2014 Results." July 24.

Rosenberg, Barry, 2014. "C4ISR Winners and Losers for 2014." *Defense News*, January 27.

Shanker, Thom, 2013. "Pentagon Releases Strategy for Arctic." *New York Times*, November 23, p. A7.

Schmidt, Kathrine, 2010. “Shipyard Shutdown Will Affect Local Workers, State Economy.” *DailyComet.com*, July 18.

Status of the Navy website, 2014. Available at: http://www.navy.mil/navydata/nav_legacy.asp?id=146 [accessed February 21, 2014].

“Tech Titans See Opportunity in Cybersecurity.” 2013. *San Jose Mercury News*, September 18.

US Department of Energy, 2013. “Energy Department Announces New Investments of over $30 Million to Better Protect the Nation’s Critical Infrastructure from Cyber Attack.” September 19.

Warwick, Graham, 2009. “Making it Happen.” *Aviation Week Space and Technology*, June 8, Vol. 170, Issue 23.

Weisgerber, Marcus, 2014. “Companies Show Move Toward Increased R&D Funding.” *Defense News*, August 4, pp. 18, 26.

Whitlock, Craig, 2013. “Air Force May Trim Drone Combat Patrols.” *Washington Post*, November 14, p. A8.

Chapter 3
Defense Industry Consolidation

The question of whether mergers in various industries lead to higher prices through increased market power or to lower prices and lower costs through greater scale efficiencies, through improved management, and/or through a greater knowledge pool has been the subject of numerous studies and public policy debates. The US economy has faced a series of merger waves over the past century which have provided the context for examining the causes and effects of mergers and acquisitions in various industries. This analysis focuses on the causes and consequences of the wave of defense mergers in the US during the 1990s. The magnitude of the consolidation in the defense sector from this wave was such that by 2001, the top five defense contractors received 34% of all prime contract awards, which was the same share that the top 10 defense firms in 1985 had received.[1] Moreover, 51 separate defense firms or units as of 1980 had been folded into the top four defense firms by 2001.[2] As the number of large defense contractors has declined, key public policy questions have arisen concerning whether the mergers have led to greater efficiencies, lower costs, and improvements in quality, or whether they have led to higher costs, fewer choices, and larger firms with unwieldy organizational structures. These concerns continue due to the fiscal constraints impacting the defense budget, such that defense contractors will likely be left with excess capacity (although to a lesser degree than following the end of the Cold War) and may consider consolidation as a means of generating efficiencies and improving profitability, in addition to expanding civilian equipment sales and overseas sales.

Many studies in other industries have found positive gains from mergers through combining complementary skill sets to generate economies of scope and new products, through combining fixed costs, including rationalization of production facilities to achieve economies of scale and density, or through entering new markets. Studies which have found that mergers generated value include Healy, Palepu, and Ruback (1990), who found improvements in pre-tax returns for 50 large mergers between 1979 and 1983. Jensen and Ruback (1983) surveyed the empirical evidence and found that "corporate takeovers generate positive gains," while Hall (1988) and others have found evidence that R&D spending and capital spending were not reduced following a merger. Gaughan (1999) argued that "there is little empirical evidence that firms combine to increase their monopoly power" and cited the results of Stillman (1983) and Eckbo (1983).

1 *Annual Industrial Capabilities Report to Congress*, 2003, p. 5.

2 *Annual Industrial Capabilities Report to Congress*, 2002, p. 2.

Nevertheless, other studies across industries suggest that mergers may not be beneficial. Roll (1986) argued that "takeover gains may have been overestimated if they exist at all." Various analyses, such as Barton and Sherman (1984), found that price increased as competition fell. Mueller (1980) found "no consistent pattern of either improved or deteriorated profitability" in his study of mergers across seven countries, and that "any economic efficiency gains from the merger would appear to be small … as would any market power increases,"[3] while Meeks (1977) concluded:

> the efficiency gains, which in public policy statements have been assumed to be the saving grace of growth by takeover, cannot … be relied upon; strong evidence was reported that the efficiency of the typical amalgamation did not improve after merger … it actually appears to have declined.[4]

Did the wave of defense mergers in the 1990s lead to cost savings for DoD? According to the *Los Angeles Times* in October, 1999:

> Almost a decade of consolidation in the defense industry has failed to deliver the benefits of lower costs for the Pentagon. And the mergers of the '90s that were supposed to produce stronger and more innovative defense contractors have more often caused corporate indigestion.[5]

Industry observers argued that innovation had suffered from the mergers, and that the companies had become too big and were expending significant effort in managing themselves.[6] Indeed, DoD, in its *Annual Industrial Capabilities Report to Congress* for January, 2001, noted that "industry restructuring has not kept pace with reduced demand in several core defense sectors."[7] On the other hand, in the *Annual Industrial Capabilities Report to Congress* for February, 2000, DoD noted that: "Consolidation has produced significant efficiencies and cost savings for DoD."[8]

The purpose of this chapter is to examine:

a. the reasons behind the merger wave and the roles of defense spending and broader merger activity in the economy on the frequency and size of defense mergers;
b. the patterns of defense mergers and some of the related antitrust concerns;

3 From Mueller, 1980, p. 306; Quoted in Scherer and Ross, 1990, p. 173.
4 From Meeks, 1977, p. 66. Quoted in Scherer and Ross, 1990, p. 173.
5 "A Decade of Defense Mergers Yields Disappointments," 1999.
6 Ibid.
7 *Annual Industrial Capabilities Report to Congress*. 2001, pp. 15–16.
8 *Annual Industrial Capabilities Report to Congress*. 2000, p. 9.

c. the reasons behind some of the mergers and the qualitative evidence on the success of the consolidations; and
d. quantitative evidence on the impact of mergers of major defense contractors on the costs of weapons systems facing DoD.

The chapter assesses the degree to which the costs faced by DoD for specific weapons systems, services, and weapons systems categories increased or decreased following a merger or acquisition of its primary manufacturer.

There have been few academic studies that quantitatively assess the impact of defense industry consolidation. Gholz and Sapolsky's (1999) qualitative study discusses workforce considerations in the defense consolidation arena and suggests that workers and local officials lobbied against post-merger restructuring and that contractors were consequently less likely to rationalize plant capacity due to the reaction of Congress in restricting consolidation cost reimbursement. The authors suggest that defense mergers did not lead to many plant closures, but that non-defense mergers did.[9] Dowdy (1997) is more optimistic in his qualitative analysis and cites the efficiencies from the acquisition of General Dynamics Air Defense systems and Convair divisions by Hughes Aircraft in terms of increased plant utilization, staff reductions, and an almost 40% reduction in unit costs.[10] Other qualitative studies discuss the role of institutions in contributing to defense consolidation (Driessnak and King, 2004), provide a general discussion of competition policy (Kovacic and Smallwood, 1994), or describe the changes in the global weapons system industry and/or compare US consolidation activity with European efforts and those of the former Soviet Union (Markusen, 1999; Dowdy, 1997). Although there have been studies examining the cost growth of weapons systems over time, the impact of policies to reduce cost growth, and the sources of cost growth, none of these studies addressed the impact of defense mergers on weapons systems' costs and consequently focused on different questions than this analysis. Examples of these types of studies focusing on cost growth include: Arena, Leonard, Murray, and Younossi (2006), Tyson, Nelson, Om, and Palmer (1989), Tyson, Harmon, and Utech (1994), Drezner, et al. (1993), and McNicol (2004).

This chapter is organized into several sections. The first section examines the relative importance of reduced defense spending and broader merger activity on the wave of defense consolidation. The second section discusses the patterns in defense merger activity across sectors and the associated antitrust concerns, as well as examines the reasons behind some of the mergers and qualitative evidence on the success of the consolidation. The third section examines quantitative evidence on the impact of mergers on total costs and cost per unit of output borne by DoD for various types of weapons systems made by various contractors. The impact of consolidation on per unit costs and total costs is examined by weapons

9 Gholz and Sapolsky, 1999/2000, pp. 23–29.

10 Dowdy, 1997, pp. 96–7.

system type, by contractor, and by service (Army, Navy, and Air Force). Finally, the fourth section provides conclusions.

The Impact of Defense Spending and Broader Merger Activity on Defense Mergers

Industry observers often cite defense spending and overall merger activity as the two forces behind defense sector mergers.[11] But is defense merger activity more linked to the level of DoD spending or to the overall level of merger activity in the economy? Which one of these considerations is more significant?

Overall defense spending, as well as defense procurement spending, grew rapidly during the 1980s, declined following the end of the Cold War, increased towards the end of the 1990s, and exhibited significant growth with the War on Terrorism. Indeed, overall defense spending grew 73.5% and defense procurement spending grew 133.1% between 1981 and 1991, while, between 1992 and 1996, overall defense spending fell 10.9% and defense procurement spending fell 34.7%. Between 1997 and 2001, overall defense spending and defense procurement spending grew 12.7% and 15.3%, respectively, while between 2002 and 2006, overall defense spending and defense procurement spending grew at 49.7% and 43.6%, respectively.[12]

The defense merger wave also occurred around the same period as the overall merger wave in the economy. As is evident in Table 3.1, growth in merger activity in the defense sector, whether measured by growth in value or growth in number of transactions, was generally lower than growth in merger activity in the overall economy. Indeed, growth in merger activity in the defense sector only exceeded growth in merger activity in the industry overall (or exhibited less negative growth) in terms of the number of transaction and in terms of value in five out of the 12 years (41.67%).

The results in Table 3.2 suggest that merger activity is much more strongly linked to overall activity in the economy. Table 3.2 provides correlations between various measures of defense merger activity and merger activity in the overall economy, as well as between defense merger activity and DoD spending, The correlations use data covering the period between 1992 and 2004. The correlations between defense merger activity (regardless of how it is measured) and DoD outlays (regardless of whether it is overall levels or procurement levels, and whether it is in the current year or the previous year) are negative, as would be expected – as defense spending goes down, defense merger activity goes up. Correlating previous year DoD overall outlays and procurement outlays with

11 Korb, 1996.

12 These growth rates were calculated by the author from the raw data in the Historical Tables (Table 3.2) for the *United States Budget for Fiscal Year 2008*, pp. 56–60. The growth rates are not annualized nor adjusted for inflation.

Table 3.1 Annual Growth Rates in Merger Activity in the Defense Sector and in the Overall Economy

Time Period	Annual growth rates for merger activity (number of transactions) in the defense sector	Annual growth rates for merger activity (number of transactions) in the overall economy	Annual growth rates for merger activity ($ value) in the defense sector	Annual growth rates for merger activity ($ value) in the overall economy
1992–1993	-44.83%	4.008%	-82.37%	45.41%
1993–1994	-6.25%	12.66%	268.1%	80.63%
1994–1995	-33.00%	17.37%	-94.13%	30.94%
1995–1996	100.0%	66.51%	8571.4%	110.8%
1996–1997	50.00%	33.32%	-46.96%	35.68%
1997–1998	70.00%	0.154%	-59.25%	83.41%
1998–1999	0.00%	18.94%	169.0%	19.16%
1999–2000	-29.4%	3.28%	392.8%	832.9%
2000–2001	-5.5%	-13.37%	-97.03%	-94.72%
2001–2002	26.47%	-12.06%	164.7%	-37.42%
2002–2003	-34.88%	9.573%	-55.97%	15.14%
2003–2004	-10.7%	22.66%	50.50%	48.78%

Note: These annual growth rates were calculated by the author from raw data found in the *Mergerstat Review for 2005*, the *Mergerstat Review for 2002*, the *Mergerstat Review for 1997*, and the *Mergerstat Review for 1996*. The defense sector, as defined by Mergerstat, encompassed firms in Standard Industry Classification (SIC) codes 3761-3769, 3721-3728, and 3795.

current year merger activity (in terms of either transactions or value) yields a stronger relationship than correlating current year outlays with current year merger activity. This suggests that, since the merger process requires time, mergers are a delayed response to spending levels in previous years.

The correlations are strongly positive between merger activity in the defense sector and merger activity in the overall economy in a given year (excluding defense mergers) – as one increases, the other increases. The correlation is strongly positive between number of defense mergers and number of mergers in the economy overall (excluding defense mergers) at 0.6498, while the correlation is very strongly positive between the dollar value of mergers in the overall economy (excluding defense mergers) and the dollar value of defense mergers at 0.9399.

In summary, Table 3.2 suggests that although the wave of defense mergers was driven by both DoD spending and by overall economic merger activity, overall economic merger activity was much more strongly correlated. Consequently, the

Table 3.2 Correlations between DoD Outlays, Merger Activity in the Economy, and Merger Activity in the Defense Sector

Correlation between ...	Number of defense merger transactions in a given year	Dollar value of defense merger transactions in a given year
Level of overall DoD outlays in a given year	-0.0269	-0.2058
Level of DoD procurement outlays in a given year	-0.3591	-0.3783
Level of overall DoD outlays in the previous year	-0.1929	-0.2947
Level of DoD procurement outlays in the previous year	-0.6097	-0.3916
Number of mergers in the overall economy in a given year	0.6498	
Dollar value of mergers in the overall economy in a given year		0.9399

Note: The statistical correlations were calculated by the author from raw data found in the Historical Tables (Table 3.2) for the Budget for Fiscal Year 2008, pp. 56–50, and from data in the raw data found in the *Mergerstat Review for 2005*, the *Mergerstat Review for 2002*, the *Mergerstat Review for 1997*, and the *Mergerstat Review for 1996.*

decline in Cold War spending and its impact on excess capacity was less important than overall economic growth, stock market conditions, and the need for defense firms to defensively merge as their rivals merged so that they would not be left out in the cold as a relatively smaller firm facing larger, consolidated competitors.

Patterns of Defense Consolidation and Antitrust Concerns

In July, 1993, Deputy Defense Secretary William Perry met with representatives of the major defense contractors at a summit known as the "Last Supper" and encouraged significant defense sector consolidation to rationalize the excess capacity left in the industry following the end of the Cold War and the consequent decline in the defense budget.[13] As is evident in Table 3.3, the wave of consolidation which ensued led to substantive changes in the industry. Between 1990 and 1998, the number of prime contractors decreased substantially due to consolidation in 10 of the 12 key defense sectors identified by DoD. The percentage reduction in

13 "Jumping the Gun: How Lockheed Martin Misread the Radar on the Northrop Merger," 1998; "War of Attrition: Defense Consolidation Rushes Toward an Era of Only 3 or 4 Giants," 1996.

Table 3.3 Reduction in Prime Contractors in Various Weapons Systems Sectors between 1990 and 1998

Sector	Number of prime contractors in 1990	Number of prime contractors in 1998	Percentage reduction
Tactical Missiles	13	4	-69.2%
Fixed Wing Aircraft	8	3	-62.5%
Expendable Launch Vehicles	6	2	-66.7%
Satellites	8	5	-37.5%
Surface Ships	8	5	-37.5%
Tactical Wheeled Vehicles	6	4	-33.3%
Tracked Combat Vehicles	3	2	-33.0%
Strategic Missiles	3	2	-33.0%
Torpedoes	3	2	-33.0%
Rotary Wing Aircraft	4	3	-25.0%

Note: Data on the sectors and the number of contractors in 1990 and 1998 are derived from the *General Accounting Office (GAO) Report to Congressional Committees on the Defense Industry: Consolidation and Options for Preserving Competition,* Washington, DC: April, 1998.

contractors exceeded 60% in three of the 10 sectors (tactical missiles, fixed wing aircraft, and expendable launch vehicles) and varied between 25% and 37.5% in the remaining seven of the 10 sectors. The major giants which emerged out of this consolidation across these sectors were Boeing, Lockheed Martin, and Northrop Grumman, and, to a lesser degree, Raytheon and General Dynamics.

With the increasing numbers of defense mergers in the mid to late 1990s, the Antitrust Division of the Department of Justice (DOJ) and the Federal Trade Commission (FTC) became more concerned that consolidation was leading to a reduction in competition and an increase in anticompetitive activity. As Joel Klein, Assistant Attorney General of the Antitrust Division of the DOJ, noted in his address before the Senate Judiciary Committee in June, 1998: "A number of defense mergers proceeded unchallenged over the last five years, which rationalized capacity, but if that rationalization goes too far, it can harm competition." [14] Indeed, DOJ had challenged two mergers in 1997 – Raytheon's acquisition of Hughes Aircraft (the aircraft subsidiary of General Motors) and Raytheon's acquisition of the defense electronics division of Texas Instruments – but then allowed both of them to go through provided that divestitures of certain key divisions occurred prior to the merger, in order to protect competition. In 1998, however, DOJ blocked the merger between Lockheed Martin and Northrop Grumman, since DOJ

14 Klein, 1998, p. 7.

believed that the merger would lead to a reduction in competition and innovation in submarine sonar systems, military aircraft radar, and various electronic warfare systems. This proposed $11.6 billion acquisition was the largest acquisition that DOJ had challenged in its history up to that point,[15] and the challenge was supported by the Pentagon since Defense Secretary Cohen also thought that the merger would be anticompetitive.[16] Lockheed and Northrop called off the merger in July, 1998, prior to their September trial date.[17] Similarly, DoD opposed the acquisition of Newport News Shipbuilding by General Dynamics due to concerns about reductions in competition.[18]

Consolidation could contribute to a negative outcome at either the vertical or the horizontal level. Vertical mergers might lead to foreclosure to competitors of key input suppliers or distributors along the vertical supply chain. For example, one of the concerns about the proposed Lockheed Martin-Northrop Grumman merger had been that Lockheed Martin would have control of a key supplier of electronics which supplied Boeing's planes and its own planes. This could enable it to limit Boeing's access to the supplier. On the other hand, Lockheed argued that the Pentagon could monitor the selections of equipment from outside suppliers and that the process was sufficiently transparent that this would not be an issue. Indeed, Lockheed argued that the mission computers in its F-16 planes came from Raytheon.[19] The announcement by the CEO of McDonnell Douglas in April, 1996 that McDonnell Douglas would stop buying parts from Loral for its jet fighters once Lockheed Martin acquired Loral is another example of concerns over vertical integration. Paul Kaminski, the chief of procurement at the Pentagon, wrote to McDonnell Douglas, stating that this could "increase the cost or lower the quality of the products you supply" and that if the best product is offered by a given supplier, which "happens to be Loral, then McDonnell Douglas should continue to buy from that company." [20]

Horizontal mergers, in the absence of viable international competition or entry by new companies, could lead to increased market power and higher prices in certain sectors. For example, one of the concerns with Raytheon's acquisition of Hughes Aircraft and the defense divisions of Texas Instruments in 1997 was that these acquisitions would provide Raytheon with a near monopoly position in spy satellite sensors, night vision equipment, and air-to-air missiles. Hughes and Raytheon had previously been strong competitors for missile contracts and,

---

15 Klein, 1998, pp. 7–8.

16 "Jumping the Gun: How Lockheed Martin Misread the Radar on the Northrop Merger," 1998.

17 "L6bn US Defense Merger Called Off," 1998.

18 *Annual Industrial Capabilities Report to Congress*, 2002, p. 2.

19 "Jumping the Gun: How Lockheed Martin Misread the Radar on the Northrop Merger," 1998.

20 "War of Attrition: Defense Consolidation Rushes Toward an Era of Only 3 or 4 Giants," 1996.

according to the chief of acquisitions at the Pentagon, Paul Kaminiski, "their competition saved taxpayers hundreds of millions of dollars, shaving 70% from Hughes' original price." Raytheon, however, had argued that other companies had competed in missile competitions and had won, citing McDonnell Douglas' and Lockheed Martin's success in bidding for the JASSM missile contract.[21]

On the other hand, vertical and horizontal consolidation can lead to efficiencies. For example, more innovative or less costly weapons systems could be developed due to greater pooling of knowledge between consolidating contractors. One instance was when Boeing, which had acquired Rockwell and McDonnell Douglas, succeeded over Lockheed in winning a $5 billion contract for a National Reconnaissance Satellite in 1999. At the time, some argued that the combination of knowledge and talent between McDonnell Douglas, Rockwell, and Boeing enabled the unified entity to win the contract and that this would not have been possible without consolidation.[22] A second example is when the Navy in early September, 1997 thought that the proposed merger between Lockheed Martin and Northrop Grumman would have actually enabled Lockheed, which had a weaker background in building naval aircraft, to compete more effectively against Boeing in the competition for the new Joint Strike Fighter.[23] The merger, as discussed earlier, did not take place.

Vertical or horizontal consolidation could also lead to the development of new divisions and a greater focus of the acquiring company on an area in which it had not previously had much expertise. For example, the acquisition of Loral by Lockheed Martin was driven by a desire for greater vertical consolidation in platforms and a greater focus on the creation of a separate communications and space group – Loral Space and Communications, in which Lockheed would have a 20% share and which would become part of the tactical systems sector within Lockheed Martin.[24] A second example was evident in the merger between Boeing and McDonnell Douglas.[25] The efficiencies from that merger were not driven by consolidation in work forces and layoffs since the overlap in military programs between the two companies was minimal. The merger between the two resulted in a greater presence of Boeing in the military sector, such that it was involved with many of the Pentagon's development programs in tactical aviation and helicopters, had a greater involvement in the missile area, and continued its expansion in the space area, which was also enhanced by Boeing's purchase of Rockwell's defense and space units.[26]

---

21 "Raytheon Deal Raises Antitrust Concerns," 1997.

22 "A Decade of Defense Mergers Yields Disappointments," 1999.

23 "Jumping the Gun: How Lockheed Martin Misread the Radar on the Northrop Merger," 1998.

24 "Business Focus, Mergers Becoming a Business Imperative," 1996.

25 Rogers, 1997.

26 Evers and Starr, 1997.

Consolidation activity could also lead to improved cost efficiencies from reduced overhead costs – combining duplicative facilities and corporate headquarters, rationalizing and reducing the workforce, pooling R&D funds, and more effectively using pre-existing capacity. Indeed, when the Lockheed-Martin Marietta merger took place in 1995, it was estimated that merging telecommunications operations, research divisions, and headquarters would save $3 billion over the following five years.[27] A second example was in the consolidation of some of Boeing's defense and space work in St. Louis and in southern California from Puget Sound to achieve overhead efficiencies from the Boeing/McDonnell Douglas merger and from the Boeing/Rockwell merger. For example, the B1/B2 program management moved to Long Beach with support work being shifted from Puget Sound to Oklahoma City and to Long Beach.[28]

Nevertheless, some of the mergers clearly failed to yield their projected savings. For example, according to the GAO,[29] Martin Marietta's 1993 acquisition of General Electric Aerospace had only yielded half of the expected cost savings three years later. Two years after the union of Hughes Aircraft and General Dynamics' missile division in 1992, the Inspector General could not verify that the consolidation had saved the projected $600 million for the Pentagon.[30]

The issue of whether DoD recognized cost savings from the wave of consolidation was further complicated by its decision to pay the restructuring costs of consolidation beginning in July, 1993, provided that certain conditions from the consolidation were met, such as that the projected savings from the restructuring would exceed the costs. Without recovering its restructuring costs from the government, the CEO of Martin Marietta, Norman Augustine, told the House of Representatives that Martin Marietta would not have paid $208 million for General Dynamics Space Systems. Martin Marietta had believed that, by consolidating the Atlas and the Titan IV space launch programs, they could save $500 million.[31]

Under the 1997 DoD Appropriations Act, projected savings needed to exceed costs by a ratio of two to one for business combinations occurring after September 30, 1996, in order for restructuring costs to be reimbursed.[32] In 1997, DoD calculated that, through September 30, 1996, for every $1.00 that it paid in restructuring costs, it estimated $1.93 in savings because it had paid $179.2 million in restructuring costs and realized savings of $346.7 million.

Nevertheless, it is still unclear whether the mergers and the reimbursement of restructuring costs saved money for the government. In several of the five

27 "FTC Staff Supports Big Defense Merger: Conditions Placed on Martin-Lockheed Deal," 1994.

28 "Boeing Restructures Manufacturing Operations," 2000.

29 "GAO: Savings from Defense Merger are Less Than Predicted," 1996.

30 Korb, 1996.

31 "Recouping the Cost of Defence," 1994.

32 Cooper, 1997, pp. 1, 3.

business combinations reviewed, savings was much less than the contractors had actually estimated. For Lockheed Martin, the estimated savings used to certify the Lockheed-Martin Marietta merger as eligible for restructuring, as of September 30, 1996, was less than half of the original savings estimate which had originally been projected.[33]

Analysis of Cost Data on Weapons Systems by Type and by Defense Contractor

The preceding sections examined qualitative evidence on the impact of the defense mergers on efficiencies. This section examines quantitative evidence on whether weapons systems, whose manufacturers merged or were engaged in an acquisition, exhibited lower per unit costs or lower total costs following the merger/acquisition. To the extent that horizontal and vertical mergers between defense contractors occurred, to what degree were efficiencies realized in the form of lower weapons systems costs?

The analysis used cost data from the summary tables in the Selected Acquisition Reports (SARS) which are submitted to Congress by DoD and which report the acquisition costs of Major Defense Acquisition Programs (MDAPS).[34] Each SAR contains various items on the mission of the weapons system and the contractors involved, as well as data on the costs of the weapons system, including baseline cost estimates and quantity estimates, current cost estimates and quantity estimates, and a decomposition of cost changes into quantity cost changes, schedule cost changes, engineering cost changes, support cost changes, estimating cost changes, and other cost changes. The period covered in the SAR data used in this analysis encompassed March, 1981 until June, 2006.[35]

---

33 Cooper, 1997, p. 8.

34 MDAP (Major Defense Acquisition Program) – "Defined in 10 USC § 2430 as a Department of Defense (DoD) acquisition program that is not a highly sensitive classified program (as determined by the Secretary of Defense) and that is designated by the Secretary of Defense as a major defense acquisition program, or that is estimated by the Secretary of Defense to require an eventual total expenditure for research, development, test, and evaluation of more than $365,000,000 (updated to FY 2000 constant dollars) or an eventual total expenditure for procurement of more than $2,190,000,000 (updated to FY 2000 constant dollars)." *Fiscal Year 2006 Major Defense Acquisition Program (MDAP) Lists*, 2006.

35 Paul G. Hough's 1992 RAND note "Pitfalls in Calculating Cost Growth from Selected Acquisition Reports" describes many of the difficulties in using SAR data, many of which don't apply to the type of data used in this study nor the analyses undertaken. The RAND note focuses on the problems in using SAR data to calculate cost growth and which estimate should be used as the baseline, while this analysis does not calculate cost growth at all. It also discusses the problems stemming from SARS not reporting base-year dollars prior to March, 1974. This analysis uses data after that period. The RAND note

Data on 304 systems with 6,840 observations was analyzed in this chapter. Systems which had six observations or fewer (i.e. they only had data for six quarterly SAR reports) were eliminated because the data was insufficient to undertake statistical analyses. Systems whose contractors did not merge during the period covered by the data were also eliminated. Finally, systems which had few or no pre-merger or post-merger data points were eliminated. This left the 45 systems used in this study. Eight of the weapons systems experienced at least two mergers during the period covered by the data, and the impact of each merger was counted separately, resulting in 53 systems. Careful attention was paid to whether the system was re-designed or the costs were re-baselined. If there were changes in the base year of the weapons system or adjustments to the name of the weapons system, then only data points for the identical name of the system with the identical base year were used, since any changes in the name or the year in the data usually indicated a major re-design or a new generation of the system which could have impacted costs separately. Consequently, for all of the systems examined in this analysis:

a. the primary contractor was involved in a merger with a major defense contractor during the period covered;[36]
b. there was enough time series data to examine the pre-merger and the post-merger period; and
c. there was no major system re-design during the period covered.

The effective dates that the mergers were completed were used in this analysis, not the announcement dates of the mergers, because it was assumed that the impact of the merger would be felt after the merger was completed, not after it was merely announced. The dates of completion were verified using Thomson Financial's SDC Platinum database, Bloomberg's database, and various corporate news releases.

Several measures were examined over time for each weapons system/program: the current year cost estimates in base year dollars of each weapons system/program over time, the quantity estimates of each weapons system, and the cost per unit of output (computed by the author as the current cost estimates in base year dollars divided by the current quantity estimates). Current cost estimates in base

notes that "when OSD escalation indexes are lower than actual inflation (the norm), cost ratios calculated from SAR base year dollars are higher" (27), but that "OSD inflation rates do capture the bulk of price level changes in the economy" (29). This is not a significant concern for this analysis because it does not calculate cost ratios and compare current estimates in base year dollars with current estimates in current year dollars. The RAND note points out that program managers reporting the costs of their programs could make errors or have inconsistent preparation techniques. This is a minimal problem here because this analysis does not compare different categories of costs.

36 Note that the CVN76 and the CVN77 were eliminated from the analysis because, during the chronological period, their primary manufacturer, Newport News, did not merge with anyone. Rather, Newport News was spun-off in 1996 from its owner, Tenneco.

year dollars capture overall pre and post-merger effects better than other variables in the SARS which decompose the cost change into quantity changes, schedule changes, engineering changes, etc. A merger could impact cost estimates through any of these avenues, so year-to-year changes in overall current year cost estimates in base year dollars provided a useful measure. Current year cost estimates in base year dollars minimize the impact of inflation since they are indexed to a base year. Nevertheless, these total cost estimates can be higher or lower following a merger based on changes in estimated quantity. Consequently, cost per unit of output is also an important metric to examine in evaluating efficiency since it provides information on the costs faced by DoD on a per unit basis for each type of weapons system.[37] The data were logged to minimize heteroskedasticity.

This analysis used two types of regression models for each of the 53 weapons systems/programs. The first model regressed the natural log of current year cost estimates in base year dollars for a given weapons system on a time trend variable and on an indicator variable that took on the value of "1" after the merger of its primary contractor and "0" before the merger. The time trend controlled for the natural increases in costs and quantity over time. The first regression model appears below:

1. *Ln(Current year cost estimates in base year dollars)i* $= \alpha + \beta 1$ *(time trend) i* + $\beta 2$ *(post-merger indicator variable)i*

The second regression model regressed the natural log of current cost per unit output in base year dollars for a given weapons system on a time trend variable and on an indicator variable that took on the value of "1" after the merger of its primary contractor and "0" before the merger. Both of the regression models enable one to see if there is a statistically significant break in the trend in the data over time – does the pattern in the data before the merger differ statistically significantly from the pattern following the merger, controlling for time?

2. *Ln(Estimated current cost per unit output in base year dollars)*$_i = \alpha + \beta_1$ *(time trend)*$_i + \beta_2$ *(post-merger indicator variable)*$_i$

Both regression models were run over the time series data for each weapons system and the author did several robustness checks. First, other versions of the models were tested, including ones with an interaction term between the time trend and the post-merger indicator variable, with little change in the results. Second, the results reported throughout this period assumed that the post-merger effect began to take place beginning with the report date of the SAR nearest chronologically to

37 DoD Instruction 7000.3g, as well as DoD *Cost Variance Handbook* and the training slides for the CARS software program used by program managers in their SAR preparations were very helpful to the author in explaining the various types of data found in SAR reports, how they are calculated, and how they are reported by program managers.

Table 3.4 Overall Impact of Defense Consolidation on Total Costs and Per Unit Costs

	Percentage of systems experiencing a positive and statistically significant change	Percentage of systems experiencing a negative and statistically significant change	Percentage of systems experiencing a positive or a negative, statistically significant change
Total cost	17.54%	26.32%	43.86%
Cost per unit	18.37%	20.41%	38.78%

Note: Note that statistical significance is at the 5% level.

the effective date of the merger, however other lags were tested as well, with little change in the results.[38]

Table 3.4 summarizes the overall results of the regression models, across weapons systems and manufacturers, on whether the wave of consolidation in the defense sector led to greater efficiencies in the form of statistically significantly (at the 5% level) lower per unit costs and lower total costs.

Several observations emerge from Table 3.4. First, a number of systems were unaffected by mergers. Between 39% and 44% of systems experienced some statistically significant change in either total costs or per unit costs following a merger, which means that between 56% and 61% of the systems were unaffected by mergers. Second, more systems exhibited a statistically significant decrease in total costs and in per unit costs than an increase, suggesting that consolidation was more likely to lead to efficiencies, although the percentages were somewhat close. These results are consistent with the overall literature on mergers and acquisitions across industries which suggests that consolidation can be welfare-enhancing or generate positive gains; examples include Healy, Palepu, and Ruback (1990), Jensen and Ruback (1983), and Gaughan (1999). Nevertheless, the results also suggest that many systems did not manifest significant efficiencies following consolidation.

Tables 3.5a and 3.5b examine the pre- and post-merger patterns in the cost data by contractor. Since many of the manufacturers of these weapons systems served as both targets and acquirers in various consolidations, the first three columns examine whether there was a statistically significant increase (at the 5% level), a statistically significant decrease, or a statistically significant change in total costs (Table 3.6a) or in cost per unit (Table 3.6b) if the manufacturer of the weapons

38 This specification was helpful in examining the impact of mergers on total costs and per unit costs on a weapons system by weapons system basis. The author also pooled all the data and ran regression specifications which provided results consistent with the results reported in this analysis, but which lacked the degree of detail on a per weapons system basis that the specifications reported in this study had.

Table 3.5a Impact of Defense Consolidation on Total Costs by Contractor and by Acquirer or Target

Manufacturer Name	Percentage of systems with a positive statistically significant change following mergers in which the manufacturer was an acquirer	Percentage of systems with a negative statistically significant change following mergers in which the manufacturer was an acquirer	Percentage of systems with a positive or negative statistically significant change following mergers in which the manufacturer was an acquirer	Percentage of systems with a positive statistically significant change following mergers in which the manufacturer was a target	Percentage of systems with a negative statistically significant change following mergers in which the manufacturer was a target	Percentage of systems with a positive or negative statistically significant change following mergers in which the manufacturer was a target
Raytheon	33.33%	11.11%	44.44%	NA	NA	NA
Lockheed	11.11%	33.33%	44.44%	0%	0%	0%
Boeing	9.09%	9.09%	18.18%	NA	NA	NA
General Dynamics	0%	50%	50%	0%	50%	50%
Northrop Grumman	0%	33.3%	33.3%	NA	NA	NA
Northrop	0%	100%	100%	NA	NA	NA
Hughes Aircraft	0%	0%	0%	37.5%	25%	62.5%
Martin Marietta	0%	100%	100%	14.29%	57.14%	71.43%
Loral	NA	NA	NA	25%	25%	50%
Rockwell	NA	NA	NA	0%	0%	0%
McDonnell Douglas	0%	0%	0%	33.3%	0%	33.3%
E-Systems	NA	NA	NA	25%	0%	25%
Texas Instruments	NA	NA	NA	50%	0%	50%
Other	20%	20%	40%	20%	30%	50%

Note: Note that statistical significance is at the 5% level.

Table 3.5b Impact of Defense Consolidation on Per Unit Costs by Contractor and by Acquirer or Target

Manufacturer Name	Percentage of systems with a positive statistically significant change following mergers in which the manufacturer was an acquirer	Percentage of systems with a negative statistically significant change following mergers in which the manufacturer was an acquirer	Percentage of systems with a positive or negative statistically significant change following mergers in which the manufacturer was an acquirer	Percentage of systems with a positive statistically significant change following mergers in which the manufacturer was a target	Percentage of systems with a negative statistically significant change following mergers in which the manufacturer was a target	Percentage of systems with a positive or negative statistically significant change following mergers in which the manufacturer was a target
Raytheon	12.5%	37.5%	50%	NA	NA	NA
Lockheed	11.11%	22.22%	33.33%	100%	0%	100%
Boeing	9.09%	18.18%	27.27%	NA	NA	NA
General Dynamics	0%	50%	50%	0%	0%	0%
Northrop Grumman	33.33%	0%	33.33%	NA	NA	NA
Northrop	0%	0%	0%	NA	NA	NA
Hughes Aircraft	0%	0%	0%	0%	33.33%	33.33%
Martin Marietta	0%	0%	0%	42.86%	42.86%	85.71%
Loral	NA	NA	NA	0%	0%	0%
Rockwell	NA	NA	NA	0%	0%	0%
McDonnell Douglas	0%	100%	100%	16.67%	16.67%	33.33%
E-Systems	NA	NA	NA	33.33%	33.33%	66.67%
Texas Instruments	NA	NA	NA	0%	50%	50%
Other	20%	20%	40%	10%	10%	20%

Note: Note that statistical significance is at the 5% level.

Table 3.6 Impact of Selected Defense Mergers on Weapons Systems Cost Estimates

	Percentage of systems made by the defense contractors involved in a specific merger which experienced a statistically significantly higher cost estimate post-merger	Percentage of systems made by defense contractors involved in a specific merger which experienced a statistically significantly lower cost estimate post-merger	Percentage of systems made by the defense contractors involved in a specific merger which experienced a statistically significantly different estimate post-merger (higher or lower)
Lockheed/Martin Marietta (March 16, 1995)			
ANSQQ892			
DMSP2			
Trident1			
ASAS			
F-22			
Longbow Hellfire			
Titan IV			
Total cost estimates	14.29%	57.14%	71.43%
Cost per unit output	42.86%	42.86%	85.71%
Boeing/McDonnell Douglas (August 1, 1997)			
Commanche2			
FA-18 E/F			
AV-8B Remanufactured			
C-17A			
Longbow Apache			
Minuteman			
T45TS			
AWACS RSIP2			
Total cost estimates	12.5%	12.5%	25%
Cost per unit output	14.29%	28.57%	42.85%

Note: Note that statistical significance is at the 5% level.

system was an acquirer in consolidations. The second three columns examine this when the manufacturer was a target in the consolidations.

Raytheon, Boeing, and Northrop Grumman always served as acquirers in the consolidations examined in this study. About one-third of Raytheon's cost estimates experienced a statistically significant increase after consolidations in which it acquired a company. Its cost per unit output, on the other hand, was more likely to have a significant decrease (37.5%) than an increase. Regardless whether total costs or cost per unit are examined, approximately 50% of Raytheon's systems were significantly impacted following mergers.

In contrast, between 18% and 27% of Boeing's systems were significantly impacted by mergers (depending on if one examines total costs or per unit costs). Boeing's systems were more likely to experience a statistically significant decrease in per unit costs than an increase, which suggests efficiency. By 1998, Boeing was one of the prime contractors in six of the 10 markets, while Raytheon was one of the prime contractors in two of the 10 markets. The results show that despite the increased market power, both Raytheon and Boeing experienced improvements in efficiency through lower per unit costs, although Raytheon's systems were more affected than Boeing's.

The data on Northrop Grumman systems and Northrop systems was limited. Nevertheless, based on the systems examined, Northrop Grumman was more likely to experience inefficiencies since its cost per unit output experienced more significant increases than decreases, although its total costs fell. By 1998, Northrop Grumman was a prime contractor in three of the 10 sectors, suggesting that increased market power could lead to some possible inefficiencies.

About half of the systems made by Lockheed and General Dynamics experienced a significant change in total costs and about half of the systems made by General Dynamics (and one-third of the systems made by Lockheed) experienced a significant change in cost per unit following a merger in which the companies served as acquirers. This suggest that, although by 1998, Lockheed was the primary contractor in five of the 10 sectors, and General Dynamics was the primary contractor in two of the 10 sectors, this increased market concentration actually led to greater efficiencies.

Lockheed and General Dynamics are the two large defense contractors which were both targets and acquirers, depending on the merger, in this dataset. In the cases where a division of Lockheed served as a target in an acquisition, the systems that the division made experienced a statistically significant increase in per unit costs after being acquired by another company, suggesting that Lockheed units which were sold tended to experience increased inefficiency in per unit costs afterwards. In the cases where a division of General Dynamics was acquired by another company, per unit costs were not affected. Nevertheless, there were only a few systems to analyze for these companies.

In summary, the results of the preceding analysis across contractors suggest that, despite the increased market concentration in key sectors, the five major defense firms became more efficient as they acquired other firms, with the possible exception of Northrop Grumman. General Dynamics and Raytheon were most affected by their acquisitions in the sense that they had a greater percentage of weapons systems experiencing either a positive or negative statistically significant change, followed by Lockheed, Northrop Grumman, and then Boeing. Lockheed and General Dynamics were the most likely to have significant reductions in total costs and in per unit costs, indicating efficiency. Northrop Grumman was the only contractor whose cost per unit output was more likely to experience an increase, indicating possible inefficiencies in per unit costs. Nevertheless, the data on Northrop Grumman systems was limited to only a few systems.

Prior to their acquisition by another large company, did the targets experience lower weapons systems costs when they acquired other companies? In other words, were firms that later became targets originally good acquirers themselves? The mergers in which Hughes Aircraft and Martin Marietta served as acquirers had no significant change in per unit costs, while McDonnell Douglas experienced significantly lower per unit cost estimates. Thus, most of these companies' acquisitions had little effect on efficiency, which may be a contributing explanation for why they were later taken over by other firms, with the exception of McDonnell Douglas.

Table 3.6 focuses on the impact of two specific acquisitions on weapons system costs – the acquisition of Martin Marietta by Lockheed and the acquisition of McDonnell Douglas by Boeing. Three key observations emerge. First, a much greater percentage of weapons systems were impacted by the Lockheed-Martin Marietta merger than the Boeing-McDonnell Douglas merger. Second, systems involved in the Boeing-McDonnell Douglas merger were more likely to have lower per unit costs than higher per unit costs. This suggests that, to the extent that systems were impacted by this merger, they tended to manifest efficiencies. Third, systems involved in the Lockheed-Martin Marietta merger were as likely to have per unit costs increase as decrease, suggesting a mixture of improvements in efficiency and reductions in efficiency.

Table 3.7 categorizes the weapons systems on a sector by sector basis to examine which sectors exhibited efficiencies and which sectors exhibited inefficiencies following the mergers. Did increased industry consolidation lead to higher costs in DoD industry sub-sectors? The sectors are based on the type of weapons system classification found in the 1998 GAO report,[39] although this analysis added the strategic electronics sector. The classification of the weapons systems into these broader categories from the 1998 GAO report was accomplished by the author through examining the description of the weapons systems in the SARS, consulting *Jane's*, analyzing materials written by the defense contractors, examining *The 2007–2008 Weapons Systems*,[40] and examining in detail each system written by the Federation of American Scientists.[41]

39 General Accounting Office, 1998.

40 Office of the Assistant Secretary of the Army for Acquisitions. Logistics, and Technology, 2007.

41 See http://www.fas.org.

Table 3.7 Impact of Defense Consolidation on Total Costs and Per Unit Costs by Weapons System Category

Category	Percentage with a positive and statistically significant increase in total cost	Percentage with a negative and statistically significant decrease in total cost	Percentage with a positive or negative statistically significant change in total cost	Percentage with a positive and statistically significant increase in cost per unit	Percentage with a negative and statistically significant decrease in cost per unit	Percentage with a positive or negative statistically significant change in cost per unit
Strategic Electronics	21.43%	14.29%	35.71%	16.67%	25%	41.67%
Rotary Aircraft	0%	0%	0%	0%	33.33%	33.33%
Tactical Missiles	25%	33%	58.3%	11.11%	33.33%	44.44%
Fixed Wing Aircraft	11.1%	44.44%	55.55%	11.11%	11.11%	22.22%
Tracked Combat Vehicles	50%	0%	0%	50%	50%	100%
Surface Ships	0%	50%	50%	0%	25%	25%
Satellites	100%	0%	100%	50%	0%	50%
Strategic Missiles	0%	40%	40%	40%	0%	40%

Note: Note that statistical significance is at the 5% level.

The strategic electronics sector, the rotary aircraft sector, the tactical missile sector, and the surface ships sector all experienced lower per unit costs, and hence greater efficiencies following consolidation. The rotary aircraft sector experienced a decline of one-quarter in the number of contractors between 1990 and 1998 (Table 3.3) and the Army was the primary beneficiary of the improved efficiency. The surface ships sector exhibited a one-third decline in the number of contractors between 1990 and 1998 (Table 3.3) and the Navy was the primary beneficiary of the improvements in efficiency from lower per unit costs. The tactical missile sector was one in which there had been a two-thirds reduction in the number of contractors between 1990 and 1998, although the increased concentration actually led to lower per unit costs. The Navy, and to a lesser degree the Army, were the primary beneficiaries of the improved efficiencies. The strategic electronics sector had much less of a reduction in the number of contractors since it is a broader sector. The Air Force was a clear beneficiary, with lower per unit costs, Navy systems were evenly split between higher and lower per unit costs, while the Army was more likely to have higher per unit costs, suggesting inefficiency.

Results for the other sectors suggested that increased concentration could lead to inefficiencies, or to a mixture of efficiencies and inefficiencies following consolidation. For example, the satellite sector, which experienced a one-third reduction in the number of contractors between 1990 and 1998, exhibited inefficiencies in the form of higher per unit costs, as did the strategic missile category. Nevertheless, the data on the satellite sector was limited. The fixed wing aircraft sector (which experienced a two-thirds reduction in the number of contractors) and the tracked combat vehicle sector (which experienced a one-third reduction in the number of contractors) experienced a mixture of efficiencies and inefficiencies.

Conclusion

The outcome of the last wave of consolidation in the US defense industry continues to be an important issue since the defense sector is currently facing challenges due to fiscal constraints and shifting defense priorities. These constraints, which are likely to continue in the coming years, may result in defense contractors facing excess capacity, providing an impetus for defense contractors to merge to rationalize costs. Nevertheless, it is unlikely that there will be mergers between larger firms, as was the case during the 1990s, due to the level of consolidation that exists from that period. Consequently, it is more likely that larger firms will merge with smaller firms to gain further knowledge in designing new equipment in cybersecurity, UAVs, etc., since the smaller firms individually may have difficulty in financing through capital markets. Moreover, from the perspective of the larger firms, these equipment areas are rapidly expanding and have uses in the civilian market as well.

The findings of this analysis are generally consistent with studies on mergers and acquisitions which found positive gains from consolidation in other industries, such as: Healy, Palepu, and Ruback (1990), Jensen and Ruback (1983), and Gaughan (1999). The overall results suggest that, although the defense sector consolidation did yield more efficiencies than inefficiencies, it is likely that even more efficiencies could have been realized. Although this analysis is by no means exhaustive, it does suggest several key findings.

First, correlations between defense merger activity and overall merger activity in the economy are strongly positive. On balance, the correlations between defense merger activity and overall merger activity are much stronger than the correlations between defense merger activity and DoD outlays. This suggests that merger activity was driven less by declines in spending following the Cold War, and more by a stronger economy and a vibrant financial market.

Second, in examining the SAR cost data on 53 weapons systems, between 39% and 44% of systems experienced some statistically significant change in either total costs or per unit costs following a merger, which means that between 56% and 61% of the systems were unaffected by mergers. Nevertheless, to the extent that the weapons systems were impacted by mergers, more systems exhibited a statistically significant decrease in total costs and in per unit costs than an increase, suggesting that consolidation was more likely to lead to efficiencies, although the percentages were somewhat close.

Third, it is likely, based on the qualitative evidence in section III and the discussion of why many of the mergers occurred that, to the extent that consolidation led to efficiencies, many of the efficiencies may have been due to the pooling of knowledge sets and rationalization of plant capacity. For example, the pooling of the knowledge sets of Boeing, Rockwell, and McDonnell Douglas may have contributed to Boeing's success in winning the contract for the National Reconnaissance Satellite over Lockheed in 1999. Similarly, Boeing's consolidation of defense and space work in St. Louis and in southern California from Puget Sound after its acquisition of McDonnell Douglas and Rockwell led to plant rationalization and overhead efficiencies.

Fourth, the results of this analysis suggest that despite the increased market concentration in key sectors, the five major defense firms became more efficient as they acquired other firms, with the possible exception of Northrop Grumman. This may, again, be a function of pooling information sets to develop a better, cheaper weapons system, or rationalization of plant capacity. General Dynamics and Raytheon had the greatest percentage of their systems affected by their acquisitions, followed by Lockheed, Northrop Grumman, and Boeing. Boeing (one of the prime contractors in six of the 10 markets by 1998, Lockheed (one of the primary contractor in five of the 10 sectors, by 1998), Raytheon (one of the prime contractors in two of the 10 markets) and General Dynamics (the primary contractor in two of the 10 sectors) experienced reductions in per unit costs following acquisitions. Northrop Grumman (a prime contractor in three of the 10 sectors by 1998), was the only contractor whose cost per unit output was more likely to experience an increase, suggesting possible inefficiencies in per unit costs. Nevertheless, the data on Northrop Grumman systems was limited to only a few systems. A closer examination of the Lockheed-Martin Marietta merger and the Boeing-McDonnell Douglas merger indicated that, although a much greater percentage of weapons systems were impacted by the Lockheed-Martin Marietta merger than the Boeing-McDonnell Douglas merger, the Boeing merger showed clearer evidence of lower per unit costs after the merger, whereas the Lockheed merger yielded a mixture of higher per unit costs and lower per unit costs across various system. The evidence also showed that, with the exception of McDonnell Douglas, most of the firms which later became the targets of the five large contractors, had previously conducted mergers as acquirers which had little effect on efficiency, which may be a contributing explanation for why they were later taken over by the larger defense firms.

Fifth, the results of the analysis indicated greater efficiencies following consolidation for many sectors through reduced per unit costs. These sectors included the rotary aircraft sector (experienced a decline of a quarter in the number of contractors by 1998), the tactical missile sector (experienced a two-thirds reduction in the number of contractors by 1998), the surface ships sector (exhibited a one-third decline in the number of contractors by 1998), and the strategic electronics sector. The Army was the primary beneficiary of the improved efficiency in the rotary aircraft sector. The Navy was the primary beneficiary of the improvements in efficiency from lower per unit costs in the surface ships sector and in the tactical missile sector. The Air Force was a clear beneficiary in the strategic electronics sector, which had much less of a reduction in the number of contractors since it is a broader sector. Other sectors, such as the fixed wing aircraft sector (which experienced a two-thirds reduction in the number of contractors) and the tracked combat vehicle sector (which experienced a one-third reduction in the number of contractors), experienced a mixture of efficiencies and inefficiencies following consolidation. Finally, the satellite sector had a one-third reduction in the number of contractors indicating that increased concentration can lead to inefficiencies, with the Air Force being the most impacted of the services. Nevertheless, the data on the satellite sector was limited.

In conclusion, the analysis suggests that, although market concentration levels in certain sectors increased due to the wave of defense mergers, DoD's costs across weapons systems tended to be lower in the post-merger period for many systems; however, there is still room for achievement of additional efficiencies on a cost per unit output basis. While further research is necessary to more fully inform the public policy discourse, this study indicates that increases in market power do not necessarily lead to an anticompetitive outcome in pricing. Additional research on innovation cycles within the weapons systems is necessary, as well as a greater assessment of the degree to which international competition or the possibility of entry of smaller competitors in some of these sub-sectors constrained cost increases. Many of the questions and concerns in the earlier rounds of consolidation may emerge if a second round begins following the continuation of defense budget cuts. If so, the next merger wave could possibly occur at a more global level and/or between larger firms and smaller firms, so an assessment of the strengths and weaknesses of the most recent round in the US during the 1990s is crucial.

References

Annual Industrial Capabilities Report to Congress. 1998. Washington, DC: Department of Defense, February.

——. 1999, Washington, DC: Department of Defense, February.

——. Washington, DC: Department of Defense, February.

——. 2001. Washington, DC: Department of Defense, January.

——. 2002. Washington, DC: Department of Defense, March.

——. 2003. Washington, DC: Department of Defense, February.

——. 2004. Washington, DC: Department of Defense.

——. 2005. Washington, DC: Department of Defense.

——. 2006. Washington, DC: Department of Defense.

——. 2007. Washington, DC: Department of Defense.

Annual Report to the President and Congress by William S. Cohen, Secretary of Defense in 2000. 2000. (Appendix B-1). Available at: http://www.dod.gov/execsec/adr2000/index.html

Arena, Mark V., Leonard, Robert S., Murray, Sheila E., and Younossi, Obaid, 2006. *Historical Cost Growth of Completed Weapon System Programs*. Santa Monica, CA: RAND Corporation, TR-343-AF.

Barton, David M. and Sherman, Roger, 1984. "The Price and Profit Effects of Horizontal Merger: A Case Study." *Journal of Industrial Economics*. Vol. 33 (December 1984), pp. 165–78.

"Boeing Restructures Manufacturing Operations." 2000. *Jane's Defence Industry*, October 1.

Budget of the United States Government for Fiscal Year 2008. 2000. Historical Tables (Table 3.2), pp. 56–60. Available at: http://www.whitehouse.gov/omb/budget/fy2008

"Business Focus, Mergers Becoming a Business Imperative." 1996. *Jane's Defence Weekly*, January 17.

Cooper, David E., 1997. *Statement of David E. Cooper, Associate Director, Defense Acquisitions Issues, National Security and International Affairs Division on Defense Industry Restructuring: Cost and Savings Issues.* Testimony before the Subcommittee on Acquisition and Technology, Committee on Armed Services, United States Senate, General Accounting Office, April 15, pp. 1, 3, 8.

"A Decade of Defense Mergers Yields Disappointments." 1999. *Los Angeles Times*, October 17.

Department of Defense Cost Variance Handbook, Washington, DC.

Department of Defense Instruction 7000.3g, 1980. Washington, DC: May.

Department of Defense, Selected Acquisition Report Summary Tables, 1981–2006. Available at: http://www.acq.osd.mil/ara/am/sar/index.html

Department of Defense, 1981–2006, *Selected Acquisitions Reports*, Washington, DC.

Dowdy, John, 1997. "Winners and Losers in the Arms Industry Downturn." *Foreign Policy*, No. 107, Summer 1997, pp. 88–101.

Drezner, Jeffrey A., Jarvaise, Jeanne M., Hess, Ronald Wayne, Hough, Paul G., and Norton, D., 1993. *An Analysis of Weapons System Cost Growth*. Santa Monica, CA: RAND Corporation, MR-291-AF.

Driessnack, John D. Lt. Col., and King. David R. Maj., 2004. "An Initial Look at Technology and Institutions on Defense Industry Consolidation." *Defense Acquisition Review Journal*, January–April, 2004, pp. 63–77.

Eckbo, B. Epsen, 1983. "Horizontal Mergers, Collusion, and Stockholder Wealth." *Journal of Financial Economics*, Vol. 11, Issue 1 (April, 1983), pp. 241–73.

Evers, Stacey and Starr, Barbara, 1997. "Business, Boeing-MDC Fusion to Inspire More Mergers." *Jane's Defence Weekly*, January 8.

Factset Mergerstat LLC, 1996. *The Mergerstat Review for 1996*. Santa Monica, CA: Factset Mergerstat LLC.

——. 1997. *The Mergerstat Review for 1997*. Santa Monica, CA: Factset Mergerstat LLC.

——. 2002. *The Mergerstat Review for 2002*. Santa Monica, CA: Factset Mergerstat LLC.

——. 2005. *The Mergerstat Review for 2005*. Santa Monica, CA: Factset Mergerstat LLC.

Federation of Amcrican Scientists. *United States Weapons Systems*. Available at: http://www.fas.org

Fiscal Year 2006 Major Defense Acquisition Program (MDAP) Lists, 2006. Available at: http://www.acq.osd.mil/ap/mdap [accessed August 3, 2006].

"FTC Staff Supports Big Defense Merger: Conditions Placed on Martin-Lockheed Deal." 1994. *The Washington Post*, December 30.

"GAO: Savings from Defense Merger Are Less Than Predicted," 1996. *Defense Daily*, September 6.

General Accounting Office (GAO), 1998. *GAO Report to Congressional Committees on the Defense Industry: Consolidation and Options for Preserving Competition.* Washington, DC: April.

Gholz, Eugene and. Sapolsky, Harvey M., 1999/2000. "Restructuring the US Defense Industry." *International Security*, Vol. 24, No. 3, Winter, pp. 5–51.

Hall, B.H., 1988. "The Effect of Takeover Activity on Corporate Research and Development." In A.J. Auerbach (ed.), *Corporate Takeover: Causes and Consequences*. Chicago, IL: University of Chicago Press.

Healy, P., Palepu, K., and Ruback R., 1990. "Does Corporate Performance Improve after Mergers?" NBER Working Paper no. 3348.

Hough, Paul G., 1992. *Rand Note: Pitfalls in Calculating Cost Growth from Selected Acquisition Reports*. Santa Monica, CA: RAND Publishing.

Jane's Information Group. *Jane's Online*. Available at: http://www.janes.com.libproxy.nps.navy.mil

Jensen, M.C. and Ruback, R.S., 1983. "The Market for Corporate Control: The Scientific Evidence." *Journal of Financial Economics*, Vol. 11, April 1983, pp. 5–50.

"Jumping the Gun: How Lockheed Martin Misread the Radar on the Northrop Merger." 1998. *Wall Street Journal*, June 19.

Klein, Joel I., 1998. *Statement of Joel I. Klein, Assistant Attorney General, Antitrust Division, US Department of Justice, Before the Committee on the Judiciary, United States Senate, Concerning Mergers and Corporate Consolidation*. Washington, DC, June 16, 1998, pp. 7–8.

Korb, Lawrence J., 1996. "Merger Mania: Should the Pentagon Pay For Defense Industry Restructuring?" *The Brookings Review*, Summer, Vol. 14, Issue 3.

Kovacic, William E. and Smallwood, Dennis, 1994. "Competition Policy, Rivalries, and Defense Industry Consolidation." *Journal of Economic Perspectives*, Vol. 8, No. 4, Fall, pp. 91–110.

"L6bn US Defense Merger Called Off." 1998. *Financial Times*, July 17.

Markusen, Ann, 1999. "The Rise of World Weapons." *Foreign Policy*, Spring, pp. 40–51.

McNicol, D.L., 2004. *Growth in the Costs of Major Weapon Procurement Programs*. Institute for Defense Analyses, IDA Paper P-3832.

Meeks, Geoffrey, 1977. *Disappointing Marriage: A Study of the Gains from Merger*. Cambridge: Cambridge University Press.

Mueller, Denis (ed.), 1980. *The Determinants and Effects of Mergers*. Cambridge: Oelgeschlager, Gunn & Hain.

Office of the Assistant Secretary of the Army for Acquisitions. Logistics, and Technology, 2007. *The 2007–2008 Weapons Systems*. Washington, DC.

"Raytheon Deal Raises Antitrust Concerns." 1997. *The Washington Post*, January 28.

"Recouping the Cost of Defence."1994. *Jane's Defence Weekly*, August 27.

Rogers, Marc, 1997. "Contracting Issues, Europe Must Decide Industry Strategy in Face of US Giants." *Jane's Defence Contracts*, August 1.

Roll, Richard, 1986. "The Hubris Hypothesis of Corporate Takeovers," *Journal of Business*, No. 59. pp. 198–216.

Scherer, F.M. and Ross, David, 1990. *Industrial Market Structure and Economic Performance.* 3rd edition. Boston, MA: Houghton Mifflin.

Stillman, Robert S., 1983. "Examining Antitrust Policy Toward Mergers," *Journal of Financial Economics*, 11, no. 1 (April 1983), pp. 225–40.

Tyson, K.W., Harmon, B.R. and Utech, D.M., 1994. *Understanding Cost and Schedule Growth in Acquisition Programs*, Institute for Defense Analyses, IDA Paper P-2967.

Tyson, K.W., Nelson, J.R., Ohm, N.I., and Palmer, P.R., 1989. *Acquiring Major Systems: Cost and Schedule Trends and Acquisition Initiative Effectiveness*. Institute for Defense Analyses, IDA Paper P-2201.

"War of Attrition: Defense Consolidation Rushes Toward an Era of Only 3 or 4 Giants." 1996. *Wall Street Journal*, December 6.

Younossi, Obaid, Arena, Mark, Leonard, Robert S., Roll, Charles Robert, Jr., Jain, Arvind, and Sollinger, Jerry M., 2007. *Is Weapon System Cost Growth Increasing? A Quantitative Assessment of Completed and Ongoing Programs.* Santa Monica, CA: RAND Corporation, MG-588-AF.

Chapter 4
Defense Industrial Alliances

The landscape of the global defense industry in the post-Cold War period has evolved in a number of ways, which has led to the development of alliances. First, the greater linkages within the overall world economy have promoted the development of alliances between defense contractors in various countries. Second, the role of allied forces with the emergence of the terrorist threat in the post-9/11 period, as well as the shifting focus to the Asia-Pacific region, has encouraged the interoperability of equipment and synergistic compatibility in computer systems for military allies. This has further motivated alliances between defense contractors in research, development, and technology transfer, both within countries as well as across countries. Third, with the shrinkage of defense budgets in the US and Europe, there is a greater emphasis on technologically innovative products at competitive costs, which promotes the formation of alliances between domestic defense contractors. As weapons systems become increasingly complex, it can be cost-effective to spread the R&D costs across different defense companies. Fourth, alliances are also important as US and European defense contractors are involved in overseas competitions and form alliances in an effort to provide strong bids. Moreover, in an effort to expand their defense industrial bases, these overseas countries also form alliances in product development between domestic and international defense contractors.

Mergers vs. Alliances

Against this backdrop, there have been a number of studies across industries on mergers, as well as on alliances. Mergers have the benefit of leading to the formation of more permanent relationships between the merging companies. With the absorption of one company into another, there are greater opportunities for cost-cutting in eliminating duplicative workforces and in reorganizing the corporate hierarchy to better internalize and reduce the transactions costs which would have been present in an arm's-length relationship. Mergers can have permanent or long-lasting effects on the market power of various companies, the ability of new firms to enter the industry, and market concentration levels. Absorption can, however, also lead to substantive integration costs and cultural/communication difficulties, which can postpone or altogether eliminate the benefits of the merger. As a consequence, the regulatory scrutiny from the antitrust authorities is important in concluding the deal.

When mergers occur between companies from different countries, the magnitude of the opportunities for benefits relative to the costs changes. Absorption costs for an international merger can increase relative to a domestic merger, especially if there are cultural or communication incompatibilities between the merging parties. The issue of which country loses jobs to the other country is often magnified by the popular press and government officials. Although the impact on market power and market concentration may be less with an international merger than a domestic merger because the definition of the relevant market is geographically larger, the regulatory review process can become more complicated since regulatory authorities from multiple countries are involved.

Mergers or acquisitions involving the US defense market have historically been problematic. Although there may have been benefits from the acquisition, Congressional representatives are often concerned about job loss, as well as the national security issues inherent in technology transfer. The acquisition is often formally disallowed, or the foreign entrant withdraws its bid in anticipation that the acquisition will be blocked if it proceeds further. This is further discussed in Chapter 5 in the context of the case study of the tanker competition in the US.

Alliances can often be a good alternative to mergers. The parties involved in the alliance can obtain some of the benefits of a merger – joint investments in R&D expenses and production equipment, knowledge transfer and technology transfer, and access to new markets. Alliances can be easier to disassemble than mergers because less integration of operations is required. As a result, integration costs are lower and the potential for cultural or communication clashes is less. Nevertheless, as discussed by Doz and Hamel (1998), in alliances in which the generation of economies of scale is a motivation, the costs of exiting the alliance can be high due to the sunk costs of investment in equipment. Since alliances may lack the depth of integration found in mergers, there could be less of an incentive for parties to invest in relationship-specific assets and to produce the types of benefits and efficiencies which would be possible in a merger. Finally, although alliances may raise fewer regulatory concerns, the degree of technology transfer, etc. is still subject to review. Government officials can also protest ensuing job loss if combined production facilities from the alliance result in a loss of jobs in one country.

Alliances have become increasingly prevalent in a variety of industries; indeed, a number of studies on alliances have been cross-sectional, rather than focused on a specific industry, such as Yoshimi and Rangan (1995) and Liedtka (1998). The importance of global competition as an impetus for alliance formation is discussed in Yoshimo and Rangan (1995). Strategic alliances can even be a defensive strategy in that, as Gomez-Casseres (1994) discussed, as more alliances are formed, there are fewer possible partners available for firms that wish to form new alliances, and "strategic gridlock" can develop.

Alliances are helpful in the defense industry for several reasons. First, the R&D costs for development of a product can be high, which is why it is more cost-effective not to duplicate efforts. Second, the primary buyers are governments, who are increasingly cash-constrained. Collaboration between companies can be more cost-effective and successful than the competition between many different companies to chase a few contracts. Third, firms benefit from each other's skills without paying the integration costs and financial costs of a merger, which can lead to a higher return on investment from the collaboration because the costs of the investment are lower. This chapter analyzes the reasons behind the formation of alliances between US contractors and between US and European defense contractors, as illustrated by several case studies of alliances, as well as assesses some of the patterns in alliance formation. Furthermore, the chapter discusses the potential for alliances in the future.

Patterns in Alliances between US Defense Contractors and Foreign Defense Contractors

Alliances between US defense firms and foreign firms are exposed to some of the same concerns as mergers, such as concerns over the potential of US jobs going overseas and national security concerns over technology transfer. Nevertheless, although alliances undergo some scrutiny, it can be easier for the parties involved in the alliance to limit the degree of their involvement with each other, at least initially, than would be the case in a merger. As the alliance deepens and trust is built, both between the two parties concerned and between the two governments involved in the alliance, the degree of involvement can increase.

Butler, Kenny, and Anchor (2000) discuss strategic alliances within the European defense industry, as well as many of the changes to the defense sector. They describe how certain defense sub-sectors have more alliances than others, and note that the electronics sector, the land vehicles sector, and the naval vessel sector have more alliances than the small arms and ordnance sector. Although they do not discuss why this might be the case, one possibility is that the sectors with more alliances are more R&D-intensive, and it is more cost-effective for the partners to share the costs than to bear the costs alone. They also discuss how cultural compatibility has not been necessary for the success of defense alliances, although 70% of UK contractors are in an alliance with a US firm. They find that many of the alliances are actually agreements for subcontracting, where the US firm is the lead contractor, or licensing agreements, where the US firm is the licenser.

An assessment of data on the number, type, and details of joint ventures and alliances between 2002 and 2005 involving US defense contractors with both other US defense contractors, as well as foreign defense contractors, indicated some interesting conclusions. Analysis of the data suggested that Lockheed Martin and Boeing had the greatest number of alliances with foreign defense contractors. Northrop Grumman, General Dynamics, and Raytheon had between one-quarter and one-third of the number of alliances with foreign contractors as Lockheed Martin and between one-third and one-half of the number of alliances with foreign contractors as Boeing. The fact that Boeing and Lockheed Martin had more alliances with foreign defense contractors during this period than other large US defense contractors may be due to:

a. the opportunities for shared R&D in the weapons systems sub-sectors in which these alliances focused; as well as
b. the success of previous alliances made by these companies, which made them more likely to be willing to enter into additional alliances, creating a positive, self-reinforcing cycle.

Assessment of the data, which divided the foreign defense contractors involved in alliances with a US defense contractor by region – Europe, the UK/Australia/Canada, Asia, and the Middle East – also yielded interesting findings. Lockheed Martin contracted half of its alliances and joint ventures involving foreign contractors with UK, Australian and Canadian contractors and the other half with Asian contractors. Northrop Grumman contracted two-thirds of its foreign alliances with UK, Australian, and Canadian contractors, and one-third with Middle Eastern contractors. General Dynamics contracted half of its foreign alliances with European contractors and half with UK, Australian, or Canadian contractors. Raytheon had 100% of its foreign alliances with European contractors. Boeing had one-third of its foreign alliances with European contractors and two-thirds with Asian contractors. Lockheed Martin and Northrop Grumman did not form an alliance with a European contractor at all during this period, while Raytheon and Boeing, which did form alliances with European contractors, did not form any alliances with UK, Australian, or Canadian contractors over this period. The dominance of UK, Australian, Canadian, or European firms as foreign partners in these alliances suggests the importance of:

a. common language;
b. geographic proximity;
c. a prior history of successful alliances with firms in that country, leading to a positive, self-reinforcing cycle; and
d. the importance of these partner countries as allies and the need for interoperability of equipment, especially in joint operations.

US and European defense contractors partially entered into alliances due to the need for more synergistic and interoperable equipment among NATO members. In 1999, Alfred G. Volkman, US Acting Deputy Under Secretary for Defense, noted that:

> The end of the Cold War, the break-up of the Soviet empire, the emerging power of rogue nations, and equally destabilizing geopolitical events are transforming our vision of the 21st century security needs and our NATO military strategy … In order to develop and field interoperable equipment, it is necessary that stronger transatlantic ties are forged … Governments would agree on common [military] requirements, then invite defense firms to form transatlantic competitive teams of their own choosing.[1]

1 Sparaco, 1999.

Nevertheless, concerns over limitations on technology transfer on national security grounds between countries was one of the greatest impediments in deepening trans-Atlantic ties. Indeed, in 1999, General Jean-Yves Helmer, Director of the DGA French armaments agency, noted that:

> the US and Europe do not share identical [defense] concepts and [operational] requirements. Nevertheless, there is ample room for synergy, *on the condition* [emphasis added] that know-how and technology can circulate freely.[2]

Barriers on export licensing and the transfer of technology limited the development of transatlantic alliances in the late 1990s and early 2000s, but the "Declaration of Principles" signed by the US and the UK in February, 2000 was an early step to greater joint research and development, and coordination of technology transfer, military requirements, etc.[3] Some argued, however, that, with the exception of the UK and Canada, the US had a lack of trust for most other countries, especially in terms of technology-transfer issues.[4]

European defense firms were also attracted to the US market because its defense market was much larger than the defense market in Europe. For example, in 2002, the US budget was three times that of EU countries. As a result, the investments of European companies in the US were 10 times greater than the value of US acquisitions in the European defense sector. In some sub-sectors of the defense market, the gap in spending and trade between the US and Europe was less. For example, Raytheon argued that in the areas on which the Thales Raytheon Systems focused – battlefield surveillance and command and control (C2) systems – there was less of a differential in spending. Consequently, the interest of the Europeans in the US defense market was driven both by disparities in spending, as well as by the perception that US R&D might drive the next generation of weapons systems, such that an alliance would give European countries access to the technologies without having to fund their development themselves.[5]

The Role of Alliances in Creating Additional Alliances among Competitors: A Case Study of the CFM Alliance and International Aero Engines

Alliances are often formed in order to combine different knowledge pools to create a new and superior product. As the market share for this product increases, the competitors in this product space may also form alliances to share knowledge and to develop an even better product than their allied competitors. The result of this defensive alliance formation can be an improved market sector, with

2 Sparaco, 1999.

3 Sparaco, 2000.

4 Barrie and Taverna, 2002.

5 Barrie and Taverna, 2002.

several innovative and competing products for the end-user developed by multiple competing alliances. The development of the CFM International alliance and the International Aero Engine alliance is an example of this.

CFM International is a joint venture between Snecma, formerly a French state-owned enterprise, and General Electric (GE). The alliance, which is one of the most successful and long-lasting alliances in the trans-Atlantic market, was formed in 1974 because GE and Snecma intended to leverage their skills developed in the engine market in the defense sector by entering the civilian market for engines, which was heavily dominated by Pratt & Whitney at the time. One of the initial hurdles was to convince the US government to allow GE to share its military technology with Snecma. As of 2007, the engines made by CFM (especially the CFM 56 engine) could be found in over 50% of the fleet of single aisle planes with 100 seats or more and are often found in Airbus 320s and Boeing 737s. The way in which the alliance was structured was that each of the two partners would be involved with the design, production, and research of their respective modules/ components within the engine. GE and Snecma's relationship was not based on equity holdings. The two firms split the proceeds from the engines in half, based on notional costs, although neither company knew the true costs of its partners.[6]

During the early 1980s, Pratt & Whitney's market share began to fall in this product space. In 1983, it created an alliance – International Aero Engines (IAE) – with MTU (part of Daimler-Benz), Fiat, Rolls-Royce, and Japanese Aero Engines to develop an engine which would compete with CFM's engines.[7] This product alliance, like CFM International, was based around the design of an engine – in this case, the V2500 engine.

By 1995, CFM International and International Aero Engines controlled 26.6% of the aero engine sector. One benefit of the alliances within the civil engine arena has been that, although Rolls Royce, Pratt & Whitney, and GE were already involved in the civil engine market, the other members of the alliances, such as Snecma, through its development of the CFM56 engine with the CFM International alliance, and Daimler-Benz, through its development of the V2500 engine as part of the International Aero Engines alliance, were able to enhance and establish their positions in this market. The creation of the CFM International alliance allowed Snecma, which had manufactured jets for the French military, to use this expertise to enter the civilian aero engine market and to build up a significant presence through its development of the CFM56 engine. Consequently, by galvanizing Pratt & Whitney and other manufacturers to form International Aero Engines, alliance formation facilitated the development of several new engines, as well as a vibrant, competitive marketplace for the end-user.[8]

6 "Business: Odd Couple," 2007.

7 "Business: Odd Couple," 2007.

8 Smith, 1997.

Alliances as a Means of Promoting National Defense Strategy: A Case Study of Trans-Atlantic Cooperation in Missile Defense

In 2002, Boeing entered into separate agreements at the Farnborough Air Show with BAE, EADS, and Alenia Spazio to cooperate on ballistic missile defense. The alliance planned to be an informational exchange in which Boeing would discuss its approach to missile defense with its European partners and they would discuss the technologies that they could incorporate in the missiles.[9] Part of the purpose of the agreements was to galvanize the interest of European governments in larger ballistic missile defense programs. It could help the US convince the Europeans that larger missile defense programs could cover NATO's European members and to show the Europeans that there would be jobs involved in it.[10]

The various European partners in the alliance were chosen due to the contributions that their expertise would provide to the project. Alenia Spazio, part of Finnmeccanica, would add their expertise in supercomputers/ data fusion, synthetic aperture radar satellites, and wideband secure telecommunications to the joint missile defense architecture discussions. EADS would add expertise in the space area from its affiliate, Astrium, as well as its knowledge of early warning satellite systems, which could locates the sites where the ballistic missiles were launched, the zone of potential impact, and the trajectory of the missile in the boost phase.[11]

Part of the reason why there was an impetus for transatlantic alliances in the missile area was that there had been several previous alliances in the missile product area.

For example, Boeing and EADS had collaborated on a study for NATO on tactical missile defenses,[12] and Boeing had worked in marketing the Meteor missile, made by EADS and BAE. Most of the previous arrangements between Boeing and EADS had involved subcontracting or marketing, while this alliance involved sharing the product development responsibilities.[13]

Consequently, this alliance was partially motivated by the need to familiarize and convince the Europeans regarding the US perspective toward larger missile defense programs. It involved sharing of knowledge between the members and a fusion of their different capabilities to produce innovative products. Alliances, particularly between Boeing and EADS, had previously existed in the missile defense arena, and the positive momentum from these previous alliances had helped in building trust and thus helped to promote the development of subsequent alliances.

9 Asker, Barrie, and Taverna, 2002.

10 "Business: Hands Across the Sea," 2002.

11 Asker, Barrie, and Taverna, 2002.

12 Asker, Barrie, and Taverna, 2002.

13 "Business: Hands Across the Sea," 2002.

Alliances Focused on Specific Product Areas: A Case Study on the Alliance between Raytheon and Thales

Many of the successful alliances between defense contractors have been focused on a specific product area. CFM International and International Aero Engines, discussed previously, are examples of successful alliances which concentrated on developing systems in a specific product area. Another example of this type of alliance is the alliance between Thales (formerly Thomson-CSF, a French company) and Raytheon, which was completed in early 2001 as Thales Raytheon Systems. This alliance was created so that the two contractors could collaborate on ground-based battlefield radar programs and air defense command/control (C2) programs.[14] By the end of 2001, Thales and Raytheon had collaborated on 17 projects.[15]

The alliance was a horizontal combination in which firewalls were built to protect against leakage of sensitive information. Thales Raytheon Systems was divided into two subsidiaries where Raytheon would have a 51% share in the US subsidiary, and Thales would have a 51% share in the European subsidiary. The revenues of Thales Raytheon Systems were split between France and the US.[16]

As was the case in the alliance formed between Boeing and EADS in the missile defense area, Thales Raytheon Systems was formed partially because the two companies involved had successfully collaborated on other fronts, thus building trust between the two parties and increasing their tendency to invest in relationship-specific assets, despite the more arms-length nature of an alliance relative to a merger. Thales and Raytheon collaborated on the Air Command Systems International (ACSI), which was a venture established in 1997 to work on the Florako air defense radar project in Switzerland, and NATO's Airborne Command and Control System (LOC1). ACSI continued to be a separate entity, but was attached to Thales Raytheon Systems.[17]

The Role of Alliances in Sharing Knowledge and R&D Costs, and Developing New Technologies

Mergers and alliances are often valuable in enabling the participating firms to generate economies of scale in both R&D costs and in production costs by sharing these costs or by spreading them over a greater number of units of output to lower per unit costs. As weapons systems have become more complex, R&D has continued to be an important and costly phase of the product development cycle.

14 Taverna, 2001.
15 "US-Euro Strategic Alliances," 2001.
16 Taverna, 2001.
17 Taverna, 2001.

One example of an alliance which was formed to share R&D costs was an alliance, led by Boeing, and including the Airbus companies of Aerospatiale SA (France), British Aerospace, and Daimler-Benz, to develop a "super-jumbo" jet. The R&D costs to develop this jet, which would have carried between 600 and 800 passengers, were $15 billion. This was too much for one contractor to sustain, and was more affordable when spread over an alliance of contractors.[18]

The project first began development in January, 1992, but collapsed in 1995 due to uncertainty in demand. Only Singapore Airlines and British Airways were willing to place orders. This underscores the importance of the need to share R&D costs, and hence the risk of product development, in an environment of uncertain demand. A second reason for the collapse of the project was the concern that it would consume so much capital that it would limit the development of the next generation of supersonic planes.[19] Consequently, although alliances are important in sharing R&D costs, the placement of the product being developed has to be evaluated in the context of the costs of the estimated future trajectory in product development.

The development of new technologies has led to the formation of alliances, as well as acquisitions, between firms. In an effort to handle the potential for reduced government spending, the larger defense contractors have deployed their capital not through investment, but through increasing dividends and share repurchases, which has resulted in a decrease in industry R&D spending. This has led to younger firms, such as Palantir or SpaceX, obtaining larger contracts, since they have engaged in more innovation and provides a greater incentive for large defense contractors to work with the government in joint government investment projects.[20]

In many cases, larger firms acquire smaller firms in growing industries in an effort to expand their knowledge base in developing new technologies. Acquisitions can be more beneficial to the smaller firms than alliances because the joint development of new technologies results in smaller firms providing larger firms with knowledge, which, in the event of a collapse of the alliance, could hurt the smaller firms, which have less capital to sustain themselves alone. Consequently, in an acquisition, larger firms gain knowledge and the smaller firms obtain the funds to continue the R&D, as well as the opportunities for profitable sales, due to the client relationships of large firms.

For example, a number of larger companies have acquired smaller companies in an effort to enter the cybersecurity industry, which has worldwide sales of over $67 billion. Indeed, companies such as Cisco, Intel, and Google have acquired smaller companies to develop their role in the cybersecurity market. For example, Cisco purchased Sourcefire, Apple purchased AuthenTex, Google bought VirusTotal, and Twitter purchased Dasient.[21]

18 Cole, 1995.

19 Cole, 1995.

20 Fryer-Biggs, 2013.

21 "Tech Titans See Opportunity in Cybersecurity," 2013.

A second example is the UAV market, which has provided opportunities for smaller, innovative, younger firms, as well as for more established defense contractors, which are expanding into the product space partially through acquisitions. For example, as discussed in Chapter 2, Northrop Grumman, which produces the Global Hawk, acquired some of its capabilities from its acquisition of Ryan Aeronautics, which had expertise in target drone production and design. Northrop also bought Swift Engineering, which has expertise in designing blended wing UAVs. Similarly, Rockwell Collins has become a significant supplier of avionics for unmanned and manned aircraft because it acquired Athena Technologies, a pioneer in flight control systems for UAVs, in 2004.[22]

Defense contractors are concerned about the sunk costs of investing in new technologies and products due to the uncertainty in future demand from DoD for the products, as a result of fiscal constraints and shifting defense priorities. Diversification through developing products with both government and civilian uses helps to mitigate risk.[23] Moreover, it enables R&D costs to be spread across military and civilian products. For example, Textron is developing a light-attack surveillance aircraft, the Scorpion, for the military, through a partnership of 22 vendors, which is based on the design of a commercial Cessna business jet and which can carry spy sensors. This reduces the costs and enables greater diversification of the risk of uncertainty in the defense marketplace.[24]

The Role of Alliances in Developing Interoperable Equipment between Allied Forces: A Case Study of the Joint Strike Fighter

The development of the Joint Strike Fighter (JSF) is an example one of the most extensive alliances in the defense sector, involving nine different contractors from various countries, led by Lockheed Martin. The JSF not only allows the various contractors to contribute their expertise to provide a better product, but it also provides a strong basis for understanding the challenges facing global defense alliances in the future, ranging from cost allocation issues, to technology transfer security issues, to global supply chain integration issues.

One of the main benefits of the creation of the JSF is that the new product, created by the sharing of technology between the various allied nations, will allow greater synchronization of subsequent operations of coalitions of these countries and the development of more similar capabilities. The intention has been for the F-35 to replace 13 different types of aircraft across 11 different countries.[25] Nine nations are participating in the JSF program, according to their levels of financial involvement. While the US is the primary customer, the UK is a Level I partner since

22 Warwick, 2009.
23 Erwin, 2013a.
24 Erwin, 2013a.
25 "Lockheed Martin," 2008.

it contributes 10% of the development costs, followed by Level II partners – the Netherlands and Italy, and then followed by Level III partners – Canada, Turkey, Australia, Norway, and Denmark.[26]

The international structure of the relationships between the US contractors and the foreign contractors on the JSF, however, has drawn criticism. By mid-2003, one of the concerns was linked to the fact that the foreign contractors on the JSF did not have to share the growing development costs, which had already increased $3 billion since the start of the system development phase. The US defense representatives argued that they could ask for assistance from their foreign allies in handling cost overruns. A second concern, voiced by the chairman of the House Government Reform Panel, Representative Shays, was that too many US jobs on the JSF were going overseas; as of that point, 18% of the contracts on the JSF had gone overseas, valued at $2.2 billion. A third concern came from the partners on the other side of the Atlantic: the Chairman of Alenia Aeronautica noted his disappointment in the return on investment in the JSF. On the US side, there were concerns that program decisions might have to be made to increase the return on investment to partner countries, but which could also lead to delays or higher costs.[27] Finally, a fifth concern arose surrounding technology transfer issues. The UK threatened to exit the JSF program unless the US shared information on the stealth technologies, etc. related to the plane.[28] Britain had invested $2 billion in the plane as of the spring of 2006, when the discussions began about their concerns over the US not sharing this technology.[29] This disagreement was subsequently resolved.

Multinational military operations require a degree of synergy between the technologies of the various allied powers. Compatibility in equipment, computer systems and communications systems is important in coordinating forces. Indeed, operations and maintenance for equipment used in joint operations is easier if the equipment between the allied forces is similar and compatible. Trans-Atlantic alliances can promote this compatibility, not only in the case of the JSF, but for subsequent products. The intention of Secretary of Defense Robert Gates to purchase more JSFs, as announced in April, 2009, emphasized not only the high quality of the collaboratively produced plane, but also the commitment of the US to systems which are compatible with its allies and which are developed through global alliances.

Alliances between Firms in an Era of Fiscal Constraints

With the tightening of fiscal constraints, as well as the shifting demands for new technologies as defense priorities evolve, larger firms tend to form alliances with

26 "F-35 Lighting II," 2009.

27 Wall, 2003.

28 "Politics and Economics," 2006.

29 "Strains in the Alliance," 2006.

each other in an effort to share knowledge and skills. Moreover, as the market for defense equipment potentially contracts and fewer competitions could occur in the future, large firms form alliances to win the competitions. In many cases, the competitions are held to replace older equipment, which make them a powerful source of revenue for defense contractors.

For example, as of the fall of 2013, the US Army wanted to acquire a new vertical lift aircraft – the Joint Multirole Demonstrator – which would ultimately replace the Army's current helicopters, especially the Apache attack helicopter and the Blackhawk utility chopper by the mid-2030s. As a result, the Army awarded technology investment agreements to Karem Aircraft, AVX Aircraft Co., Sikorsky Aircraft, and Bell Aircraft. The funding agreements matched investments from the companies over six years with funding from AMRDEC (the Army's Aviation and Missile Research, Development, and Engineering Center). As a result, Bell Helicopter teamed with Lockheed Martin to produce the V-280 Valor, which would be the successor to the V-22 Osprey, while Sikorsky teamed with Boeing to provide a helicopter closely modeled on Sikorsky's X-2 demonstrator. Neither AVX Aircraft nor Karem Aircraft had built operational aircraft,[30] which provides an advantage to the alliances between larger firms.

A second example is the alliance formed by Boeing and Lockheed Martin, announced in October, 2013, for the Air Force's Long Range Strike Program, which is the second time that they have formed an alliance on the next-generation bomber. The Long Range Strike Bomber would replace the B-52s and B-1s.[31]

Boeing and Lockheed formed an earlier alliance in 2008, which ended in 2010 when the requirements for the bomber changed. As of the fall of 2013, the Air Force's top priorities are the bomber program, the F-35, and the KC-46 tanker. Consequently, Boeing and Lockheed Martin's involvement in the strike bomber program, should they succeed in obtaining the contract, further strengthens their role in the defense industrial base, since Lockheed Martin is the lead contractor on the F-35 program and Boeing is the primary contractor for the KC-46 tanker. Indeed, Boeing was also the contractor for the B-52s, which, on average, are over 50 years old. As of 2013, Northrop Grumman is a significant contender, partially due to its prior experience with the stealthy B-2 Spirit. The sequestration program, as of 2013, had not impacted the funding for the Long Range Strike Program. The Air Force is likely to purchase 100 of these bombers, at $550 million each, and the bombers are expected to enter service in the mid-2020s.[32]

New bomber programs are begun by the Air Force every three decades, which makes the winner of the competition likely to be in a significant position for a long time. Northrop Grumman, which built the B-2 in the late 1970s, now has a niche in unmanned systems and plans to explore whether the design for the new bomber can be adjusted to be flown unmanned. The Air Force, which provided

30 Parsons and Insinna, 2013.

31 Barnes and Cameron, 2013.

32 Mehta, 2013; Barnes and Cameron, 2013.

$8.7 billion in R&D over the next five years for the bomber, hopes to seek bids on the potential contract.[33]

Expanding overseas sales has also been important for the profitability of US defense contractors, with the shrinking defense market. As a result, US defense contractors have formed alliances with each other in order to submit competitive bids in overseas competitions. For example, the US bid (which ultimately lost to China's bid) in the Turkish competition in 2013 to provide it with a missile defense system involved an alliance. The US bid involved Lockheed Martin and Raytheon teaming, such that Raytheon's Patriot air defense batteries would fire Lockheed Martin's Patriot PAC-3 interceptor missiles. Nevertheless, in 2012, the United Arab Emirates spent $1.1 billion in purchasing 48 missile batteries made by Lockheed Martin – Terminal High Altitude Area Defense (THAAD) – which uses Raytheon's AN/TPY-2 radar.[34]

International sales can also lead to alliances between US/EU defense contractors and the defense contractors in a given country, as the country, in addition to buying equipment from the western defense contractors, also tries to obtain knowledge to build its defense industrial base. Countries in Asia, Latin America, and the Middle East which import defense equipment from US, EU, and other foreign defense contractors are trying to build their defense industrial bases. As a result, their domestic defense contractors often engage in knowledge-sharing alliances with US, EU, and foreign defense contractors to learn more about designing and building key defense equipment. Indeed, India has engaged in alliances with Western defense contractors to gain knowledge; examples include: Mahindra Defense systems, which engaged in a $20 billion joint venture beginning in January 2009 with BAE to develop mine protected vehicles; Larsen & Toubro, which teamed with Raytheon in February 2010 to upgrade the Indian Army's Russian origin T-72 tanks; and Hindustan Aeronautics, which has been building the Hawk 132 Advanced Trainer Jet under license from BAE.[35] Moreover, India and Russia have considered developing alliances: in 2010, Russia's Sukhoi Desin Bureau and India's state-owned Hindustan Aeronautics Ltd. considered forming an alliance to jointly develop the FGFA, but, as of the fall of 2013, a final contract had not been reached.[36]

Conclusion

As seen in this chapter, alliances between domestic and international defense manufacturers have become increasingly important due to the emergence of the common global threat of terrorism, the greater price sensitivity of governments concerning weapons systems costs, and the shrinkage of defense budgets. As a

33 Barnes and Cameron, 2013.

34 Erwin, 2013b.

35 Misquitta, 2009; Krishna, 2010.

36 Raghuvanshi, 2013.

result of national security concerns and integration costs, alliances can often be easier to develop than mergers and can ultimately provide a prelude to an eventual merger between the parties if the alliance is successful. They can provide many of the benefits of mergers, such as sharing R&D costs and/or allowing access into new markets, without many of the costs of mergers – difficulty in exiting, substantive integration costs, etc. Nevertheless, development of new technologies can lead to mergers of larger and smaller firms in an effort to exchange knowledge for capital on a more permanent basis, as in the cybersecurity market and the UAV market. Alliances between larger firms to develop new products are more likely than mergers of large firms due to concerns about increasing market power.

The case studies in this analysis highlighted the role of alliances in achieving various outcomes: spurring alliances between competitors to ultimately create a market with several new products (CFM International and International Aero Engines); promoting national defense strategies (Boeing's alliance with EADS and other manufacturers in the missile arena); sharing R&D costs (the failed alliance between Boeing and other manufacturers to build a "super-jumbo" jet); and developing interoperable equipment between allied nations (the JSF). Moreover, the tendency for larger defense contractors to form alliances to win large competitions in an era of potentially fewer future competitions is evident in the alliance formed between Boeing and Lockheed for the Long-Range Strike Bomber and the alliances formed by other large firms for the Army's Joint Multirole Demonstrator. Moreover, alliances between US/EU defense contractors and overseas defense contractors often occur as the overseas nations try to build their defense industrial bases.

As countries are increasingly faced with budgetary constraints as a result of the current financial crisis, the fiscal strains imposed by an ageing population, and other areas, such as education, infrastructure, etc., defense budgets will likely be under more pressure. Moreover, there will be a greater emphasis on obtaining innovative weapons systems products at low costs and in a timely manner. Consequently, there will increasingly be significant opportunities for global alliances in the defense sector to play a valuable role in helping governments meet the challenges of the new millennium.

References

"Analysts Assess Damage to Boeing in Aftermath of Contract Loss." 2009. *Marketwatch*, March 3.

Asker, James R, Barrie, Douglas, and Taverna Michael A., 2002. "US, European Firms Join on Missile Defense." *Aviation Week and Space Technology*, July 29.

Barnes, Julian E. and Cameron, Doug, 2013. "Boeing, Lockheed Pursue Bomber." *Wall Street Journal*, October 26–27, B4.

Barrie, Douglas and Taverna, Michael, 2002. "Allure of Pentagon Purse Sustains European Interest." *Aviation Week and Space Technology*, July 22.

Beattie, Alan, 2008. "US Issues Warning on Stoking Protectionism." *Financial Times*, May 5.

"Business: Hands Across the Sea: Defense Alliances." 2002. *The Economist*, July 27.

Butler, Colin, Kenny, Brian and Anchor, John, 2000. "Strategic Alliances in the European Defense Industry." *European Business Review*,Vol. 12, Issue 6.

"Buy America." 2006. *Government Executive*, April 15.

Cole, Jeff, 1995. "Boeing-led Alliance Halts 'Super-Jumbo' Jet. *Wall Street Journal*, July 10.

Dallmeyer, Dorinda, 1987. "National Security and the Semiconductor Industry." *Technology Review*, November/December, pp. 47–55.

Doz, Y. and Hamel, G., 1998. *Alliance Advantage: the Art of Creating Value through Partnering*. Boston, MA: Harvard Business School Press.

Drawbaugh, Kevin, 2008a. "Congress in turmoil over Air Force tanker decision." *Reuters*, February 29.

——. 2008b. "U.S. Congress Roiled by Air Force Tanker Decision." *Reuters*, March 3.

Erwin, Sandra I., 2013a. "Companies See Bright Spots in Bleak Market." *National Defense*, November.

——. 2013b. "Missile Defense Market Getting Tougher for U.S. Firms." *National Defense*, October 29.

"F-35 Lightning II." *Wikipedia*. Available at: http://en.wikipedia.org [accessed April 12, 2009].

Fryer-Biggs, Zachary, 2013. "Industry Girds for Extended US Sequester." *Defense News*, October 20.

Gomez-Casseres, B., 1994. "Group versus Group: How Alliance Networks Compete." *Harvard Business Review*, July/August, Vol. 72, No. 4.

Hepher, Tim, 2008. "Eads Shares Soar after Big US Defense Deal." *Reuters*, March 3.

Hinton, Christopher, 2008. "Boeing Files Protest over the Air Force Tanker Award." *Marketwatch*, March 11.

Krishna, E. Jai, 2010. "Larsen Eyes Revenue from Defense Nuclear Power." *Wall Street Journal Online*, February 16.

Liedtka, J.M., 1998. "Synergy Revisited: How a 'Screwball Buzzword' Can Be Good for the Bottom Line." *Business Strategy Review*, Vol. 9, No. 2. pp. 45–55.

"Lockheed Martin Aeronautics Company: F-35: Delivering on the Promise to Redefine National Strategic Capabilities." 2008. *Economics and Business Week*, October 4.

Lynch, David J., 2006. "Some Would Like to Build a Wall Around the US Economy." *USA Today*, March 15.

Mehta, Aaron, 2013. "Boeing, Lockheed Team on Long Range Strike Bomber." *Defense News*, October 25.

Misquitta, Sonya, 2009. "Defense Contractors Target Big Jump in India's Military Spending." *Wall Street Journal*, July 17, p. B1.

"Northrop Group Wins $35B Air Force Deal." 2008. *CNNMoney.com*, February 29, 2008.

"Northrop Grumman Fires Back on Tanker Debate." 2008. *CNBC.com*, March 5, 2008.

Parsons, Dan and Insinna, Valerie, 2013. "Army Selects New Helicopter Designs." *National Defense*, October 2.

Pearlstein, Steven, 2006. "Ports Furor is Just Protectionism, with a French Accent." *Washington Post*, March 1.

Platt, Gordon, 2006. "Cross-Border Acquisitions Facing Growing Interference from Tighter Security Reviews." *Global Finance*, May.

"Politics and Economics: US, UK are Likely to End Clash That's Marred Fighter Jet Project." 2006. *Wall Street Journal*, December 12.

Raghuvanshi, Vivek, 2013. "India Extends Relations with China, Russia." *Defense News*, October 25.

Randolph, Monique, 2008. "Top Acquisition Official: Tanker Acquisition Top Priority." *Air Force Print News Today*, July 21.

Shalal-Esa, Andrea and Wolf, Jim, 2008. "Pentagon Reopening Contest to Build Aerial Tankers." *Reuters*, July 9.

Shalal-Esa, Andrea, 2008a. "Air Force Agrees to Brief Boeing on Tanker Loss." *Reuters*, March 4.

——. 2008b. "U.S. Plans Expedited Rerun of Aerial Tanker Contest." *Reuters*, July 9.

Shearer, Brent, 2006. "Raising Barriers to Inbound Deals." *Mergers and Acquisitions*, May.

Smith, David J., 1997. "Strategic Alliances in the Aerospace Industry: A Case of Europe Emerging or Converging?" *European Business Review*, Vol. 97, Issue 4.

Sparaco, Pierre, 1999. "Security Concerns Impede Alliances." *Aviation Week and Space Technology*. April 26.

"Strains in the Alliance." 2006. *Time*, April 10.

Taverna, Michale A., 2001. "Raytheon, Thales Form Defense Venture." *Aviation Week and Space Technology*, January 1.

"Tech Titans See Opportunity in Cybersecurity." 2013. *San Jose Mercury News*, September 19.

"US-Euro Strategic Alliances Will Outpace Company Mergers." 2001. *Aviation Week and Space Technology*, December 3.

Wall, Robert, 2003. "Fear." *Aviation Week and Space Technology*, July 28, 2003.

Wallace, James, 2008. "Boeing Air Tanker Bid Takes Hit." *Seattle Post-Intelligencer*, August 7.

Warwick, Graham, 2009. "Making it Happen." *Aviation Week Space and Technology*, June 8, 2009, Vol. 170, Issue 23.

Wolf, Jim and Shalal-Esa, Andrea, 2008. "Northrop-Eads Beats Boeing to Build US Tanker." *Reuters*, March 2.

Yoshimo, M. and Rangan, U., 1995. *Strategic Alliances: An Entrepreneurial Approach*. Boston, MA: Harvard Business School Press,.

Chapter 5
Globalization and the US Defense Industrial Base: A Case Study of the Tanker Competition between Boeing and Northrop Grumman/EADS

The growth in the global economy and the trend toward outsourcing have given rise to concerns over the composition and strength of the US industrial base, as well as the degree to which the US is dependent on other countries for certain goods and commodities. These concerns have appeared across a variety of industries in the dialogue between Congressional representatives and their constituents, between companies and their employees, and between policymaking representatives of different nations. The dialogue becomes particularly heated when the industries involved are deemed important to national security and to the US defense industrial base.

This chapter examines the concerns surrounding the alleged inroads of foreign manufacturers into the US defense industrial base, as well as the background behind the concerns, with a specific focus on the recent competition between Boeing and a team composed of Northrop Grumman and European Aerospace and Defense Systems (EADS) over what may be the second largest defense contract in US history to supply the USAF with a new fleet of aerial refueling tankers. The tanker competition is a case study of the difficulties in preserving the US defense industrial base with the increasing globalization of the defense industry and the challenges of:

a. potentially facing the loss of the position of the incumbent domestic contractor;
b. the international damage when foreign bids are turned down, given the perception of an open market; and
c. the conflict over which states obtain employment and economic growth through the awarding of a contract.

In the 2008 tanker competition, the potential entrant (Northrop/EADS) deployed a different strategy in an effort to enter the market than previous entrants. The foreign entrant, EADS, teamed up with a US manufacturer (Northrop) and directly invested in the US defense industrial base by creating jobs and building facilities. In many of the historical cases, on the other hand, foreign involvement in the defense sector involved the entry into the US markets by acquiring a US company or by producing the products overseas and importing them into the US.

A Historical Taxonomy of Strategies to Limit the Role of Foreign Manufacturers in the US Defense Industrial Base

National security concerns have played a significant role in a variety of different industries. The desire to limit the role of foreign manufacturers in the US defense industrial base has taken on several forms. These include:

a. blocking mergers between US and foreign companies to prevent greater dominance of foreign firms in industries deemed important to national security;
b. the introduction of tariffs, quotas, and Voluntary Restraint Agreements (VRAs) to protect the domestic industry; and
c. attempts to strengthen Buy America legislation.

In all of these cases, the vocal role of Congressional representatives from areas which potentially would lose business if foreign competitors entered the market has been important. This section provides historical examples of the deployment of these strategies.

Strategy No. 1: Prevent Foreign Entrants from Merging with Domestic Competitors

The strategy of preventing potential foreign entrants from obtaining a strong foothold in the domestic market by acquiring a domestic firm has often been used, both in the US and in other countries. Supporters of the strategy have argued that the sector concerned is of key importance to the domestic defense industrial base. The acquisition has either been formally disallowed or the foreign entrant has withdrawn its bid in anticipation that the acquisition will be blocked if it proceeds further.

Acquisitions of US companies by foreign companies are usually protested by Congressional representatives and other US industrial competitors, but not always by the Pentagon or regulatory agencies. The Committee on Foreign Investment in the United States (CFIUS) was created in 1975 to balance the costs and benefits of foreign investment in the US. It has the right to block acquisitions of US companies by foreign companies if such an acquisition would impact national security. The committee conducts a 30-day review and then undertakes a more in-depth 45-day investigation if problems arise in the 30-day probe. Between 1988 and 2006, it only rejected one out of over 1,750 potential acquisitions – the attempted acquisition by China National Aero-Technology Imports and Exports to purchase a manufacturer of aircraft parts in Seattle in 1990. Only 25 cases were fully investigated and only 12 of them were forwarded to the President for a decision.[1]

An example of the strategy of preventing a foreign company from obtaining a foothold by acquiring a domestic company occurred in the US semiconductor industry in the 1980s. The Japanese firm Fujitsu announced that it planned to purchase 80% of Fairchild Semiconductors, which was the second largest seller of chips to the US military. The US semiconductor industry was important for early warning, air-defense, and air-to-surface attack systems, naval surface warfare, tanks, and conventional artillery. Between 1978 and 1987, Japan had increased its share of the semiconductor industry from 28% to 50%, while the share of the US in semiconductors had fallen from 55% to 44%. For particular types of chips, such as DRAM chips, the share of US companies fell from 90% in 1975 to 5% by 1986. A Congressional outcry ensued following Fujitsu's proposal. Senator Howard Metzenbaum of Ohio argued that jobs would be lost, while B. Jay Cooper, the Press Secretary for the Department of Commerce, argued that the deal would place "vital national interests at stake." Several proposals were suggested, including a proposal that the merged firm would not be allowed to have military contracts and a proposal that Fairchild would not provide Fujitsu with military technology. The outcome of the protests was that, in March, 1987, Fujitsu withdrew its offer, and National Semiconductor bought Fairchild and became the sixth largest chipmaker in the world. It is doubtful, however, whether national security would have been compromised had the Fujitsu-Fairchild merger taken place since DoD could have acquired 95% of Fairchild's products through other suppliers. Moreover, although this was not often mentioned in the debate, Fairchild had been owned by Schlumberger, a French company, since 1979.[2]

Another example of a failed attempt at entering the US market occurred in 2005 in the US oil sector. As in the semiconductor case, the foreign acquirer withdrew its offer due to a substantive Congressional outcry. China National Overseas Oil Corporation (CNOOC), a Chinese state-owned company, tendered a bid to purchase Unocal Corp for $18.5 billion. Chevron, the other bidder, was offering only $17.1 billion, but it mobilized Congressional representatives to express their concerns about a Chinese firm playing a significant role in the US oil sector. In the wake of this outcry, CNOOC withdrew its bid even before the CFIUS review and Chevron acquired Unocal. Similarly, Check Point Software Technologies (Israel) decided not to pursue its acquisition of the US software company Sourcefire when national security concerns were raised.[3]

Concerns over national security can lead to some form of separation or divestiture of operating units linked to the defense sector so that the rest of the acquisition can proceed. One example was the merger of Alcatel (a French firm) and Lucent Technologies. Since Bell Labs, a division of Lucent, had undertaken a

1 "Buy America," 2006.
2 Dallmeyer, 1987, pp. 47–55.
3 Shearer, 2006.

number of projects for the US government, Bell Labs would be insulated from the new firm and would become a separate US subsidiary with an independent board. A second example was the concern over the acquisition of Peninsula & Orient Steam Navigation Co. (P&O), a British firm, by the state-owned Dubai Ports Worldwide (DPW). This acquisition provoked a Congressional outcry because it would have resulted in a foreign company managing six US ports. DPW agreed to sell the ports to a US company in the wake of strong Congressional opposition.[4]

The US is not the only country, however, which uses protectionism to block mergers. For example, Dominique de Villepin, who served as the French Premier, designated 11 sectors of the French economy as sensitive for national security and blocked the merger of PepsiCo (US) and Danone (French) under "economic patriotism." He further encouraged the merger of Gaz de France (a French gas supplier with significant state involvement) with Suez (a French power and water supplier) at the expense of a bid by the Italian company Enel. Similarly, Italy has blocked foreign takeovers of many of its banks.[5] New Zealand blocked a bid from Canada to acquire Auckland Airport, while Canada prevented Alliant Techsystems (US) from acquiring the satellite and space robotics arm of MacDonald-Dettwiler. As globalization exposes vulnerabilities, it is likely that countries will continue to promote domestic champions by preventing foreign acquisitions through protectionist concerns linked to national security.

Strategy No. 2: Tariff Protection

A second strategy to protect domestic industries is to enact tariffs or quotas. The evidence is mixed on the degree to which domestic job preservation offsets the potentially higher prices and reduced quantities facing US consumers, whether tariff/quota protection is welfare enhancing, and who wins and who loses. A number of studies have been done on the impact of tariffs and quotas in different industries. Examples include Bergstrom (1982) and Maskin and Newbery (1990), both of whom examined tariffs in the oil industry, Razin and Svensson (1983) and Gardner and Kimbrough (1990), both of whom examined the impact of trade-balance triggered tariffs, and Thompson and Reuveny (1998), who provide a broader historical perspective on the impact of tariffs on trade fluctuations.

Under section 232 of the Trade Enhancement Act of 1962, manufacturers in industries can seek tariff protection or other regulatory protection from imports for national security reasons. For example, President Carter and President Ford imposed oil import taxes to stimulate the US oil production industry to be more self-sufficient.[6] During the 1980s, the US government also attempted to protect the domestic steel and auto industries from foreign competition through VRAs, which increased tariffs or increased import quotas on foreign imports. Indeed,

4 Lynch, 2006; "Buy America," 2006; Shearer, 2006.

5 Pearlstein, 2006; Platt, 2006; Beattie, 2008.

6 Bradsher, 1993.

some industry leaders in the steel industry argued that VRAs were a response to the fact that overseas producers were often supported by their governments, which enabled them to offer lower prices in the US markets. US steel manufacturers pointed to the subsidies that British Steel, Finsider (state-owned Italian steel firm), and French, Spanish, and Brazilian steel firms received from their governments,[7] and argued that VRAs and tariffs helped to provide them with a more level playing field. Another example of tariff protection for a sector important to the defense industrial base was the imposition of tariffs on steel imports into the US in 2002.

Strategy No. 3: Strengthening the Buy America Legislation

Several studies have examined the impact of discriminatory procurement practices on foreign trade, including Collie and Hviid (2001), who examined international procurement as a signal of export quality and discussed the role of "buy British" sentiments in the UK government's choice of army ambulances in 1996. The application by the US of the Buy America Act has become a source of concern for overseas manufacturers, especially over the last several years. The legislation was originally passed in 1933 to protect the US defense industrial base and manufacturing sectors by requiring federal agencies to buy products with "substantially all" US content, where "substantially all" came to mean, over time, about 50% of the content.[8] Indeed, Lowinger (1976) examined the impact of the Buy America legislation during the 1960s and found that it had significantly reduced imports. As many US sectors waned, even prior to the 9/11 attacks, other countries felt that the US was conflating a broad definition of national security interests with a desire for protectionism. Indeed, in a July, 2001 report on barriers to trade and investment in the US, the EU argued that the broad definition of US national security interests was leading to a "disguised form of protectionism."[9]

In the fall of 2003, Duncan Hunter, chairman of the House Armed Services Committee, introduced a stronger Buy America provision in the FY 2004 Defense Authorization Bill that would have required the Pentagon to choose components for military systems which, wherever possible, were made in the US. US defense contractors opposed the bill because they feared retaliation in overseas markets, as did the White House and other agencies that feared that US trade obligations would be infringed on. The EU, the Pentagon, and John Warner, who chaired the Senate Armed Services Committee, opposed the provision. The final provision in the Defense Authorization Bill was considerably weaker, and only required the Pentagon to examine the capabilities of US producers to make systems, to develop the Defense Industrial Base Capabilities Fund to support them, to encourage US defense manufacturers to use domestic parts, and to keep a list of foreign

7 Modic, 1989, pp. 85–8.

8 Manzullo, 2003.

9 "EU Voiced Concern on US 'Protectionism,'" 2002.

manufacturers who had previously restricted sales of military goods to the US due to its military operations.[10]

Mr Hunter again introduced legislative language in the House version of the FY 2005 Defense Authorization Bill to restrict the Pentagon's purchases from foreign suppliers, but, again, this triggered opposition from the major US defense contractors, such as Lockheed-Martin, Raytheon, and Boeing, as well as the Bush administration, and several foreign governments, including the UK. Indeed, the UK defense secretary, Geoff Hoon, noted that the UK would consider retaliating and buying fewer US weapons products if the bill were passed.[11]

At times, the Buy America legislation can compromise national security. For example, at the beginning of the Iraq war, there was a shortage of body armor. Polyethylene fiber is an important component for body armor, and it was only made by two companies – Honeywell and DSM, a Dutch firm which has plants in Greenville, NC and in Europe. The US government requested both firms to produce as much fiber as they could in a short period of time. Unfortunately, demand outstripped domestic production by those two companies. DSM's European plants had plenty of fiber, but the Buy America rules prevented it from importing the fiber into the US.[12]

The Need for a Different Strategy? The Tanker Competition between Boeing and Northrop Grumman/EADS

The 2008 competition between Boeing and a team composed of Northrop Grumman and European Aerospace and Defense Industries (EADS) to supply the US Air Force (USAF) with a new fleet of aerial refueling tankers posed a different challenge and required a different set of strategies. This contract, which may be the largest defense contract in history with the exception of the F-35 Joint Strike Fighter, involved a competition between two US firms, one of which was allied with a European firm. EADS, through its Airbus subsidiary, was a competitor to Boeing in the commercial market. Unlike some of the earlier examples, this situation did not involve an acquisition, nor did it involve EADS serving as the primary contractor and attempting to import its products into the US. The Congressional reaction here was split due to EADS' plans to invest in the US industrial base by building facilities in the US and creating US jobs, as well as because it allied itself with another US firm. Consequently, this is a landmark case not only in terms of the size of the defense contract, but also in terms of the relationship of the US with the broader European defense market and the impact of the US reaction to the tanker competition on global perceptions concerning the openness of the US defense market.

10 Brun-Rovet, 2003; Steffes, 2004.

11 Speigel, 2004.

12 Wayne, 2005.

The 2008 tanker competition was very important to the USAF because the average age of the existing KC-135 tankers was 47 years and the planes were first put into service in 1957. The Air Force had 531 tankers from the Eisenhower period and 59 tankers built by McDonnell Douglas in the 1980s, prior to its merger with Boeing (1997). In an effort to replace the tankers, the USAF attempted lease and buy 100 modified KC-767 tankers from Boeing for $23.5 billion, but this was overturned in 2004 following a Pentagon procurement scandal, in which one of the key Air Force procurement officials, Darleen Druyen, and the CFO of Boeing, Michael Sears, went to jail. Consequently, this competition between Boeing and Northrop Grumman/EADS represented the second attempt at replacing the USAF's aging tanker fleet. The 2008 competition was initially over a $1.5 billion contract, covering four test aircraft. The intent was then to buy 175 more planes, for a total value of $35 billion. The Air Force hoped to operate the new tankers in 2013. While the $35 billion amount would stretch over 10–15 years, an additional $60 billion in revenue could come from maintenance and parts, such that the overall contract would be worth approximately $100 billion.[13]

In this competition, Boeing displayed the behavior of a traditional incumbent. It had been the provider of refueling tankers to the USAF for almost 50 years and had what was often referred to as a "monopoly." Consequently, it was strongly favored to win the competition. Indeed, prior to the announcement of the winner, a poll was conducted of 10 industry analysts in which all of them predicted a win by Boeing.[14] As a result, when the Air Force announced that the Northrop Grumman/EADS team had won the contract on February 29, 2008, Boeing indicated shortly afterward that it was dissatisfied with the decision.

There was, however, some indication prior to the announcement, that the Air Force was concerned about a protest from the losing competitor. This could have been because the contract was so lucrative and important and they felt that the loser would be disappointed. In addition, some officials may have anticipated that if Boeing, the incumbent tanker manufacturer, lost the contract, it would be more likely than Northrop Grumman or EADS to launch a protest. As early as February 22, 2008, it was reported that "the Air Force has said it expects a protest and has been extra careful in documenting its decision-making process." General Mosely (USAF Chief of Staff at the time), in a February 28 statement, noted that he hoped that whoever lost the contest would not challenge the result by lodging a protest with General Accounting Office, which would delay the timeline for the delivery of the tankers to the USAF.[15]

Boeing did, in fact, launch a protest with the GAO on March 10. It argued that the Air Force, during the course of the competition, had changed its requirements on the amount of ramp space and how closely the tankers could be parked to

13 "Analysts assess damage to Boeing in aftermath of contract loss," 2009; "Northrop Group Wins $35B Air Force Deal," 2008; Hinton, 2008c.

14 "Northrop Group Wins $35B Air Force Deal," 2008; Wolf, and Shalal-Esa, 2008.

15 Wolf, 2008a.

each other and that "the changes were designed to keep them [Northrop] in the competition." [16] Boeing felt that the process was "replete with irregularities," which "placed Boeing at a competitive disadvantage." Many analysts felt that Boeing's protest would not succeed. For example, Myles Walton, an analyst at Oppenheimer & Co, stated "given the initial judgment by the Air Force combined with the Northrop team's better score on four out of five criteria, we anticipate Boeing's protest will be denied."[17] Moreover, complaints have often been unsuccessful with the GAO. Only 249 of the 1,327 bid complaints lodged with GAO in 2006 received an official decision, in 71% of which the GAO denied the complaint and supported the government's earlier decision. In FY 2007, of the 1393 cases filed and closed, 16% of them were ruled to have merit by GAO. [18] Nevertheless, on June 19, the GAO recommended that the competition be reopened and upheld eight of Boeing's 100 protests, although they stated that they found no evidence of "intentional wrongdoing" by USAF procurement officials.[19]

In evaluating the competition, it is important to examine the two proposals since the differences between the two proposals and the specifications of the Air Force's Request for Proposal (RFP) partially drove Boeing's protest, although it played a much smaller role in the Congressional protest (as will be discussed). The two proposals differed from each other in size and in fuel offloading capacity. The Northrop/EADS proposal, the KC-45a, was a modification of the Airbus A330. Boeing, on the other hand, offered a variant of its traditional 767 model. Air Force General Arthur Lichte noted that the KC-45a provided "more passengers, more cargo, more fuel to offload" and that the bigger capacity of the Northrop/EADS tanker had been an important consideration in awarding the contract.[20] The Northrop tanker carried more fuel – 250,000 pounds – than the Boeing tanker at 202,000 pounds. On the other hand, Boeing's 767 model had some advantages over the Airbus 330 model in that it could land on narrower, shorter airstrips, such as those in developing countries in Africa or in Afghanistan.[21] Boeing argued that:

> As the requirements were changed to accommodate the bigger, less capable Airbus plane, evaluators arbitrarily discounted the significant strengths of the KC-767, compromising operational capabilities, including the ability to refuel a more versatile array of aircraft such as the V-22 and even the survivability of the tanker during the most dangerous missions it would encounter.[22]

They further noted that:

---

16 Hinton, 2008c.
17 Rigby, 2008b.
18 "Boeing to Protest $35B Tanker Deal," 2008; Crown, and Epstein, 2008.
19 Randolph, 2008.
20 "Northrop Group Wins $35B Air Force Deal," 2008.
21 "Tanker Deal: Why Boeing Shouldn't Protest," 2008; Hinton, 2008a.
22 "Boeing Protests U.S. Air Force Tanker Contract Award," 2008.

> Initial reports also indicate there may well have been factors beyond those stated in the RFP, or weighted differently than we understood they would be, used to make the decision.[23]

Jim Albaugh, CEO of Boeing's Integrated Defense Systems, argued:

> In our reading of the RFP, it wasn't about a big airplane. If they'd wanted a big airplane, obviously we could have offered the 777. And we were discouraged from offering the 777.[24]

Boeing also emphasized the benefits of their tanker's lower fuel requirements in an environment of rising fuel prices and released a report on March 17 stating that, over the next 40 years, it would cost the Air Force an extra $30 billion in fuel costs to operate the 179 Airbus A330–200 refueling tankers relative to a similar number of Boeing tankers since the A330–200 required 24% more fuel than the 767–200ER.[25]

The GAO's report that was issued in mid-June and upheld eight of Boeing's 100 protests focused on procedural errors in the competition, rather than an assessment of the two competing proposals. For example, the GAO argued that, although the RFP requested competitors to meet as many of the technical requirements as they could, the USAF did not take into account that Boeing had satisfied more of the non-mandatory technical requirements than Northrop/EADS. In addition, the GAO stated that although the RFP had noted that "no consideration will be provided for exceeding [key performance parameter] objectives," the Northrop proposal was deemed superior partially because it did exceed a key performance parameter linked to aerial refueling capacity. Although the USAF informed Boeing at one point that it had satisfied a key performance parameter objective, it later decided that Boeing had not met that objective, but did not inform Boeing. Finally, the GAO identified a series of errors in the USAF's calculation of the lifecycle costs of the two proposals and felt that, when these errors were corrected, Boeing's proposal had a lower lifecycle cost than the Northrop/EADS proposal.[26]

Was Boeing's original protest rooted in its perception that the competition had operationally been run incorrectly or was Boeing concerned, as a traditional incumbent, that its turf was being encroached upon? First, without the tanker contract, the 767 production lines would be likely to wind down in the absence of additional orders. In 2007, Boeing only sold 36 Boeing 767s and, having sold 1,000 over the past 30 years, only had 51 left to deliver. On the other hand, the 787 Dreamliner, which is a successor to the 767, received 369 orders in 2007, although,

---

23 "Boeing Request Immediate KC-X Tanker Briefing," 2008.

24 Carpenter, 2008.

25 "Boeing: Study Projects That as Oil Prices Climb, 767 Tanker Most Cost Efficient," 2008.

26 General Accounting Office (GAO), 2008a.

due to substantive delays in the 787, preservation of the 767 line would enable 767s to provide 787 customers with interim lift capacity. Second, this was not the first time that Boeing and EADS competed against each other in tanker competitions and that EADS had won. Between 2001 and 2008, Boeing and EADS faced each other in competitions for tankers in six countries, of which EADS had won four of the competitions to supply a total of 25 planes (Saudi Arabia, UAE, Australia, and Britain), while Boeing had won in Italy and Japan to supply eight planes.[27]

The reasons why EADS triumphed over Boeing in some of the other competitions was that the contract was awarded to the bidder which seemed the most sensitive to the needs of the client, the most flexible, and the most willing to make investments in the relationship. EADS, as a newcomer in the tanker business, manifested the traditional behavior of a successful entrant in terms of being innovative and absorbing risk, while Boeing played the role of the traditional incumbent. For example, Boeing and BAE Systems in the UK competed against EADS for a $26 billion contract to replace the UK's fleet of military refueling tankers in 2004 and lost the contract to EADS, which had not built a tanker before and which had proposed a modification of the commercial Airbus A330. The UK felt that EADS was more willing to make concessions and assume the financial risk in constructing the planes and then leasing them to the government, whereas Boeing did not offer such terms. Although Boeing's C-17 transport plane had been successful there, the tanker business had been handled by a different division of Boeing than the C-17. The competition in Australia provides another illustrative example in that the Australian government was impressed by EADS' willingness to use its own R&D money to develop and test a boom, whereas Boeing used an older boom in its proposal and suggested that it would build a newer type of boom only if it won the large US contract.[28]

Boeing's repeated loss in tanker competitions prior to 2008 suggested the possibility towards future erosion of its defense businesses. Its delays on the 787, combined with the reputational loss from the earlier tanker procurement debacle in 2004, were not helpful to its image at the time. If the trend had continued in which Boeing lost tanker competitions and Japan and Italy ended up with "orphan fleets" – i.e. they were the only countries with Boeing tankers – then their fleets could have cost more to maintain than if Boeing had developed the scale economies in costs to maintain the parts through obtaining other contracts, especially the US contract. As a result, other potential customers would have been less likely to choose Boeing in the competition when they saw these higher maintenance costs and the cycle would become self-reinforcing. The loss of the tanker contract in itself would not have affected Boeing on an annual basis, in that it only would have led to 12–18 additional tankers per year, which would have been a small number in comparison to the 450 commercial aircraft that it made each year.[29] But, since

27 Rigby, 2008a; "Boeing's Trouble with Tankers," 2008; Hinton, 2008b.

28 "Boeing's Trouble with Tankers," 2008.

29 Tessler, 2008.

it was a very large contract in the long run, at a time when defense expenditures could have plateaued, it could have had long-range significance.

Boeing's strategy was to focus its protest on alleged irregularities in the course of the tanker competition. Unlike the Congressional representatives, it did not focus its arguments as much on problems associated with EADS being a European firm.

This could be because it realized that it was a foreign entrant into the defense sectors of many other countries, which made it much harder for it to focus its arguments on the unfairness of a government outsourcing a contract to a foreign supplier. Boeing has traditionally sold a number of aircraft to foreign countries: it sold C-17 planes to the UK, Australia, and Canada; F-15 jets to Japan, Korea, and Singapore; and aerial refueling tankers to Italy and Japan. Of the $66.4 billion which comprised Boeing's 2007 revenue, about $27.1 billion came from overseas sales (commercial and military). Sales to Europe comprised $6.3 billion, of which 16% of that came from sales to the military.[30]

The Congressional representatives, on the other hand, focused their arguments on the dangers inherent in allowing a foreign firm to obtain one of the largest defense contracts in US history. Consequently, in effect, a 2-pronged strategy was followed in which Boeing attacked the process and Congressional representatives attacked the issues linked to EADS being a foreign contractor. Boeing could not focus on the strategy of attacking EADS as a foreign contractor because Boeing was itself a foreign contractor in the defense markets of other countries. The Congressional representatives from the regions in Washington, Kansas, and Connecticut which would have benefited if Boeing had received the contract strongly protested the decision and argued that the Northrop/EADS proposal would send US jobs overseas. They were assisted by the labor unions. The AFL-CIO and the United Steelworkers Union echoed concerns about sourcing the contract to a foreign manufacturer.[31] In a statement reported on March 3, the General Vice President of the International Association of Machinists and Aerospace Workers, Rich Michalski, said, "President Bush and his administration have today denied real economic stimulus to the American people and chosen instead to create jobs in Toulouse, France."[32]

Congressional representatives also used additional tools which were available to them based on the committees on which they served. For example, Senator Maria Cantwell from Washington stated that she would block President Bush's nomination of the new Secretary of the Air Force, Michael Donley, until she became convinced that "the Air Force and Pentagon will conduct a fair tanker competition."[33] The House appropriations defense committee introduced a provision attached to the FY 2009 military spending bill. This provision, if enacted, would have forced

30 Tessler, 2008.

31 Shalal-Esa, 2008a.

32 "Boeing Calls for Air Force Review of Tanker Contract Awarded to Northrop," 2008.

33 Wallace, 2008.

the Pentagon to consider the impact of the tanker contract on the defense industrial base, which would have strengthened the control of Congressional representatives in defense procurement issues.[34] Representative Norm Dicks from the state of Washington noted that: "If the Air Force doesn't get it right, I am going to reserve all my options as a member of the Appropriations Committee to offer amendments and do anything I can to stop this thing from going forward."[35] Moreover, some Congressional representatives also suggested that they would block the funding for the tanker if the award were given to Northrop/EADS.

Unlike many other situations involving foreign acquisitions or importation of foreign goods into the US, however, Congressional representatives were divided in their desire to protest the award of the initial contract to Northrop Grumman/EADS. This was because the Northrop Grumman/EADS tanker would have created 48,000 jobs in the US, especially in Alabama, and the Alabama Congressional delegation was very supportive of the results. Senator Richard Shelby (Alabama) noted that the contract would bring 7,000 jobs to Alabama and that "Any assertion that this award outsources jobs to France is simply false."[36] About 60% of the Northrop/EADS tanker would have been made in the US. Some of the parts would be manufactured in Germany, France, Spain, and Great Britain, but assembly of the tanker would occur in Mobile, AL, where EADS planned to build the third largest manufacturing facility in the world and where it also planned to assemble a commercial freighter version of the A330.[37]

The Boeing tanker, on the other hand, would have benefited different areas of the country than those areas which would have reaped economic gains from the Northrop/EADS presence. About 85% of Boeing's tanker would have been made in the US, although the tail would have been made in Italy and the fuselage in Japan. Boeing argued that 44,000 new and existing jobs would have been assisted by the contract, across 40 states and 300 suppliers. Wichita, Kansas and Everett, Washington would have been major locations for tanker production and the engines in the tanker, made by Pratt & Whitney, would have been made in Connecticut.[38] Nevertheless, despite the fact that the Northrop/EADS tanker would create 48,000 jobs in the US, Kansas Representative Tiahrt continued to argue that "I cannot believe we would create French jobs in place of Kansas jobs."[39]

From the perspective of a foreign manufacturer attempting to establish a foothold in the US market, the decision of EADS to team with a US manufacturer and move production into dollar-zone countries was a sensible strategy. If it had instead attempted to acquire a US company in order to enter the US market, it

34 Sevastopulo, and Kevin, 2008; Wolf, 2008b.

35 "Northrop Grumman Had Originally Clinched the USAF Contract in February, with Boeing's Objections," 2008.

36 Drawbaugh, 2008a; Drawbaugh, 2008b.

37 Wolf, and Shalal-Esa, 2008.

38 Tessler, 2008.

39 Drawbaugh, 2008a.

would have run the risk of the merger being blocked on national security grounds. Similarly, EADS' strategy of teaming with a US manufacturer and attempting to shift production into the US avoided the problems inherent in producing the goods outside the US, including the distortions of tariffs or quotas being placed on its goods when they are imported into the US. Finally, EADS was exposed to the weak dollar, because it paid its suppliers in euros, but sold airliners in dollars. Moving production into dollar-zone countries, as it would have with the plant in Alabama, would have helped to mitigate its exchange rate risk.

The dialogue again raised the question of whether the US wanted the best product at the lowest cost or was attempting to create/maintain jobs. Some of the Congressional representatives simplified the debate into one of US jobs versus foreign jobs, but, as this chapter discussed earlier, US jobs would have been created through both proposals. The difference, however, lay in where the jobs were located. The debate of US jobs versus foreign jobs was more appropriately held in industries which faced a flood of cheaper and/or better-quality imports, rather than in this teaming arrangement between a US manufacturer and a European manufacturer. Senator McCain noted: "I've never believed that defense programs should be – that the major reason for them should be to create jobs. I've always felt that the best thing to do is to create the best weapon system we can at cost to taxpayers."[40] These thoughts were echoed in the comments of Pentagon acquisition chief John Young, who noted: "I don't think anybody wants to run the department as a jobs program," further arguing that lawmakers usually focused on asking him to reduce the costs of weapons systems and that a decision by Congress to ban sourcing of contracts to foreign companies could lead to reciprocal retaliation on the part of the Europeans.[41]

The initial award of the contract to the team of Northrop/EADS reinforced the perception of many that, as Defense Secretary Robert Gates had stated, "defense manufacturing is a global business,"[42] particularly as the US had allied with many other countries in combating the War on Terror. Many perceived this initial award as the harvest of improved relations with France and that it would have been much harder for European manufacturers to claim that US markets were closed to them. French President Nicholas Sarkozy stated on March 3: "If Germany and France had not shown from the beginning that we were friends and allies of the United States, would it have been possible to have such a commercial victory?"[43] Significantly, EADS failed in a similar competition in 2003, at the time when the then-President of France, Jacques Chirac, was opposing the US involvement in Iraq. Other defense firms also saw the award of the contract to Northrop/EADS as a positive move toward open markets. Bob Stevens, CEO of Lockheed, said, in response to the award of the initial contract, that this "should put an end to

40 Drawbaugh, 2008b.

41 Shalal-Esa, 2008a.

42 Northrop Grumman Fires Back on Tanker Debate," 2008.

43 Hepher, 2008.

well-worn laments that the US markets are closed to European interests or that the US is unwilling to partner for the long term with industry in NATO countries." Nevertheless, after the GAO handed down its ruling, Stevens tried to emphasize that the ruling was rooted in process-related issues and noted that "I don't think that the tanker should be viewed as a trade issue as much as an acquisition issue."[44] This may partially have been because of concerns of retaliation on the part of European manufacturers if Northrop/EADS lost the contract. Indeed, the EU's lucrative $118 billion defense market had recently opened up to more cross-border competition through a newly drafted European Commission directive on defense procurement.

Although Congressional representatives and the popular press noted that concerns about national security could be key in involving a foreign contractor in manufacturing US weapons systems, Boeing did not emphasize this point as much in its complaints over the awarding of the initial contract. This may have been because there have been other instances in which foreign contractors had worked on key US defense projects. For example, EADS worked on a $2 billion Army contract to replace 345 "Huey" helicopters, in addition to providing the Coast Guard with radar systems and search and rescue aircraft.[45] The presidential helicopter was partially built by Italy's Finmeccanica, while Britain's BAE Systems had been involved in a number of US DoD projects, since it purchased United Defense Industries in 2005.[46] Moreover, all classified military technology on the KC-45a would have been installed by Northrop after the aircraft was assembled, so that EADS would not have handled it.[47]

The USAF announced on July 9, 2008 that it would reopen the competition and would focus it on the eight areas of protest sustained by the GAO. Unlike the previous competition, which was overseen by the USAF, this competition would be overseen by John Young, the chief of weapons procurement and the Undersecretary of Defense for Acquisition, Technology, and Logistics at the Pentagon. Following the issuance of its new RFP, the Pentagon planned to provide the bidders with 45 days to present new proposals, with the winner of the new competition being selected by early January, 2009. The Air Force stated explicitly that, in the new competition, it would provide extra credit for a larger plane with additional fuel offload capacity.[48] Loren Thompson from the Lexington Institute noted that: "It appears that the Pentagon's default process is to fix the process and then award the contract to Northrop Grumman again."[49] The USAF planned to extend the definition of the time period over which lifecycle costs were calculated from 25 years to 40 years, which favored the Boeing proposal, due to its lower

44 Wardell, 2008.
45 Wingfield, 2008.
46 Lagorce, 2008.
47 Hinton, 2008e.
48 Shalal-Esa, 2008b; Shalal-Esa, and Wolf, 2008; Wallace, 2008.
49 Hinton, 2008b.

fuel costs. Nevertheless, it also planned to separate lifecycle costs from acquisition costs and to give greater weight to the nearer-term acquisition costs, which could reduce the advantages of Boeing's proposal.[50]

Boeing, faced with the opportunity to propose the larger 777 in light of this "extra credit" suggested in the draft RFP, claimed that it would pull out of the competition if it were not provided with more time to develop a modified 777. This would lead to a delay in the process and would push the decision of the tanker winner into the next Administration and Congress.[51] Boeing again deployed its Congressional tools to ultimately influence the process by creating a procedural delay to enable a potentially more favorable political landscape. Northrop/ EADS, however, attempted to counter that strategy by strengthening their base of Congressional support in the southeast. This is because they were considering the possibility of presenting a modification of the A330–200 freighter tanker, rather than the earlier proposal which had involved a modification of the A330 passenger aircraft. This would have enabled the development of greater economies of scale and cost efficiencies in its Mobile, AL plants and would have further enhanced the value of that operation and the consequent support of Congressional representatives in that area. Nevertheless, in early September, 2008, the USAF canceled the competition and announced that it would re-open it during the new administration.

The third competition for the KC-X tanker began in September, 2009, with 373 requirements.[52] The winner of the competition would deliver, over 12 years, 179 aircraft for $35 billion. A $3.5 billion fixed price contract was the initial contract, which would result in the development and delivery, by 2017, of 18 aircraft. All 415 of the KC-135Rs would be retired after two existing competitions took place in the subsequent decades.[53] The Air Force's focus on "warfighter effectiveness" would be included in judging the competition. Moreover, "the unit cost of each aircraft would be adjusted to reflect lifecycle cost over 40 years."[54]

Boeing and EADS were the competitors in this tanker competition, while Northrop Grumman withdrew from the competition. Initially, it appeared that Boeing would be the only competitor, however, EADS decided to enter after DoD extended the deadline to enter the competition by 60 days. EADS, as in the prior competition with Northrop Grumman, planned to provide tankers based on the A-330, supporting 48,000 jobs in assembly in Mobile, AL. As in the prior competition, Boeing planned to provide tankers based on the 767, which would support 50,000 jobs and which would be assembled in Seattle, WA with modifications in Wichita, KS.[55]

50 Hinton, 2008b; Wallace, 2008.

51 Epstein, 2008.

52 "Boeing Wins KC-X Tanker Battle," 2011; Garamone, 2009.

53 Hebert, 2010.

54 "Boeing Wins KC-X Tanker Battle," 2011; Garamone, 2009.

55 Hebert, 2010.

Issues were raised by the World Trade Organization (WTO) regarding government support received by Airbus (the parent of EADS). The WTO ruled in June, 2010 that Airbus had received illegal "launch aid" from governments in Europe in developing aircraft and that EADS would have a competitive advantage over Boeing. DoD, however, argued that the rulings of the WTO could not be included in determining the winner of the competition.[56]

A bid was also submitted in the competition from an alliance between a US firm and a foreign firm – US Aerospace (US firm) and Antonov Aeronautical (Ukrainian firm). Antonov had manufactured and designed a variety of different types of aircraft since 1946, including commercial airliners (the AN-148 and the AN-158) and strategic airlifters (AN-225 Mriya strategic airlifter and the AN-124 Condor strategic airlifter). Antonov, a state-owned company, had significant production facilities in Kiev, Ukraine and also operated Antonov Airlines and Antonov Airport. US Aerospace, located in southern California (Santa Fe Springs and Rancho Cucamonga, CA) supplied a number of large DoD contractors, including Lockheed Martin, Boeing, and L-3, as well as DoD and the Air Force. The bid submitted by the alliance involved supplying 179 tankers at $150 million per plane, for a total of $29.5 billion, with the planes being assembled in the US at US Aerospace, but being built by Antonov in Ukraine.[57]

The Air Force, however, disallowed the bid on the basis that it was five minutes late. US Aerospace launched a complaint with the Government Accountability Office (GAO) that the Air Force had deliberately allowed the delivery person to wait long enough to miss the deadline. GAO, in its September 16, 2010 ruling, cleared the Air Force of conspiring to block the submission, and, in a decision on October 6, 2010, denied US Aerospace's claim.[58]

In February, 2011, Boeing was announced as the winner of the competition. Although many had thought that EADS would win the contract, Boeing's success suggested that EADS' A-330, which could cost more in terms of operating and building it and which burned more fuel per flight hour than Boeing's 767, had been an issue of concern for the Air Force. Unfortunately, however, EADS' loss in the competition did not reinforce the perception that US had become more open to foreign contractors. Moreover, as discussed in the next chapter, the US has not issued a significant share of the volume or the value of DoD contracts to foreign contractors.[59]

The 2008 tanker competition differs from other types of foreign involvement in the defense industrial base in that the US was not buying military equipment and products which were assembled overseas, nor did the foreign company attempt to acquire a US company. Rather, in the case of the 2008 tanker competition, the

56 Hebert, 2010.

57 "US Aerospace and Antonov Submit $29.5 Billion Bid for KC-X Tanker Program," 2010.

58 Censer, 2010.

59 "Boeing Wins KC-X Tanker Battle," 2011; Garamone, 2009.

foreign manufacturer formed a strategic alliance with a domestic manufacturer. In both the 2008 competition and its successor competition, the foreign manufacturer planned to assemble the product on US soil by making substantive investments in the US industrial base by building factories and creating jobs. While Boeing could not focus its protests directly on the problems with a foreign manufacturer making domestic military equipment because it had been the foreign manufacturer doing this in other countries, it did argue that the procedure by which the contract was initially awarded to Northrop/EADS was flawed. In the meantime, the Congressional representatives were divided due to Northrop/EADS' involvement in the defense industrial base. Nevertheless, the Congressional representatives which were most affected by the Boeing contract held key positions on Congressional committees, and, consequently, were able to suggest threats in the event that Boeing did not win the competition. They further focused their arguments on issues relating to US jobs allegedly going overseas and the resulting shrinkage of the US defense industrial base. These arguments ran the risk of alienating the European market and potentially even leading to a backlash against Europeans purchasing US products, which would particularly have impacted Boeing, due to its substantive overseas sales. Moreover, Boeing attempted to introduce additional procedural concerns to delay the decision, in the hope that the new Congress and the new Administration would be more supportive of its case.

The 2008 tanker case appeared to set important precedents for an "open market," which was not sustained by the subsequent tanker competition. The initial award of the tanker to the Northrop/EADS team suggested that the government procurement process did not always favor incumbents and that there was an increasing emphasis on obtaining the most appropriate product at the best cost. Moreover, the initial award of the tanker indicated that the transparent process was often open to the global marketplace, especially when the range of national security threats, such as the terrorist threat, were more globally focused. Nevertheless, the decision to provide Boeing with the contract, instead of EADS, in the third competition, despite the similarity in the proposals from the 2008 tanker competition, suggests that the "open market" precedent from the 2008 award to EADS was less strong than it could have been. Moreover, in the third competition, the fact that the proposal of the alliance between the US firm (US Aerospace) and the foreign firm (Antonov Aeronautical) was rejected from participation in the competition due to being five minutes late did not reinforce the perception of an "open market."

It is possible that the strength of various Congressional representatives regarding where the tanker would be made played a significant role in the decision-making process, since an unfavorable contract award, from the perspective of significant Congressional representatives, could lead to further delays, which DoD could not afford due to its immediate need for the tankers. In the coming years, this may contribute to a greater reluctance on the part of contractors to make the necessary investments to create the best product at the lowest cost to the government. Rather, the contractors may be more likely to focus on locating production in states which have powerful Congressional representatives, rather than the states which have the lowest cost or which are otherwise more appropriate for production. Indeed, if this occurs, it could lead to a reduction in innovation, since the focus will have shifted from the quality of the product to the importance of political considerations within Congress. Moreover, it reduces the importance of having a transparent and well-documented government procurement process if Congress can ultimately block the funding for the winning proposal anyway.

Conclusion

DoD has been facing challenges due to fiscal constraints, as well as the evolution of defense priorities, and has focused on the balance between the best product and the lowest price, which potentially can have tradeoffs. The production of defense equipment within the US is key because of its role in creating jobs and sustaining economic growth, as well as maintaining skillsets of defense industry employees. Nevertheless, foreign firms which are willing to assemble equipment in the US can create regional economic growth, maintain the skillsets of workers, and produce innovative designs at lower cost.

Congressional representatives have played a key role in protecting the defense industrial base from encroachment by foreign competition. Their concerns over job loss in their districts have often motivated a vocal response to attempts of foreign companies to acquire domestic companies or to a declining market share of domestic companies as foreign imports enter the market. They have used a variety of strategies, including:

a. discouraging the foreign bidder by the strength of their protests, such that the bidder withdraws its bid, or supporting the recommendation of other regulatory agencies that the merger be blocked or that key assets be divested prior to the completion of the merger;
b. imposing tariffs, quotas, or VRAs on foreign imports to reduce the attractiveness of their pricing relative to domestic products; and
c. strengthening the Buy America legislation to limit the incentives of US purchasing agents to buy goods from overseas.

The 2008 tanker competition between Boeing and Northrop/EADS has been a landmark case highlighting the challenges facing the US industrial base. This competition involved one of the largest defense contracts in US history at a time when defense spending was likely to shrink in the coming years. The ultimate outcome of the competition was unfortunate in that it did not promote the perception of the integration of the US market with the global market. Similar proposals from the 2008 competition re-appeared in the final competition, yet Boeing's proposal, rather than EADS' proposal, was chosen. This may have implications in future years for the perceived importance of key Congressional representatives in states where a given proposal indicates that the equipment will be produced.

The situation was different from the other examples discussed in the chapter because the foreign company was allied with a US company in the 2008 competition and was actually augmenting the US defense industrial base by building manufacturing facilities and creating jobs. Boeing followed a two-pronged strategy of focusing its own arguments on procedural issues, while effectively allowing Congressional representatives to focus their dialogue on concerns over the involvement of a foreign company in the US national defense sector. Boeing could not credibly focus its arguments on the involvement of a foreign company in a domestic defense industry due to its role as a foreign company in overseas defense markets. EADS, as the entrant, allied itself with a US firm and proposed to grow the domestic defense industrial base by producing and assembling the aircraft domestically, rather than overseas. Its decision to make a substantial investment in the US defense industrial base split Congressional representatives, to a degree that would not have been possible if it had followed a strategy of attempting to import its products into the US.

The tanker competition highlighted the divide in the US between a desire for open markets and for the best product at the lowest cost, and the concerns of certain US states that they might lose jobs to other US states and overseas countries. The vocal outcry of Congressional representatives and their actions in threatening to block funding for the tanker emphasized that the positions on Congressional committees held by representatives in affected states matter. Hopefully, over time, this will not alter the incentives of defense contractors to locate production in the states with the best skillsets and the lowest costs, rather than to locate production in states with powerful Congressional representatives.

The US has allied itself with a number of nations to combat the War on Terror, since the terrorist threat, unlike previous military threats, transcends the boundaries of nation-states. The new millennium has brought challenges for the US in dealing with the dual role of foreign companies and governments as economic competitors and as allies against a common enemy. Hopefully, over time, the lessons of the importance of a global marketplace will become more obvious in these industries, while, simultaneously, achieving a balance between global cooperation and protection of sensitive information impacting national security.

References

Aerospace Industries Association of America. Annual. *Aerospace Facts and Figures*, annual. Washington, DC.

"Analysts Assess Damage to Boeing in Aftermath of Contract Loss." 2009. *Marketwatch*, March 3.

Beattie, Alan, 2008. "US Issues Warning on Stoking Protectionism." *Financial Times*, May 5.

Bergstrom, Theodore C., 1982. "On Capturing Oil Rents with a National Excise Tax." *American Economic Review*, March, pp. 194–201.

"Boeing Calls for Air Force Review of Tanker Contract Awarded to Northrop." 2008. *Los Angeles Business*, March 3.

"Boeing Protests U.S. Air Force Tanker Contract Award." 2008. Press release, March 11.

"Boeing Requests Immediate KC-X Tanker Briefing." 2008. Press release from the Boeing Corporation, March 4.

"Boeing Study Projects That as Oil Prices Climb, 767 Tanker Most Cost Efficient." 2008. Press release, March 17.

"Boeing to Protest $35B Tanker Deal." 2008. *CNNMoney.com*, March 11.

"Boeing's Trouble with Tankers." 2008. *Business Week*, March 11.

"Boeing Wins KC-X Tanker Battle." 2011. *Air Force Times*, February 24.

Bradsher, Keith, 1993. "Industries Seek Protection as Vital to US Security." *New York Times*, January 19.

Brun-Rovet, Marianne, 2003. "Buy America' Provision Dropped from US Weapons Bill." *Financial Times*, November 8.

"Buy America." 2006. *Government Executive*, April 15.

Carpenter, Dave, 2008. "Boeing Says its Tanker Proposal Superior." *Associated Press*, March 5.

Censer, Marjorie, 2010. "GAO Clears Air Force in Tanker Case." *Washington Post*, October 7.

Collie, David R. and Hviid, Morten, 2001. "International Procurement as a Signal of Export Quality." *The Economic Journal*, Vol. 111, April, pp. 374–90.

Crown, Judith and Keith Epstein, Keith, 2008. "Boeing Files Tanker Protest." *Business Week*, March 11.

Dallmeyer, Dorinda, 1987. "National Security and the Semiconductor Industry." *Technology Review*, November/December, pp. 47–55.

Drawbaugh, Kevin, 2008a. "Congress in Turmoil over Air Force Tanker Decision." *Reuters,* February 29.

Drawbaugh, Kevin, 2008b. "U.S. Congress Roiled by Air Force Tanker Decision." *Reuters,* March 3.

Epstein, Keith, 2008. "Boeing's Tanker Challenges Mount." *Business Week*, August 28.

"EU Voiced Concern on US 'Protectionism.'" 2002. *Financial Times*, April 6.

Garamone, Jim, 2009. "DoD Announces Requirement for New Aerial Tanker Competition." *American Forces Press Service*. Available at: http://www.defense.gov [accessed September 24, 2009].

Gardner, Grant W. and Kimborough, Kent P., 1990. "The Effects of Trade-Balance-Triggered Tariffs." *International Economic Review*, Vol. 31, No. 1, February.

General Accounting Office (GAO), 2008a. *Statement of Daniel I. Gordon, Deputy General Counsel on Air Force Procurement: Aerial Refueling Tanker Protest*, July 10.

——. 2008b. *Statement Regarding the Bid Protest Decision Resolving the Aerial Refueling Tanker Protest by the Boeing Company, B-311344*, et al., June 18.

Hebert, Adam J., 2010. "A Tanker for the Air Force." *Air Force Magazine: The Online Journal of the Air Force Association*, September.

Hepher, Tim, 2008. "EADS Shares Soar after Big US Defense Deal." *Reuters*, March 3.

Hinton, Christopher, 2008a. "Air Force Set to Award $40 Bn Air Tanker Contract." *Marketwatch*, February 22.

——. 2008b. "Air Force Still Leans Toward Northrop Tanker Plane." *Marketwatch*, July 21.

——. 2008c. "Boeing Files Protest over the Air Force Tanker Award." *Marketwatch*, March 11.

——. 2008d. "Boeing Might Deliver Just 45 Dreamliners in 2009." *Marketwatch*, March 12.

——. 2008e. "US Air Force Debriefs Northrop Grumman on Tanker Award." *Marketwatch*, March 10.

Lagorce, Audrey, 2008. "Northrop-EADS Tanker Wins a Boon for Europe." *Marketwatch*, March 3.

Lowinger, Thomas C., 1976. "Discrimination in Government Procurement of Foreign Goods in the US and Western Europe." *Southern Economic Journal*, Vol. 42, No. 3, January, pp. 451–60.

Lynch, David J., 2006. "Some Would Like to Build a Wall Around the US Economy." *USA Today*, March 15.

Manzullo, Donald A., 2003. "Goal is US Security." *USA Today*, August 11.

Maskin, Eric and Newbery, David, 1990. "Disadvantageous Oil Tariffs and Dynamic Consistency." *The American Economic Review*, Vol. 80, No. 1, March, pp. 143–56.

Modic, Stanley J., 1989. "Steel VRAs." *Industry Week*, September 18, pp. 85–8.

"Northrop Group Wins $35B Air Force Deal." 2008. *CNNMoney.com*, February 29.

"Northrop Grumman Fires Back on Tanker Debate."2008. *CNBC.com*, March 5.

"Northrop Grumman Had Originally Clinched the USAF Contract in February, with Boeing's Objections." 2008. *CNNMoney.com*, June 18.

Office of the Deputy Under Secretary of Defense for Industrial Policy, 2004. *Report Required by Section 812 of the National Defense Authorization Act for Fiscal Year 2004: Foreign Sources of Supply*, November.

——. 2005. *Report Required by Section 812 of the National Defense Authorization Act for Fiscal Year 2004: Foreign Sources of Supply, Addendum Incorporating Fiscal Year 2004 Contract Information*, March.

——. 2006. *Report Required by Section 812 of the National Defense Authorization Act for Fiscal Year 2004: Foreign Sources of Supply*, April.

——. 2007. *Report Required by Section 812 of the National Defense Authorization Act for Fiscal Year 2004: Foreign Sources of Supply*, November.

Pearlstein, Steven, 2006. "Ports Furor is Just Protectionism, with a French Accent." *Washington Post*, March 1.

Platt, Gordon, 2006. "Cross-Border Acquisitions Facing Growing Interference from Tighter Security Reviews." *Global Finance*, May.

Randolph, Monique, 2008. "Top Acquisition Official: Tanker Acquisition Top Priority." *Air Force Print News Today*, July 21.

Razin, Assaf, and Svensson, Lars E. O., 1983. "Trade Taxes and the Current Account." *Economics Letters*, Vol. 13, pp. 55–7.

Rigby, Bill, 2008a. "Boeing Faces Questions after Tanker Loss." *Reuters*, March 3.

——. 2008b. "Boeing says Air Force Tanker Award Flawed." *Reuters*, March 11.

Sevastopulo, Demetri, and Done, Kevin, 2008. "Congress Move Could Force Pentagon Rethink on Tankers." *Financial Times*, July 31.

Shalal-Esa, Andrea. 2008a. "Air Force Agrees to Brief Boeing on Tanker Loss." *Reuters*, March 4.

——. 2008b. "US Plans Expedited Rerun of Aerial Tanker Contest." *Reuters*, July 9.

Shalal-Esa, Andrea and Wolf, Jim, 2008. "Pentagon Reopening Contest to Build Aerial Tankers." *Reuters*, July 9.

Shearer, Brent, 2006. "Raising Barriers to Inbound Deals." *Mergers and Acquisitions*, May.

Spiegel, Peter, 2004. "Weapons Hawk Makes American Defense Contractors Nervous." *Financial Times*, July 31.

Steffes, Peter M., 2004. "The Reality of the 'Buy America' Provisions." *National Defense*, February.

"Tanker Deal: Why Boeing Shouldn't Protest." 2008. *Business Week*, March 4.

Tessler, Joelle, 2008. "Air Force Tanker Can't Be All American." *Associated Press*, March 6.

"US Aerospace and Antonov Submit $29.5 Billion Bid for KC-X Tanker Program." 2010. *Examiner.com*, July 12.

US Bureau of Economic Analysis. 2007. *Survey of Current Business*, April

US Census Bureau, 2004. *U.S. International Trade in Goods and Services*. Series FT-900 (07–04), July.

——. 2007. "Related Trade Party – 2006." May 10.

Wallace, James, 2008. "Boeing Air Tanker Bid Takes Hit." *Seattle Post Intelligencer*, August 7.

Wardell, Jane, 2008. "Lockheed Martin Urges Transparency." *Associated Press*, July 13.

Wayne, Leslie, 2005. "Pentagon Looks Outside Borders to Equip Troops." *New York Times*, September 27.

Wingfield, Brian, 2008. "Boeing Fights Back." *Forbes.com*, March 11, 2008.

Wolf, Jim, 2008a. "US Air Force Chief Extends Tanker Suspense." *Reuters*, February 28.

Wolf, Jim, 2008b. "US Lawmakers Urge Broader Pentagon Tanker Review." *Reuters*, July 30.

Wolf, Jim and Shalal-Esa, Andrea, 2008. "Northrop-Eads Beats Boeing to Build US Tanker." *Reuters*, March 2.

Chapter 6
The Role of the US in the International Defense Market

To what degree is the US involved in overseas trade in areas which are important to national defense? This chapter examines the role of the US as an importer and as an exporter of defense equipment. Defense contractors are facing fiscal constraints and potentially reduced orders from DoD for certain types of equipment. Moreover, with the shift in defense priorities away from Afghanistan and toward the Asia-Pacific region, demand has shifted away from various types of equipment and toward other types. Consequently, defense contractors are expanding toward overseas markets; nevertheless, many of these countries also face budgetary constraints. The diversification of defense contractors into the civilian sector can also be profitable and can lead to increased sales in overseas markets. Nevertheless, slower economic growth in overseas countries can impact the ability of overseas buyers to purchase civilian equipment due to reduced demand for their services. Finally, the chapter concludes with an examination of the aerospace industry as a case study assessing the role of the US as an importer and exporter, in both the military and the civilian sectors.

Table 6.1 Percentage of DoD Procurement Budget Expended on Purchases from Foreign Suppliers

Fiscal Year	Percentage of DoD Procurement Budget Expended on Purchases from Foreign Suppliers
FY 2009	6.8%
FY 2010	7.7%
FY 2011	6.4%
FY 2012	6.1%
FY 2013	6.4%

Sources: Report to Congress on Department of Defense Fiscal Year 2013 Purchases from Foreign Entities. May 2014, Section 1; *Report to Congress on Department of Defense Fiscal Year 2012 Purchases from Foreign Entities*. July 2013, Section 1; *Report to Congress on Department of Defense Fiscal Year 2011 Purchases from Foreign Entities*. May, 2012, Section 1; *Report to Congress on Department of Defense Fiscal Year 2010 Purchases from Foreign Entities*. July 2011, Section 1; *Report to Congress on Department of Defense Fiscal Year 2009 Purchases from Foreign Entities*. August 2010, Section 1.

The US is not a significant importer of military equipment or related civilian equipment due to its strong industrial base in many of those areas. For example, in examining world imports of ships – both military and civilian – between 2008 and 2012, the US was the twelfth largest importer in 2008, when it imported 2.6% of the global market for ship imports, and was the fourteenth largest importer by 2012, when it imported 2.7%. In examining the world imports of aircraft – both military and civilian – between 2008 and 2012, the US varied between being the second largest importer and the third largest importer. Nevertheless, due to the fragmented market, US imports of aircraft only varied between 12% and 13% of the global market for aircraft imports, despite being one of the most significant importers.

Purchases by DoD from foreign suppliers – defense imports – have been relatively low. Table 6.1 shows the share of the DoD procurement budget spent on purchases from foreign suppliers. Between 6.1% and 7.7% of the DOD procurement budget between 2009 and 2013 was spent on purchases from foreign suppliers.

Table 6.2 Percentage of Total Purchases of DoD from Foreign Suppliers in Various Categories: FY 2009–FY 2013

Category	FY 2009	FY 2010	FY 2011	FY 2012	FY 2013
Air Frames and Spares	1.283%	0.995%	2.27%	2.336%	4.38%
Aircraft Engines and Spares	0.589%	0.246%	0.462%	0.244%	0.28%
Other Aircraft Equipment	1.040%	1.292%	0.858%	1.081%	1.99%
Missile and Space Systems	0.093%	0.231%	0.32%	0.123%	0.15%
Ships	1.451%	0.764%	0.848%	0.836%	1.19%
Combat Vehicles	1.471%	10.407%	5.140%	1.807%	0.72%
Non-Combat Vehicles	0.447%	0.292%	0.70%	0.314%	0.30%
Weapons	1.525%	1.293%	1.698%	1.128%	0.49%
Ammunition	0.503%	0.661%	0.645%	0.907%	0.58%
Electronics and Communication Equipment	1.011%	0.924%	0.835%	0.802%	0.71%
Petroleum	28.026%	26.986%	27.848%	33.307%	41.54%
Subsistence	12.444%	12.404%	12.722%	15.402%	12.55%
Construction	17.394%	9.937%	12.003%	14.157%	9.77%

Sources: Report to Congress on Department of Defense Fiscal Year 2013 Purchases from Foreign Entities. May 2014, Table 2; *Report to Congress on Department of Defense Fiscal Year 2012 Purchases from Foreign Entities*. July 2013, Table 2; *Report to Congress on Department of Defense Fiscal Year 2011 Purchases from Foreign Entities*. May, 2012, Table 2; *Report to Congress on Department of Defense Fiscal Year 2010 Purchases from Foreign Entities*. July 2011, Table 2; *Report to Congress on Department of Defense Fiscal Year 2009 Purchases from Foreign Entities*. August 2010, Table 2.

A variety of different categories involve DoD as an importer; however, it imports a relatively small percentage of defense equipment. Table 6.2 shows the percentage of total purchases by DoD from foreign suppliers by equipment categories. The share of purchases of military equipment by DoD from foreign suppliers was relatively low – usually under 2.5%. Combat vehicles were an exception; they comprised 10.407% of the total purchases of DoD from foreign suppliers in FY 2010 and 5.140% in FY 2011. Moreover, airframes and spares, although usually under 2.5%, were 4.38% of DoD purchases from foreign suppliers in FY 2013. The other equipment categories with the highest shares of total purchases comprised 2% or less of total purchases from foreign suppliers. These included: weapons (between 1.698% and 0.49%); other aircraft equipment (between 1.99% and 0.858%); ships (between 1.451% and 0.764%); and electronics and communications equipment (between 1.011% and 0.71%).

Non-equipment categories, however, were the dominant share of DoD purchases from foreign suppliers. For example, petroleum purchases ranged between 41.54% and 26.986%, construction ranged between 17.394% and 9.77%, and subsistence ranged between 15.402% and 12.404%. Consequently, the shares of DoD equipment purchases from foreign entities has been relatively small; over half of the purchases have been construction, subsistence, and petroleum.

In most categories of military equipment, DoD awards contracts largely to domestic contractors rather than to foreign contractors. Table 6.3 shows the

Table 6.3 Percentage of Total Number of Contracts Awarded by DoD to Foreign Contractors in Different Categories

	FY 2003	FY 2004	FY 2005	FY 2006	FY 2007	FY 2008
Airframes and related assemblies and spares						
US	0.96	0.966	0.966	0.991	0.979	0.979
Foreign	0.04	0.034	0.034	0.009	0.021	0.021
Aircraft engines and related spares and spare parts						
US	0.955	0.945	0.917	0.965	0.964	0.973
Foreign	0.045	0.055	0.083	0.035	0.036	0.027
Other aircraft equipment and supplies						
US	0.973	0.977	0.97	0.967	0.986	0.977
Foreign	0.027	0.023	0.03	0.033	0.014	0.023
Missile and space systems						
US	0.975	0.982	0.981	0.702	0.983	0.993
Foreign	0.025	0.018	0.019	0.298	0.017	0.007

Table 6.3 Continued

	FY 2003	FY 2004	FY 2005	FY 2006	FY 2007	FY 2008
Ships						
US	0.984	0.987	0.984	0.997	0.988	0.946
Foreign	0.016	0.013	0.016	0.003	0.012	0.054
Combat vehicles						
US	0.954	0.939	0.955	0.977	0.97	0.94
Foreign	0.046	0.061	0.045	0.023	0.03	0.06
Non-combat vehicles						
US	0.937	0.937	0.922	0.927	0.936	0.902
Foreign	0.063	0.063	0.078	0.073	0.064	0.098
Weapons						
US	0.96	0.969	0.967	0.898	0.966	0.939
Foreign	0.04	0.031	0.033	0.102	0.034	0.061
Ammunition						
US	0.936	0.918	0.932	0.949	0.93	0.094
Foreign	0.064	0.082	0.068	0.051	0.07	0.061
Electronics and Communications Equipment						
US	0.981	0.982	0.98	0.981	0.987	0.99
Foreign	0.019	0.018	0.02	0.019	0.013	0.01
Total						
US	0.969	0.97	0.964	0.976	0.979	0.973
Foreign	0.031	0.03	0.036	0.024	0.021	0.027

Sources: Report Required by Section 812 of the National Defense Authorization Act for Fiscal Year 2004: Foreign Sources of Supply, November, 2004 (Office of the Under Secretary of Defense, Acquisitions, Technology, and Logistics); *Report Required by Section 812 of the National Defense Authorization Act for Fiscal Year 2004: Foreign Sources of Supply, Addendum Incorporating Fiscal Year 2004 Contract Information*, March, 2005 (Office of the Under Secretary of Defense, Acquisitions, Technology, and Logistics); *Report Required by Section 812 of the National Defense Authorization Act for Fiscal Year 2004: Foreign Sources of Supply*, April, 2006 (Office of the Under Secretary of Defense, Acquisitions, Technology, and Logistics); *Report Required by Section 812 of the National Defense Authorization Act for Fiscal Year 2004: Foreign Sources of Supply,* November, 2007 (Office of the Under Secretary of Defense, Acquisitions, Technology, and Logistics).

percentage of the total number of contracts awarded by DoD to foreign contractors in different categories of military equipment between 2003 and 2008. This table indicates that in 2003, the percentage of the total number of contracts awarded by

DoD to foreign contractors varied from 1.6% (ships) to 6.3–6.4% (non-combat vehicles and ammunition, respectively). By 2008, the percentage of total contracts awarded by DoD to foreign contractors varied from 0.7% (missiles and space systems) to 9.8% (combat vehicles). The share of the total number of contracts to foreign contractors dropped between 2003 and 2008 for airframes, aircraft engines, and missile and space systems. The other categories exhibited volatility, although non-combat vehicles, weapons and ammunition categories tended to show some growth. Across all the categories, the share of the number of contracts made with foreign contractors declined from 3.1% in 2003 to 2.7% in 2008.

Similarly, as shown in Table 6.4, the total value of contracts awarded by DoD in various military equipment categories to foreign contractors between 2003 and

Table 6.4 Percentage of Total Value of Contracts Awarded by DoD to Foreign Contractors in Different Categories

	FY 2003	FY 2004	FY 2005	FY 2006	FY 2007	FY 2008
Airframes and related assemblies and spares						
US	0.993	0.989	0.987	0.991	0.997	0.995
Foreign	0.007	0.011	0.013	0.009	0.003	0.005
Aircraft engines and related spares and spare parts						
US	0.984	0.983	0.941	0.965	0.983	0.989
Foreign	0.016	0.017	0.059	0.035	0.017	0.011
Other aircraft equipment and supplies						
US	0.969	0.962	0.979	0.967	0.975	0.978
Foreign	0.031	0.038	0.021	0.033	0.025	0.022
Missile and space systems						
US	0.992	0.996	0.998	0.702	0.999	0.999
Foreign	0.008	0.004	0.002	0.298	0.001	0.001
Ships						
US	0.996	0.998	0.996	0.997	0.998	0.988
Foreign	0.004	0.002	0.004	0.003	0.002	0.012
Combat vehicles						
US	0.976	0.928	0.98	0.977	0.988	0.953
Foreign	0.024	0.072	0.02	0.023	0.012	0.047
Non-combat vehicles						
US	0.968	0.97	0.963	0.927	0.97	0.984
Foreign	0.032	0.03	0.037	0.073	0.03	0.016

Table 6.4 Continued

	FY 2003	FY 2004	FY 2005	FY 2006	FY 2007	FY 2008
Weapons						
US	0.97	0.975	0.887	0.898	0.914	0.912
Foreign	0.03	0.025	0.113	0.102	0.086	0.088
Ammunition						
US	0.956	0.914	0.929	0.949	0.916	0.92
Foreign	0.044	0.086	0.071	0.051	0.084	0.08
Electronics and Communications Equipment						
US	0.978	0.978	0.985	0.981	0.991	0.988
Foreign	0.022	0.022	0.015	0.019	0.009	0.012
Total						
US	0.985	0.98	0.976	0.976	0.985	0.982
Foreign	0.015	0.02	0.024	0.024	0.015	0.018

Sources: Report Required by Section 812 of the National Defense Authorization Act for Fiscal Year 2004: Foreign Sources of Supply, November, 2004 (Office of the Under Secretary of Defense, Acquisitions, Technology, and Logistics); *Report Required by Section 812 of the National Defense Authorization Act for Fiscal Year 2004: Foreign Sources of Supply, Addendum Incorporating Fiscal Year 2004 Contract Information*, March, 2005 (Office of the Under Secretary of Defense, Acquisitions, Technology, and Logistics); *Report Required by Section 812 of the National Defense Authorization Act for Fiscal Year 2004: Foreign Sources of Supply*, April, 2006 (Office of the Under Secretary of Defense, Acquisitions, Technology, and Logistics); *Report Required by Section 812 of the National Defense Authorization Act for Fiscal Year 2004: Foreign Sources of Supply*, November, 2007 (Office of the Under Secretary of Defense, Acquisitions, Technology, and Logistics).

2008 was small, reinforcing the fact that DoD was not a significant purchaser of military equipment from foreign contractors. This provides similar results to Table 6.3. Table 6.4 shows the percentage of the total value of contracts awarded by DoD to foreign contractors in the different categories between 2003 and 2008. The share of the total value of contracts awarded to foreign contractors varied from 0.4% (ships) to 4.4% (ammunition). By 2008, the share of the total value of contracts varied from 0.1% (missiles and space systems) to 8.8% (weapons). Between 2003 and 2008, a declining share of the value of contracts made with foreign contractors occurred in categories, such as aircraft equipment/supplies and electronics and communications equipment. Many of the categories exhibited volatility during the period, although some of the categories did experience a rising share of the value of contracts made with foreign contractors, such as combat vehicles (in some years) and ammunition.

When it awards contracts to foreign contractors, DoD has tended to work particularly with Canada, as well as with the UK and Germany. The countries to whom the DoD has awarded over 10 contracts in one of the years during the 2003–2008 period is illustrated in Table 6.5. In 2003, the countries which received over 100 contracts that year were: Canada (857), the UK (623), Germany (157), and Israel (120). By 2008, the countries which received over 100 contracts were: Canada (937), the UK (533), Japan (474), Germany (120), and Singapore (103). For many of the countries during the period, there was significant volatility in the number of contracts. For example, Colombia went from 13 contracts in 2007 to 99 in 2008, while Japan expanded from 48 contracts in 2007 to 474 in 2008. Singapore went from 16 contracts in 2007 to 103 in 2008. Canada and Germany

Table 6.5 Countries to Whom DoD has Awarded Over 10 Contracts in One of the Years During the 2003–2008 Period

	FY 2003 Volume	FY 2004 Volume	FY 2005 Volume	FY 2006 Volume	FY 2007 Volume	FY 2008 Volume
Australia	6	18	15	9	14	18
Bahrain	5	7	1	1	7	48
Belgium	40	33	32	17	7	67
Canada	857	775	831	746	501	937
Colombia		4	9	8	13	99
Denmark	5	11	8	11	1	5
France	57	55	53	48	19	59
Gabon	12	15	10	9		1
Germany	157	161	143	112	84	120
Greece	4	5	3			27
Iraq	0	4	27	3	18	19
Israel	120	113	111	92	44	73
Italy	68	35	41	28	14	39
Japan	58	48	38	39	48	474
Korea	20	26	215	17	24	
Kuwait	87	53	35	11	12	66
The Netherlands	17	5	7	5	2	3
Norway	10	3	8	8	15	14
Romania	0	8	5	10	5	0
Saudi Arabia	14	22	39	18	3	28
Senegal	0	1	10	22		
Singapore	30	15	15	27	16	103

Table 6.5 Continued

	FY 2003 Volume	FY 2004 Volume	FY 2005 Volume	FY 2006 Volume	FY 2007 Volume	FY 2008 Volume
South Korea						79
Sweden	35	27	2	16	11	12
Switzerland	10	1	4	2	1	3
Turkey	4	2	4	3	8	16
United Arab Emirates			3	7	9	21
United Kingdom	623	649	720	498	312	533

Sources: Report Required by Section 812 of the National Defense Authorization Act for Fiscal Year 2004: Foreign Sources of Supply, November, 2004 (Office of the Under Secretary of Defense, Acquisitions, Technology, and Logistics); *Report Required by Section 812 of the National Defense Authorization Act for Fiscal Year 2004: Foreign Sources of Supply, Addendum Incorporating Fiscal Year 2004 Contract Information*, March, 2005 (Office of the Under Secretary of Defense, Acquisitions, Technology, and Logistics); *Report Required by Section 812 of the National Defense Authorization Act for Fiscal Year 2004: Foreign Sources of Supply*, April, 2006 (Office of the Under Secretary of Defense, Acquisitions, Technology, and Logistics); *Report Required by Section 812 of the National Defense Authorization Act for Fiscal Year 2004: Foreign Sources of Supply FY 2006 Report*, November, 2007 (Office of the Under Secretary of Defense, Acquisitions, Technology, and Logistics); *Report Required by Section 812 of the National Defense Authorization Act for Fiscal Year 2004: Foreign Sources of Supply FY 2007 Report*, September, 2008 (Office of the Under Secretary of Defense, Acquisitions, Technology, and Logistics); *Report Required by Section 812 of the National Defense Authorization Act for Fiscal Year 2004: Foreign Sources of Supply FY 2008 Report*, October, 2009 (Office of the Under Secretary of Defense, Acquisitions, Technology, and Logistics).

were the only countries which consistently received over 100 contracts on an annual basis. Israel dropped below 100 contracts beginning in 2006.

Although DoD does not import significant amounts of military equipment and provides foreign defense contractors with a much smaller share of contracts than US defense contractors, DoD does work with foreign defense contractors in several other ways, reinforcing interdependencies between US and foreign industrial bases. For example, the littoral combat ships are being built in foreign-owned facilities in Mobile, AL (Austal) and Marinette, Wisconsin (Marinette Marine, owned by Fincantieri). Moreover, the US is involved with the defense industrial bases of other countries through global supply chain arrangements. The development of the Joint Strike Fighter (JSF) is an example of one of the most extensive global alliances in the defense sector. Former Secretary of Defense Robert Gates emphasized the commitment of the US to systems which are compatible with its allies and which are developed through global alliances in the context of purchasing more JSFs.[1] The F-35 involves nine different contractors from various countries, led by Lockheed Martin, with Northrop Grumman and BAE as the primary subcontractors, and is supported by 600 suppliers in 30 countries.[2] The F-35 is intended to replace 13 different types of aircraft across 11 different countries.[3]

Table 6.6 DoD Foreign Military Sales (FMS) on Significant Military Equipment (SME) Exceeding $2 Million: FY 2009–FY 2011

Fiscal Year	Number of Countries	Total Value of FMS
FY 2009	44	$8,041,376,740
FY 2010	47	$9,451.945,114
FY 2011	60	$14,000,000,000

Sources: Report to Congress on Department of Defense Sales of Significant Military Equipment to Foreign Entities Fiscal Year 2009, July 2010, p. 2; *Report to Congress on Department of Defense Sales of Significant Military Equipment to Foreign Entities Fiscal Year 2010*, June 2011, p. 2; *Report to Congress on Department of Defense Sales of Significant Military Equipment to Foreign Entities Fiscal Year 2011*, May 2012, p. 2.

Not surprisingly, the US defense sector is highly involved in the global defense market as an exporter. DoD sold equipment to between 44 and 60 countries through the foreign military sales (FMS) program between 2009 and 2011, ranging from $8 billion to $14 billion (Table 6.6).

One example of a defense product with significant exports to overseas countries by the US are missile defense systems. As of the fall of 2013, there continued to be significant demand from countries, such as South Korea and Qatar. Lockheed Martin and Raytheon have been the designers of the majority of the US military's missile defense system and the increased concerns from other countries regarding the ballistic missiles in Iran and North Korea provide greater opportunities for international sales. Although manufacturers in other countries, such as China and Russia, are involved in the international competitions, they have not sold this type of equipment before to other countries and lack the historical background in sales and maintenance that US firms have. Maintenance is particularly important for missile defense systems; about two-thirds of the cost of a system, such as the Patriot, is in maintaining it and operating it.[4]

1 "Aeronautics Company; F-35: Delivering the Promise to Redefine National Strategic Capabilities," 2008, p. 331.

2 "Aerospace Production: Connecting Flights," 2009, p. 36.

3 "Aeronautics Company; F-35: Délivering the Promise to Redefine National Strategic Capabilities," 2008, p. 331.

4 Erwin, 2013a.

In the recent competition for missile defense held in Turkey, Chinese and Russian firms, as well as Raytheon's Patriot air defense batteries that would fire Lockheed Martin's Patriot advanced capability 3 (PAC-3) interceptor missiles, are contenders. The Turkish sale would likely involve a purchase of 100 PAC-3 missiles, although 200 missiles are needed to maintain annual production for Lockheed Martin. Other purchasers of PAC-3 missiles include: Japan, Germany, the Netherlands, Kuwait, Taiwan, and the United Arab Emirates. Defense contractors, such as Boeing, also benefit from the sale because Boeing provides the "seeker" for the PAC-3, which "targets incoming missiles and guides them to the interceptor."[5]

Terminal High Altitude Area Defense (THAAD) is another missile defense system and Lockheed Martin's first international customer for THAAD, which recently signed the contract, is the United Arab Emirates. Other possible buyers, as of the fall of 2013, included South Korea and Saudi Arabia. Nevertheless, Lockheed is unlikely to be successful in MEADS – the Medium Extended Air Defense Systems – which the Army had funded up to $2 billion in its development, along with the Italian and German governments. This is because the Army backed out recently, although Germany and Italy agreed to continue on with production, which could lead to delays and funding shortfalls. Following the Army's decision to withdraw from MEADS, "Congress passed legislation that requires the Defense Department to transition key components of MEADS into existing Patriot systems."[6]

Another area of possible growth in international sales of defense equipment is the expanding interest in the Navy's Aegis air-defense system, which has been purchased by NATO countries and for which Lockheed Martin is the prime contractor.[7]

In short, these and other examples illustrate the significant role played by the US in global trade for military equipment, as well as related civilian equipment. This is important in supporting the industrial base in the US due to the fiscal constraints in government budgets and, in some cases, reduced demand for certain types of equipment due to shifting defense priorities. The US is a significant exporter to a number of countries of both civilian products, as well as military products, through FMS, although it is less of an importer.

The Role of International Trade in the Aerospace Sector

The global aerospace market provides an important case study in understanding the subsectors within the aerospace market which are key to the US in terms of imports and exports in both the civilian and the military markets. In many cases, civilian designs are valuable to a given contractor in developing a military design

5 Erwin, 2013a.
6 Erwin, 2013a.
7 Erwin, 2013a.

proposal and in establishing the reliability of a given contractor in producing the military equipment. US exports of civilian equipment, especially aircraft, are based on demand for transport on airlines. The degree to which airlines require new aircraft is often based on their projections of passenger travel, which itself is often based on factors such as economic growth and unemployment. The risk of job difficulties and slow economic growth may limit passenger travel, which reduces the need for additional civilian aircraft for airlines. US exports of military equipment are often impacted by budgetary constraints faced by overseas countries, which often reduces their demand for US military equipment. Moreover, the defense threats facing the overseas countries determine the types of military equipment that they need.

Not surprisingly, US exports significantly exceed US imports in the aerospace sector, including both civilian and military markets. Figure 6.1 shows the balance of trade for the US aerospace sector. Although exports and imports declined in 2009 and 2010 with the slowdown in economic growth, they have recovered and are exhibiting growth.

US exports in the aerospace sector are heavily dependent on the exports of civilian aircraft, engines, missiles, and spacecraft, which far exceeded the imports of military aircraft, engines, missiles, and spacecraft. Figure 6.2 shows the exports

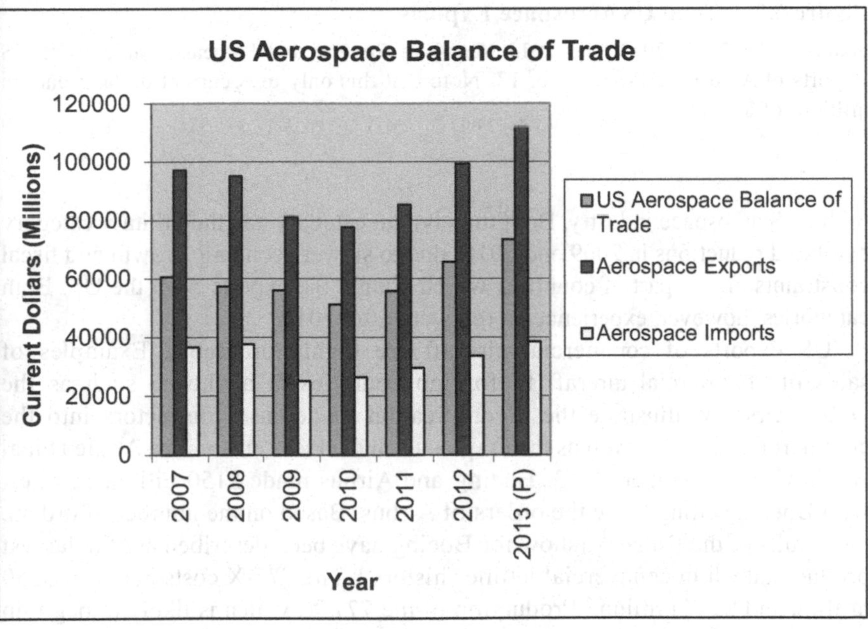

Figure 6.1 The US Aerospace Balance of Trade

Source: AIA 2012 and 2013 Year-End Review and Forecast; data in millions of dollars (current dollars; Table VI: US Aerospace Balance of Trade, p. 15).

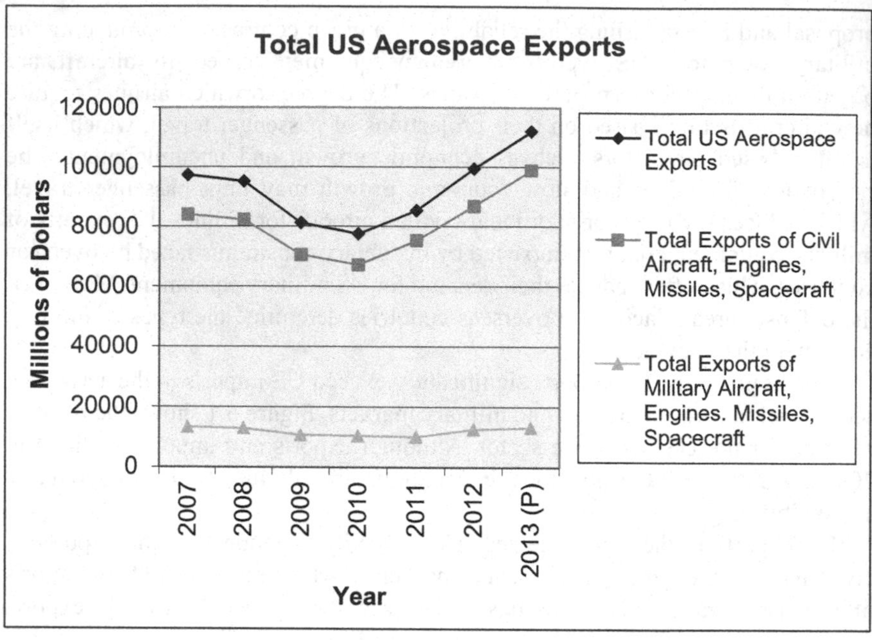

Figure 6.2 Total US Aerospace Exports

Source: AIA 2011, 2012, and 2013 Year-End Review and Forecast; Table VIII: US Exports of Aerospace Products, p. 17. Note that this only uses current dollars; data in millions of $.

in the US aerospace industry. Both the civilian category and the military category exhibited reductions in 2009 and 2010, due to slower economic growth and fiscal constraints on the part of countries which bought the exports from the US. Both categories, however, experienced a recovery after 2010.

US exports of commercial aircraft are highly profitable. Examples of sales of commercial aircraft to foreign countries at airshows, such as the Dubai Airshow, illustrate the diversification of defense contractors into the commercial sector, as well as the emphasis on overseas customers. At the Dubai Airshow in November, 2013, Boeing and Airbus made $150 billion in sales, with Boeing selling twice the orders of Airbus. Based on the number of orders, the results of the Dubai Airshow for Boeing have been described as "the largest product launch in commercial jetliner history." The 777-X costs between $350 million and $377 million.[8] Production of the 777-X, which is likely to begin in 2017, is based on two models: the 777–8X flies 9,300 nautical miles but holds

8 Jones and Cameron, 2013, pp. B1 and B2.

350 passengers, while the 777–9X flies 8,200 nautical miles and carries over 400 passengers.[9]

Middle Eastern airlines have significant demand for commercial aircraft in order to expand their capacity. Etihad Airways expanded its capacity between 2009 and 2014 by 15.4%, Qatar Airways expanded by 16.9%, Emirates Airways expanded by 13.1%, and Turkish Airlines expanded by 17.6%. Qatar Airways is twice the size of Etihad and half the size of Emirates.[10] Consequently, at the Dubai Airshow, Boeing sold several types of planes – 342 orders, which included 737s, 777s, and 787s (including the 1,000th order of the 787).[11] The batch of orders for the 777-X were worth $95 billion and they were sold to Flydubai, Qatar Airways, Etihad Airways, and Emirates Airline. Etihad and Qatar Airways made up 225 of the orders for the 777-X, followed by Deutsche Lufthansa at 34 orders. Emirates placed 150 orders for the 777-X model at $76 billion and 50 A380s at $23 billion. Flydubai placed $11.4 billion of orders for 111 Boeing 737s, while Qatar Airways spent $19 billion for 50 777-Xs, and Etihad ordered 25 planes, as well as 30 787 Dreamliners.[12]

Although US exports of military equipment are not as substantial as exports of civilian equipment, they significantly expand the global defense industrial base and the military capabilities of overseas nations. Exports of military aircraft are the most significant, followed by exports of military missiles, rockets, and parts. The preliminary data for 2013 suggest that exports of military missiles, rockets and parts will exceed the exports of military aircraft. Demand by overseas nations for this equipment is dependent not only on the equipment that they need for defense threats, but also on the fiscal strain on their defense budget. The strain on the defense budgets of overseas countries can be impacted by size of the overall federal budget for overseas countries. Figure 6.3 shows the US exports of categories of military aerospace products. All four categories experienced a decline following the financial crisis and slowing economic growth, but then began to recover.

In examining US exports of different categories of military aircraft (Figure 6.4), exports of military aircraft fighters and bombers were the leading category prior to 2010, but, beginning in 2011, exports of military aircraft transports became the most significant category of exports. The financial crisis and slowing economic growth reduced US exports of aircraft fighters, bombers, used aircraft, and aircraft transports, although they subsequently experienced recovery. Nevertheless, demand by overseas countries for fighters and bombers has dropped the most sharply and demand for military transports has grown the most rapidly.

Overseas sales of the F-35 provide an example of the challenges and opportunities facing US military sales of aircraft. Many countries have significant demand for new, innovative equipment due to the potential defense threats that they

9 Drew, 2013, p. B4.

10 Jones and Cameron, 2013, pp. B1 and B2.

11 Drew, 2013, p. B4.

12 Christopher, 2013, p. B4.

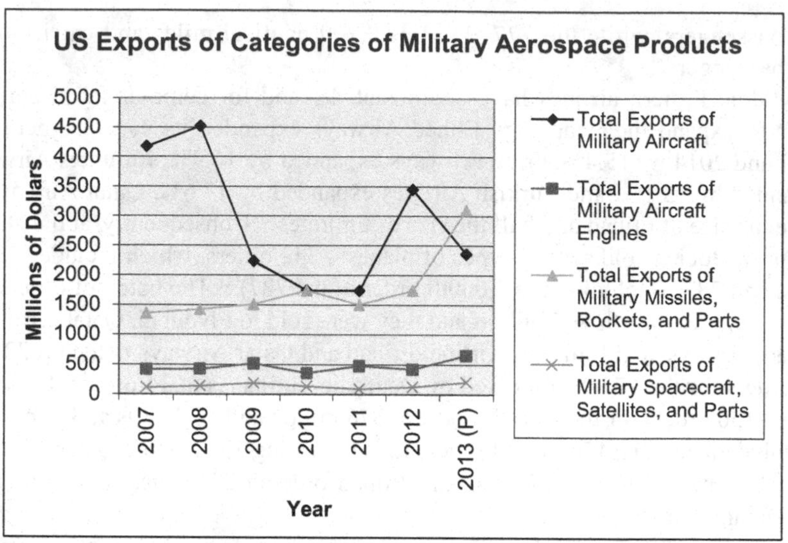

Figure 6.3 US Exports of Categories of Military Aerospace Products
Source: AIA 2011, 2012, and 2013 Year-End Review and Forecast; Table VIII: US Exports of Aerospace Products, p. 17. Note that this only uses current dollars; data in millions of $.

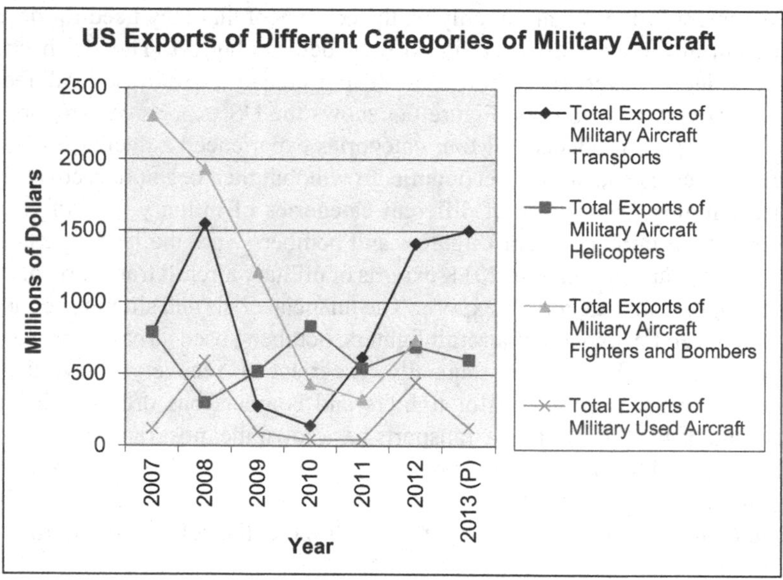

Figure 6.4 US Exports of Different Categories of Military Aircraft
Source: AIA 2011, 2012, and 2013 Year-End Review and Forecast; Table VIII: US Exports of Aerospace Products, p. 17. Note that this only uses current dollars; data in millions of $.

are facing. Rising costs, however, have made it more difficult for them to afford F-35s, due to budgetary constraints and slow economic growth. As of November, 2013, sales of F-35s comprised 15% of the revenue of Lockheed Martin, which focused on making half of its F-35 sales to overseas customers. The first country in the Asia-Pacific region to receive an F-35 is Australia, which receives its first delivery in 2014, followed by Japan in 2016, which contracted for 42 planes, most of which are being constructed in Nagoya due to the offset program. In September 2013, the Netherlands contracted to purchase 37 more F-35s at $6.1 billon.[13] Nevertheless, with budgetary constraints, some countries have had to re-evaluate their purchases. Canada had originally suggested that it would buy F-35s to fill its demand for 65 new fighters, but it has re-opened the competition; moreover, Italy is also considering how many F-35s it will buy.

The important role of the F-35 in responding to perceived defense threats for the purchasing countries has a significant impact on sales. As of the winter of 2013, South Korea was strongly considering purchasing 40 F-35s to fulfill its need to replace its aging planes and may consider a subsequent deal for an additional set of 20 F-35s, particularly given South Korea's need for radar-evading stealth capabilities in response to potential threats from North Korea. Boeing's F-15SE, which was considered in South Korea's contest, did not have a strong enough stealth capability. EADS' Eurofighter was another contender.[14]

Unfortunately, US exports of military aircraft can be hindered by high costs, despite the innovative technology and the value of these aircraft in facing defense threats. Indeed, US equipment is often at a competitive disadvantage with equipment from other countries due to high prices. One example is the F-35, which costs $100 million. The average costs of manufacturing the F-35 will likely decline, however, as more sales occur; Lockheed Martin estimates that in less than five years, the US will allow more sales to Middle Eastern countries. As of the fall of 2013, the only Middle Eastern countries to whom Lockheed Martin could sell the F-35 were Turkey and Israel.[15] Indeed, Israel's F-35s will also be delivered in 2016.[16]

Another example of the high prices of US exports of military aircraft is Boeing's F/A-18E/F Super Hornet and its F-15 Eagle. The average price of a given fighter which is purchased by overseas countries is $65 million (the cost of the Super Hornet), which is much lower than the $100 million of the F-35. Nevertheless, 30 of the 52 countries which buy fighters tend to purchase in the price range of $35 million to $50 million, which is similar to the price of an F-16 or the Mirage 2000, the latter of which is made by Dassault Aviation (France). Lockheed Martin, which makes both the F-35 and the F-16, however, focuses its international sales and marketing efforts much more on the F-35, although fewer countries can afford

13 Kwaak and Cameron, 2013, p. B4.
14 Kwaak and Cameron, 2013, p. B4.
15 Cameron, 2013, pp. B1 and B3.
16 Cameron, 2013, pp. B1 and B3.

it. As of the fall of 2013, only seven countries could afford the F-18 and another seven had possibilities of purchasing (in some cases additional) F-35s, such as South Korea, Japan, Israel, and Singapore.[17]

Due to its affordability, many countries had been disappointed when the Air Force ended production of the F-22s at 187 planes. Countries such as Bolivia, Kenya, and Argentina can only afford to purchase used planes. As a result, there is a bifurcation of the categories of countries which can afford different types of aircraft based on the size of their defense budgets in the wake of slowness in economic growth.[18] Unfortunately, from the perspective of defense contractors, there are limits to the demands of various countries for particular types of US exports of military aircraft. As of the fall of 2013, the last F-15 was scheduled to be delivered to Saudi Arabia in 2018–2019, the last exports of the F/A-18 E/F would be delivered during 2015–2016 unless other countries, such as Brazil or Kuwait, purchased it, and the production for F-16s would continue until 2017 based on the

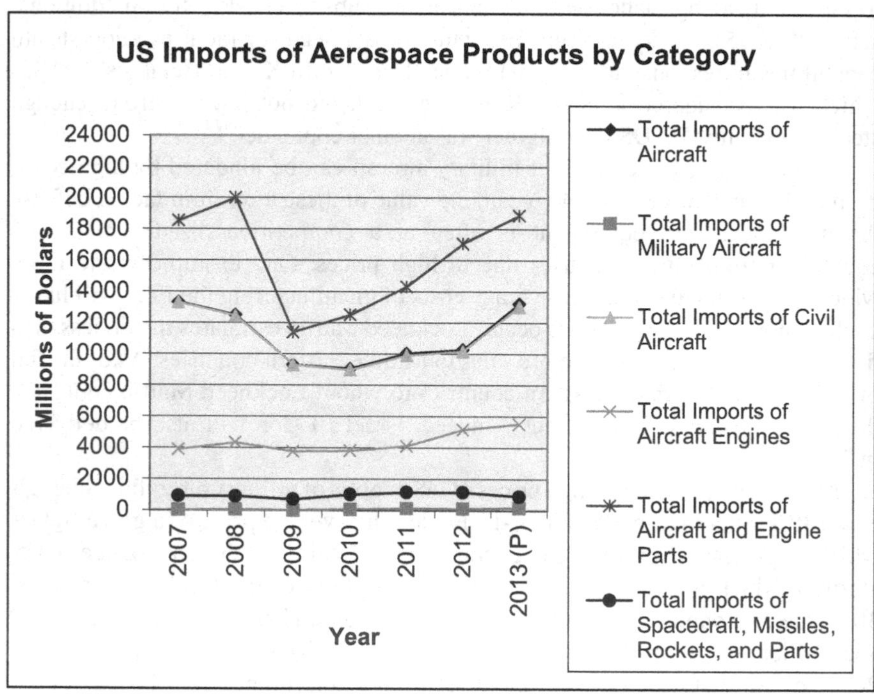

Figure 6.5 US Imports of Aerospace Products by Category

Source: AIA 2011, 2012, and 2013 Year-End Review and Forecast; Table VII: US Imports of Aerospace Products, p. 16. Note that this only uses current dollars; data in millions of $.

17 Erwin, 2013b.

18 Erwin, 2013b.

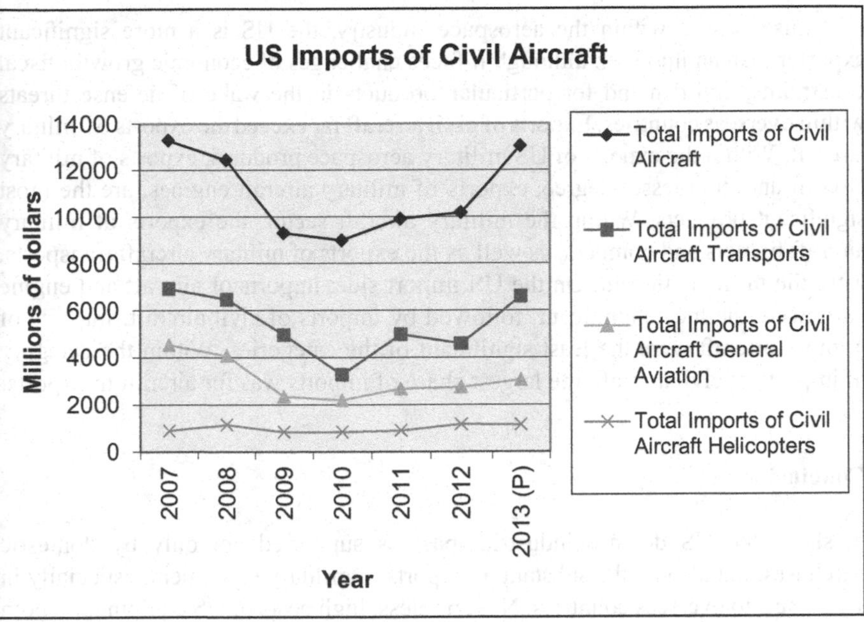

Figure 6.6 US Imports of Civil Aircraft

Source: AIA 2011, 2012, and 2013 Year-End Review and Forecast; Table VII: US Imports of Aerospace Products, p. 16. Note that this only uses current dollars; data in millions of $.

current order book (note that, since 1970, 4600 F-16 planes have been sold). The difficulty with the F/A-18 is that it has not thrived on exports to the degree that the F-16 line has and is more dependent on purchases by DoD, especially the Navy.[19]

While the US is a significant exporter of aerospace products, it is a much less significant importer. The largest category of US imports in aerospace products is aircraft and engine parts, followed by imports of civil aircraft. Figure 6.5 shows US imports of aircraft by category. Many of the categories, especially the aircraft and engine parts category and the civil aircraft category, experienced a decline in 2009 with the slowing of economic growth, but also exhibited a recovery in subsequent years.

US imports of civilian aircraft are much greater than US imports of military aircraft. Figure 6.6 examines, in more detail, imports of various subcategories of civil aircraft. As was evident in the earlier figure, imports of civil aircraft declined in 2009 and 2010 as a result of the reduction in economic growth limiting demand for civil aircraft. After 2010, the market for civil aircraft improved. Similar patterns were evident in imports of civil aircraft transports – the largest of the three subcategories – as well as civil aircraft general aviation and helicopters.

19 Erwin, 2013b.

Consequently, within the aerospace industry, the US is a more significant exporter than an importer, although it faces challenges of economic growth, fiscal constraints, and demand for particular products in the wake of defense threats within overseas countries. Exports of civil aircraft far exceed the exports of military aircraft. Within the exports of US military aerospace products, exports of military aircraft and, to a lesser degree, exports of military aircraft engines, are the most significant products. Within the military aircraft sector, the exports of military aircraft fighters and bombers, as well as the exports of military aircraft transports, were the most significant. On the US import side, imports of aircraft and engine parts were the most significant, followed by imports of civil aircraft. Imports of military aircraft were the least significant of the categories. Within the category of imports of civil aircraft, the largest share of imports was for aircraft transports.

Conclusion

In short, the US defense industrial base is supported not only by domestic purchases, but also by the substantial exports in military equipment, especially in aerospace, to overseas countries. Nevertheless, high costs of US equipment – both commercial and military – as well as slowing economic growth and budget constraints, can restrict demand from overseas countries, despite their need for civilian transport equipment and military threats. While diversification of US defense contractors is important in the commercial sector and in overseas sales, the continued sustainability of the US defense market is the backbone for stable or growing profitability. As US demand lessens, however, overseas sales of both commercial and military equipment helps to promote innovation and continuation of existing models. Finally, although the US does not award many contracts to foreign contractors, it does work with them in the global supply chain, as exemplified by the F-35, highlighting the globalization of the defense industrial base in recent decades.

References

"Aeronautics Company; F-35: Delivering the Promise to Redefine National Strategic Capabilities." 2008. *Business and Finance Week*, October 4, p. 331.

"Aerospace Production: Connecting Flights." 2009. *The Engineer*, November 23, p. 36.

Cameron, Doug, 2013. "Clipped by U.S., Lockheed CEO Aims Abroad." *Wall Street Journal*, December 7–8, pp. B1 and B3.

Drew, Christopher, 2013. "New Boeing Jet, 777X, Hits $95 Billion in Orders." *New York Times*, November 18, p. B4.

Erwin, Sandra I., 2013a. "Lockheed Martin Courts New International Buyers for its Missile Defense Systems." *National Defense*, September 10.

——. 2013b. "U.S. Tactical Aircraft Too Expensive for Most International Buyers." *National Defense*, October 28.

Jones, Rory and Cameron, Doug, 2013. "Boeing, Airbus Reel in Orders." *Wall Street Journal*, November 18, pp. B1 and B2.

Kwaak, Jeyup S. and Cameron, Doug, 2013. "Lockheed Zeroes in on Jet Contract." *Wall Street Journal*, November 23–24, p. B4.

Chapter 7
The Global Defense Market

This chapter examines the global defense industry in detail. It explores the impact of economic growth, shifting defense priorities, and fiscal constraints on the evolution of global trade in the defense market, as well as the efforts of countries to protect their own defense industrial bases. It first examines the countries with the largest defense budgets and explores the challenges facing them by patterns in economic growth and government debt. The chapter then focuses on the trade patterns for defense equipment and assesses the economic growth and government debt in the countries which are the leading exporters and the leading importers in the global defense market. The historical evolution of global trade in ships and aircraft provides a case study of the changes in the importance of various countries as importers and exporters in these markets. It then discusses the strategies for countries in preserving their defense industrial base. Finally, the case study of the European defense industrial base and fiscal constraints explores the challenges and solutions facing European countries. It examines the impact of budgetary constraints in various countries on the European defense industrial base, the role of alliances between countries in developing equipment, and the strategies of contractors in closing key facilities, redeveloping their company, and focusing on international sales.

Military Expenditures in a Global Context

US military expenditures and defense budgets have been discussed in earlier chapters and lead to a series of questions. How does US defense spending compare with the military expenditures in the other countries with the highest military expenditures? What are the trends in economic growth and government debt over the past 12 years for these countries? The economic growth (or lack thereof) in many of the countries and fluctuations in their military expenditures will impact the profitability of US defense contractors who increasingly rely on expanding their international military sales, as well as the ability of domestic industrial bases of overseas countries to develop. The following charts reveal a number of important aspects of the global defense market.

Figure 7.1 compares the countries with the highest military expenditures in 2013. The US had the highest military expenditure, distantly followed by China at $188 billion and Russia at $87.8 billion. The UK, Japan, France, Saudi Arabia, India, and Germany were in a range of $47.4–$67 billion. Countries such as Italy, Brazil, South Korea, Australia, the UAE, and Turkey had even smaller military

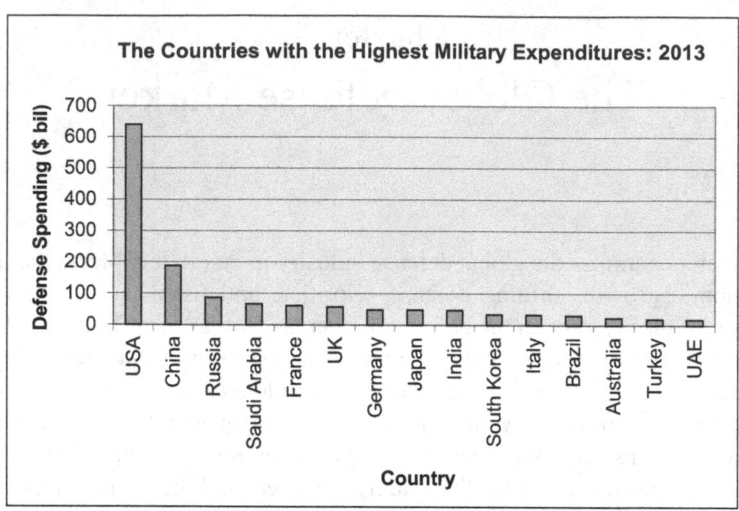

Figure 7.1 The Countries with the Highest Military Expenditures: 2013

Source: Sam Perlo-Freeman and Carina Solmirano, *SIPRI Fact Sheet: Trends in World Military Expenditure, 2013*, April 2014, Table 1, p. 2.

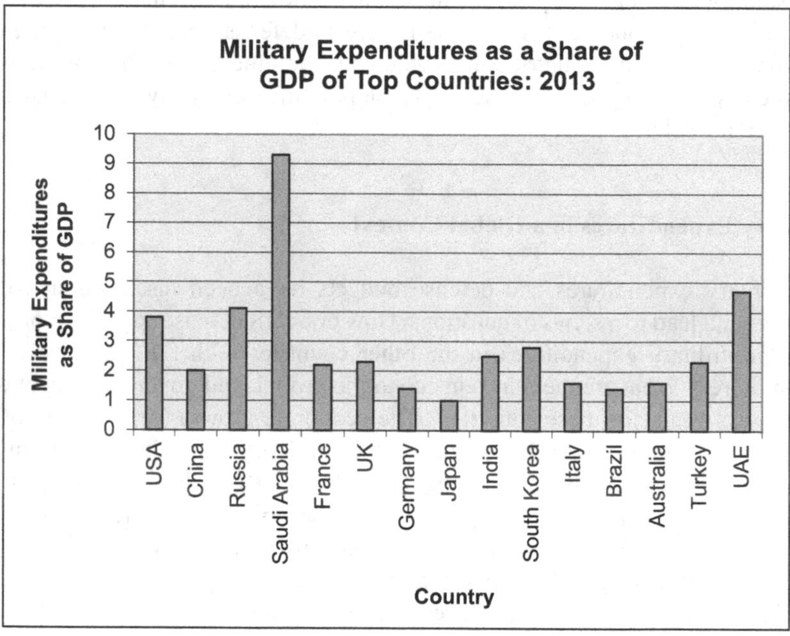

Figure 7.2 Military Expenditures as a Share of GDP of Top Countries: 2013

Source: Sam Perlo-Freeman and Carina Solmirano, *SIPRI Fact Sheet: Trends in World Military Expenditure, 2013*, April 2014, Table 1, p. 2.

expenditures, but were still among the top 15 countries. Most of the countries with highest expenditures declined in their military spending from 2012. Over the period between 2004 and 2013, some countries, such as Saudi Arabia (118%), China (170%), and Russia (108%), experienced growth in military expenditures of over 100%. Other countries experienced declining growth – Italy (-26%), France (-6.4%), the UK (-2.5%), and Japan (0.2%). For defense contractors in the US and abroad, many of the countries with smaller, but growing, defense budgets provide vibrant markets for overseas sales.

The importance of defense spending in the economies of various countries is illustrated in Figure 7.2, which shows military expenditures as a share of GDP for the top 15 countries in 2013. Although Saudi Arabia had lower military

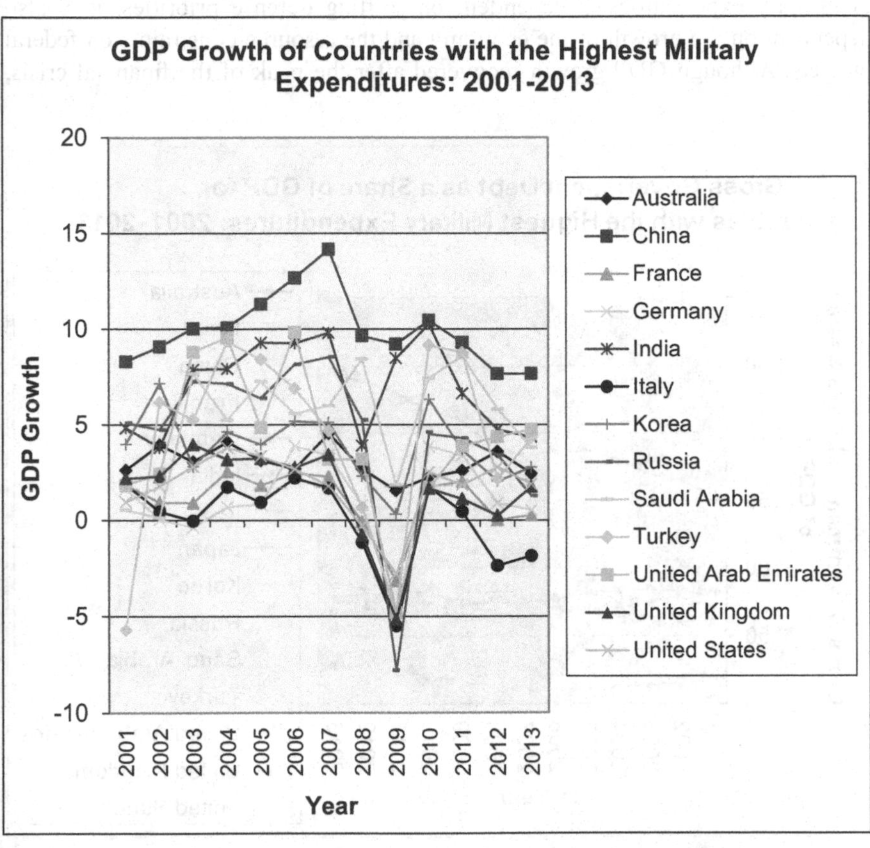

Figure 7.3 GDP Growth of Countries with the Highest Military Expenditures: 2001–2013

Source: Sam Perlo-Freeman and Carina Solmirano, *SIPRI Fact Sheet: Trends in World Military Expenditure, 2013*, April 2014, Table 1, p. 2; *IMF World Economic Outlook Database*. See https://www.imf.org/external/data.htm

expenditures than the US, China, or Russia, its military spending was 9.3% of its GDP and was much higher than the share of military expenditures for the US as a share of GDP, which was 3.8%. Similarly, the UAE, which had the fifteenth highest military expenditure, and Russia, which had the third highest military expenditures, exceeded the US in terms of its military spending as a share of GDP (4.7% and 4.1%, respectively). Countries such as the UK, France, India, South Korea and Turkey had military expenditures which ranged between 2% and 3% of GDP. Countries such as China, Germany, Italy, Brazil, and Australia have military expenditures which were at or below 2% of GDP. Of the top 15 countries, Japan's military expenditures are the smallest share of GDP at 1%.

The GDP growth of the countries with the highest military expenditures between 2001 and 2013 is evident in Figure 7.3. While the size and growth of military expenditures is dependent on shifting defense priorities, it is also dependent on the growth of the economy and the resource constraints on federal budgets. Although GDP growth recovered after the peak of the financial crisis,

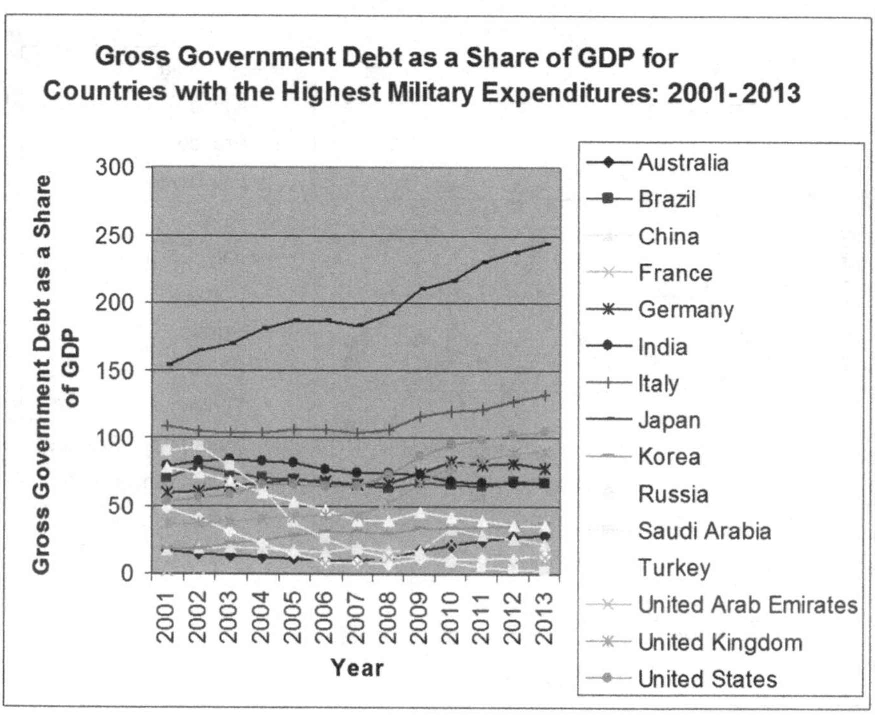

Figure 7.4 Gross Government Debt as a Share of GDP for Countries with the Highest Military Expenditures: 2001–2013

Source: Sam Perlo-Freeman and Carina Solmirano, *SIPRI Fact Sheet: Trends in World Military Expenditure, 2013*, April 2014, Table 1, p. 2; *IMF World Economic Outlook Database*. See https://www.imf.org/external/data.htm

there has been a decrease in GDP growth in many countries in recent years, which may limit defense expenditures. Between 2011 and 2012, there was a decline in GDP growth in many countries, including: Brazil, China, France, Germany, India, and Italy. Nevertheless, Japan, Australia and the US showed an improvement in GDP growth during that period. Between 2012 and 2013, many of the countries experienced similar GDP growth, although Australia, the US, Germany, Russia, and Saudi Arabia declined, and the UK, South Korea, Turkey, and Italy showed improvement.

Government debt as a share of GDP has increased since the financial crisis, which may limit defense expenditures in future years, as illustrated in Figure 7.4. This is because increasing government debt can lead to higher interest payments, which can, in turn, reduce federal budget expenditures in other areas, including defense. Japan has had the most significant increase in gross government debt as a share of GDP, ranging from just above 150% in 2000 to almost 250% in 2013. Japan's debt levels as a share of GDP not only are the highest, but also grew the most rapidly. Consequently, it is not surprising that, as suggested by Figure 7.2 and as discussed earlier, Japan's military expenditures are only 1% of GDP – the smallest of the 15 countries with the highest military expenditures. Italy is another country which, like Japan, had debt as a share of GDP at over 100% in 2001 and reached 130% by 2013. A few countries, however, have managed to reduce their debt as a share of GDP since 2001: Australia, India, Russia, Saudi Arabia, and Turkey. Between 2011 and 2013, most countries increased debt as a share of GDP; only China, Saudi Arabia, the UAE, and Turkey experienced a decrease, while countries such as Australia, Brazil, Japan, South Korea, France, Italy, Russia, the UK, and the US increased debt as a share of GDP.

Trade Patterns for Defense Equipment

As discussed in the previous chapters, defense contractors have had to develop strategies to handle challenges to the defense industrial base. One of the strategies for defense contractors who have difficulty selling defense products in their home country, due to reductions in military expenditures (which can be linked to increased interest payments, etc. in the broader federal budget), is to export their products to countries overseas. This was evident in the discussion of the profitability of overseas sales for large defense contractors. This section takes a global perspective and examines the leading exporting and importing countries of defense equipment and services, including their GDP growth and government gross debt as a share of GDP.

The leading exporters of defense equipment in 2013 were Russia and the US, followed distantly by other countries, such as: China, France, the UK, Germany, Ukraine, Italy, Spain, and Israel (Figure 7.5).

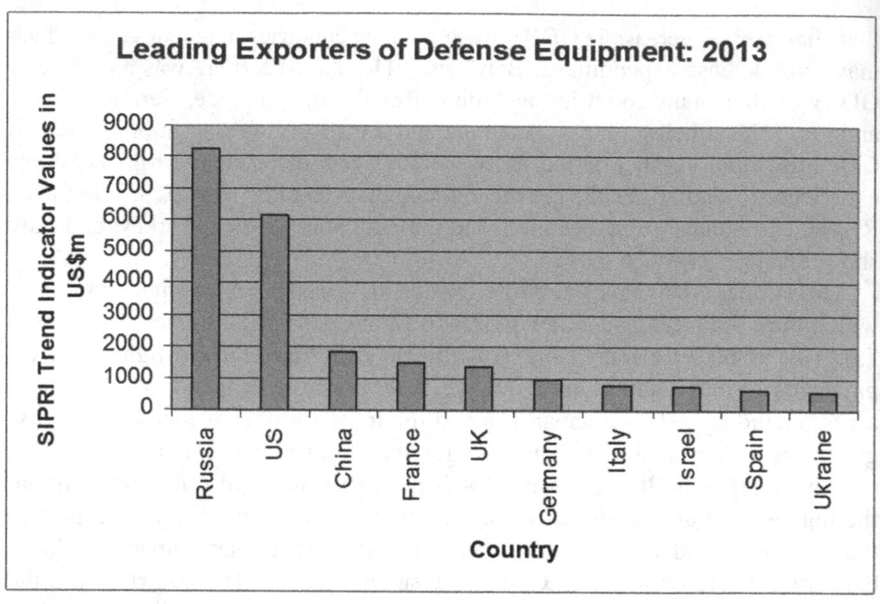

Figure 7.5 Leading Exporters of Defense Equipment: 2013
Source: SIPRI Military Expenditures Database. See http://www.sipri.org/research/armaments/transfers/measuring/recent-trends-in-arms-transfers

The GDP growth for the top exporters of defense equipment between 2001 and 2013 is illustrated in Figure 7.6. All of the countries which are top exporters of defense equipment showed a substantial decline in GDP growth at the height of the financial crisis in 2009, with China experiencing the least decline and Ukraine experiencing the most significant decline. All of the countries improved in GDP growth between 2009 and 2010, but have experienced a decline in GDP growth again since 2010. The US and the UK are the only two countries of the top exporting countries which experienced a modest increase in GDP growth between 2011 and 2013. Countries such as China, France, Germany, Israel, Italy, Russia, Spain, and the Ukraine experienced significantly lower GDP growth by 2013 relative to 2011.

Figure 7.7 shows the trends in government gross debt as a share of GDP for the top exporters of defense equipment between 2001 and 2013. For most countries, government debt as a share of GDP has increased, especially following the financial crisis, beginning in 2008. Italy has the highest debt as a share of GDP – around 130% – and has consistently followed this pattern over the past 12 years. The US has the second highest share of debt at 104% of GDP, with a sharp increase following the financial crisis such that it rose from 72% of GDP in 2008. Germany and France had debt as a share of GDP which was similar to the US between 2000 and 2007 and, like the US, their debt as a share of GDP increased following the financial crisis,

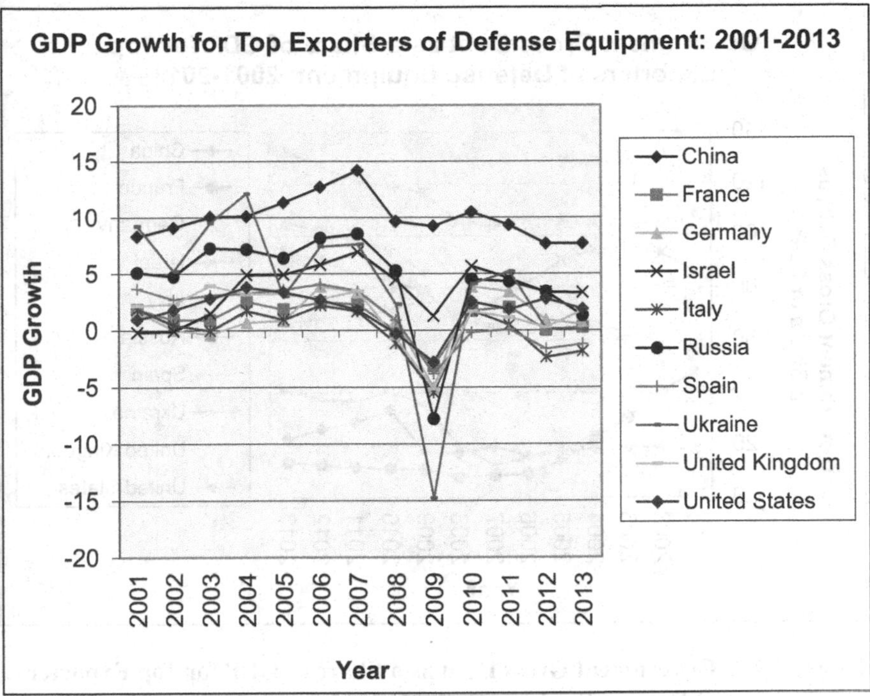

Figure 7.6 GDP Growth for the Top Exporters of Defense Equipment: 2001–2013

Source: SIPRI Military Expenditures Database. See http://www.sipri.org/research/armaments/transfers/measuring/recent-trends-in-arms-transfers; *IMF World Economic Outlook Database*. See https://www.imf.org/external/data.htm

although the US' gross debt as a share of GDP grew faster, such that it surpassed Germany and France. Several countries over the past 12 years have experienced reductions in debt as a share of GDP, including Russia and Israel. Between 2011 and 2013, China and Israel reduced their gross debt as a share of GDP, Russia and Germany were relatively flat, and the other countries (France, Italy, Spain, Ukraine, the UK, and the US) increased their gross debt as a share of GDP.

Partially as a result of the increase in government gross debt as a share of GDP, many of the countries are developing fiscal strategies to reduce the federal budget debts and deficits, which can result in a reduction in defense expenditures. As a result, many of the defense contractors in these countries are likely to export their defense equipment and services to other countries. Consequently, due to flat or declining GDP growth and rising government debt, defense contractors in the leading exporting countries in defense equipment are likely to need to depend on foreign sales (rather than domestic sales) to a greater degree than previously to sustain recent profitability and productivity levels.

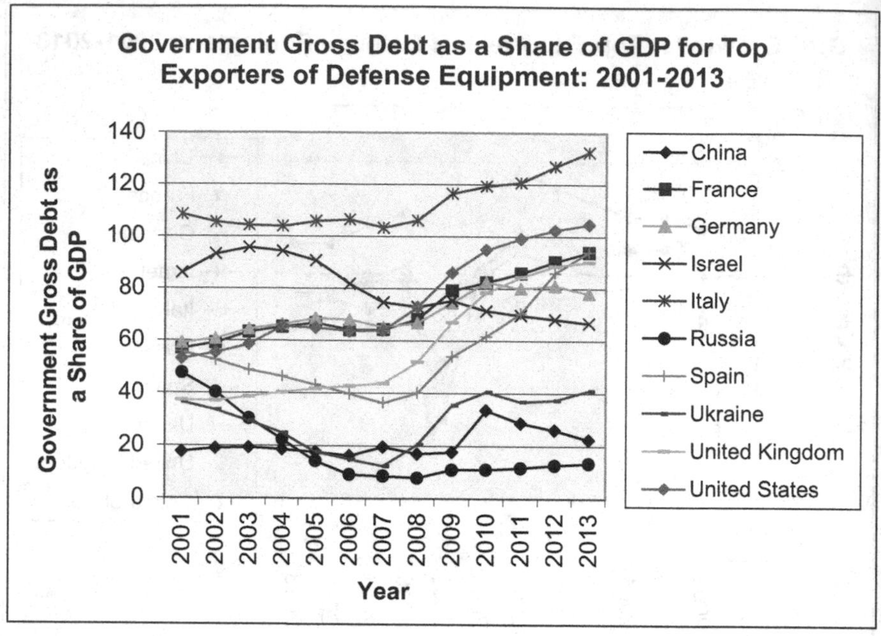

Figure 7.7 Government Gross Debt as a Share of GDP for Top Exporters of Defense Equipment: 2001–2013

Source: SIPRI Military Expenditures Database. See http://www.sipri.org/research/armaments/transfers/measuring/recent-trends-in-arms-transfers; *IMF World Economic Outlook Database*. See https://www.imf.org/external/data.htm

Which countries are the leading markets for defense equipment? The figures over the next few pages examine the economic growth and government debt of the leading importers in the global defense market.

The countries which were the leading importers of defense equipment in 2013 are depicted in Figure 7.8. India was the leading importer, followed distantly by the UAE, China, and Saudi Arabia, and then Pakistan, Afghanistan, the US, Turkey, Algeria, and South Korea.

Figure 7.9 shows the GDP growth for the top importers of defense equipment between 2001 and 2013. Most countries experienced higher GDP growth in 2013 relative to 2001 (except for Azerbaijan and South Korea), although these countries experienced a sharp decline during the financial crisis in 2009 and then significant recovery following the financial crisis. Nevertheless, most countries experienced declining GDP growth between 2011 and 2013, Indeed, only Azerbaijan and UAE had higher GDP growth in 2013 than in 2011. Countries such as China, India, South Korea, Saudi Arabia, and Turkey experienced declines, while Algeria, Pakistan, and the US had similar GDP growth rates. By 2013, the countries with the highest GDP growth were China, Azerbaijan, UAE, India, and Turkey.

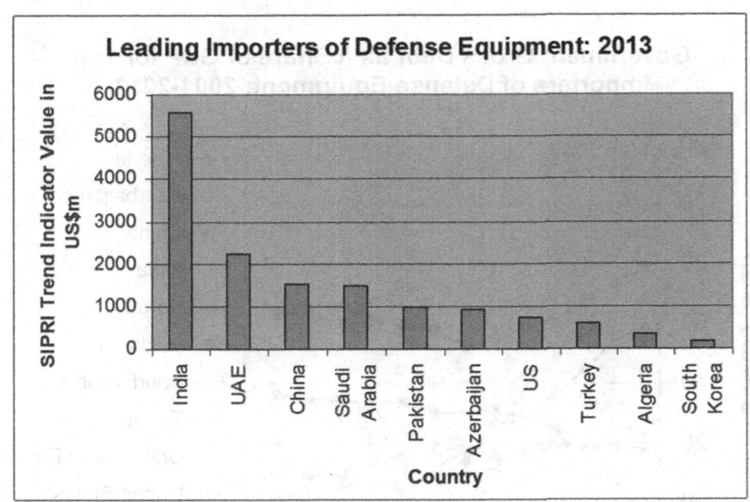

Figure 7.8 Leading Importers of Defense Equipment: 2013

Source: SIPRI Military Expenditures Database. See: http://www.sipri.org/research/armaments/transfers/measuring/recent-trends-in-arms-transfers

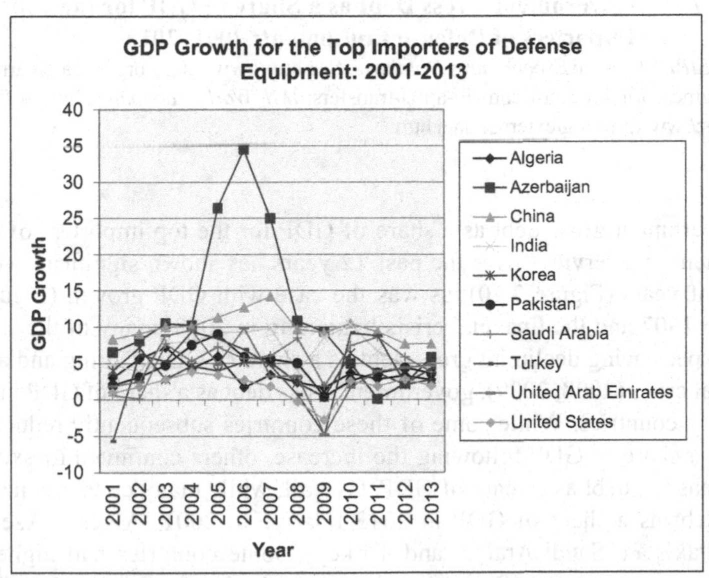

Figure 7.9 GDP Growth for the Top Importers of Defense Equipment: 2001–2013

Source: SIPRI Military Expenditures Database. See http://www.sipri.org/research/armaments/transfers/measuring/recent-trends-in-arms-transfers; *IMF World Economic Outlook Database*. See https://www.imf.org/external/data.htm

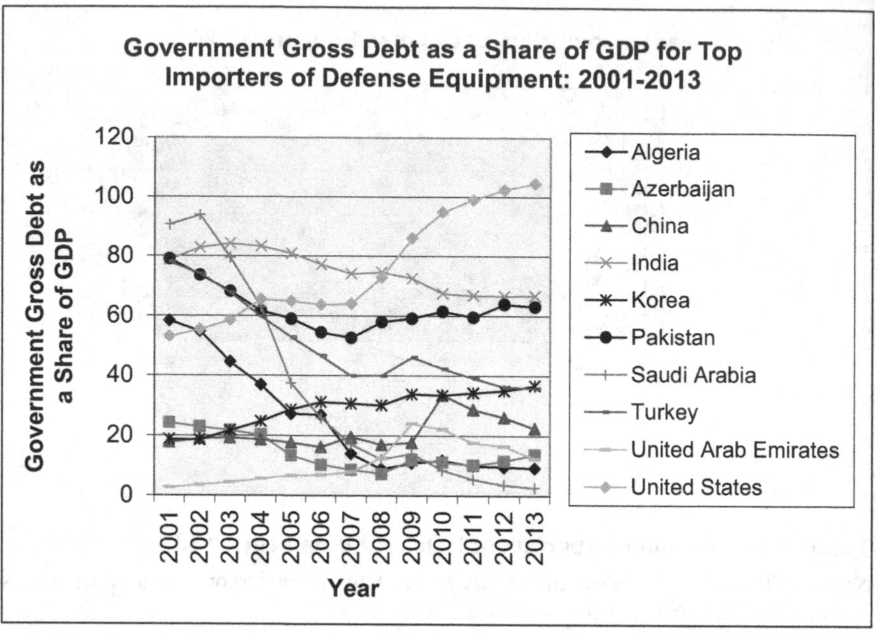

Figure 7.10 Government Gross Debt as a Share of GDP for the Top Importers of Defense Equipment: 2001–2013

Source: SIPRI Military Expenditures Database. See http://www.sipri.org/research/armaments/transfers/measuring/recent-trends-in-arms-transfers; *IMF World Economic Outlook Database.* See https://www.imf.org/external/data.htm

Government gross debt as a share of GDP for the top importers of defense equipment and services over the past 12 years has shown significant volatility in recent years (Figure 7.10), as was the case with GDP growth (Figure 7.9). Prior to 2007 and the financial crisis beginning in 2008, many of the countries were experiencing declining gross debt as a share of GDP. During and after the financial crisis (2008, 2009), government gross debt as a share of GDP increased for many countries. While some of these countries subsequently reduced their debt as a share of GDP following the increase, others continued to experience an increase in debt as a share of GDP. Overall, while many countries had lower gross debt as a share of GDP in 2013 relative to 2001 (Algeria, Azerbaijan, India, Pakistan, Saudi Arabia, and Turkey), some countries had higher gross debt as a share of GDP (China, South Korea, the UAE, and the US). Between 2011 and 2013, some of the countries experienced rising gross debt as a share of GDP (Azerbaijan, South Korea, Pakistan, and the US), while others experienced declining gross debt as a share of GDP (China, Saudi Arabia, Turkey, and the UAE) and other countries experienced little change (Algeria and India). By

2013, the US had the highest government gross debt as a share of GDP, followed by India and Pakistan.

The macroeconomic conditions for the top importing countries of defense equipment and services suggest that these countries are likely to continue to purchase defense equipment, but the markets will not exhibit the substantial growth that they experienced in prior years.

In short, due to declining GDP growth and rising/flat government debt as a share of GDP for some of the countries, defense spending will be constricted by the need to reduce the debt and to stimulate economic growth, which, in turn, can place financial constraints on the federal budget and on the defense sector. On the other hand, as defense priorities shift for many of these countries, the need to purchase new equipment and maintain existing equipment, as well as train military personnel, can be deemed to be more significant drivers for defense spending than limitations of financial constraints. Consequently, defense contractors face challenges both domestically and overseas in maintaining or increasing demand for equipment from their customers.

Case Study of Global Defense Markets: Ships and Aircraft

The global defense market for ships and aircraft provides a valuable case study of the evolution of these markets in recent years. The charts below show the recent

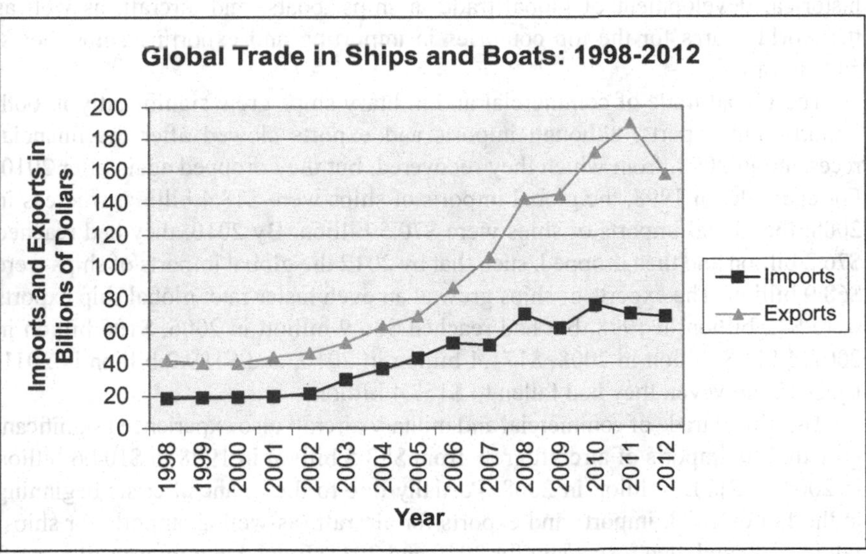

Figure 7.11 Global Trade in Ships and Boats: 1998–2012
Source: UN International Merchandise Trade Statistics, 2012, p. 793.

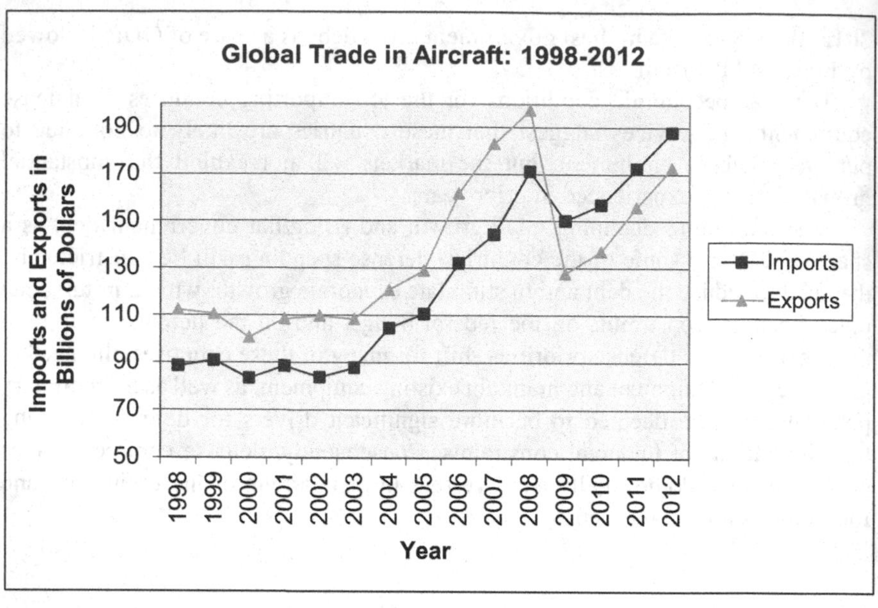

Figure 7.12 Global Trade in Aircraft: 1998–2012
Source: UN International Merchandise Trade Statistics, 2012, p. 792.

historical development of global trade in ships, boats, and aircraft, as well as the world shares for the top countries in importing and exporting ships, boats, and aircraft.

The global trade of commercial and military ships grew significantly in both imports and exports, although imports and exports slowed after the financial recession in 2008, from which they recovered, but they dropped again after 2010. For example, in 1998, the global imports of ships were $18.4 billion, whereas in 2008, the global imports of ships were $70.5 billion. By 2010, they had reached $76.2 billion and then dropped, such that by 2012 the global imports of ships were $68.9 billion. The exports of ships grew at an even faster rate: global ship exports were $42 billion in 1998, but had reached $86.9 billion in 2006, $106 billion in 2007, $142.8 billion in 2008, $171.4 billion in 2010, and $189.2 billion in 2011. By 2012, however, they had fallen to $157.4 billion.

The global trade of commercial and military aircraft also experienced significant growth. The imports of aircraft grew from $88.5 billion in 1998 to $104.6 billion in 2004 to $171.1 billion in 2008. Partially due to the financial crisis beginning in the fall of 2008, imports and exports for aircraft (as well as imports for ships) fell in 2009 and then rose again for 2010 and 2011. Indeed, the imports of aircraft fell to $150.1 billion in 2009, returning to 2008 levels by 2011 at $172.3 billion. Similarly, the exports of aircraft fell from $197.3 billion in 2008 to $127.6 billion in 2009, rising in 2010 to reach $155.9 billion in 2011. Exports of aircraft, like

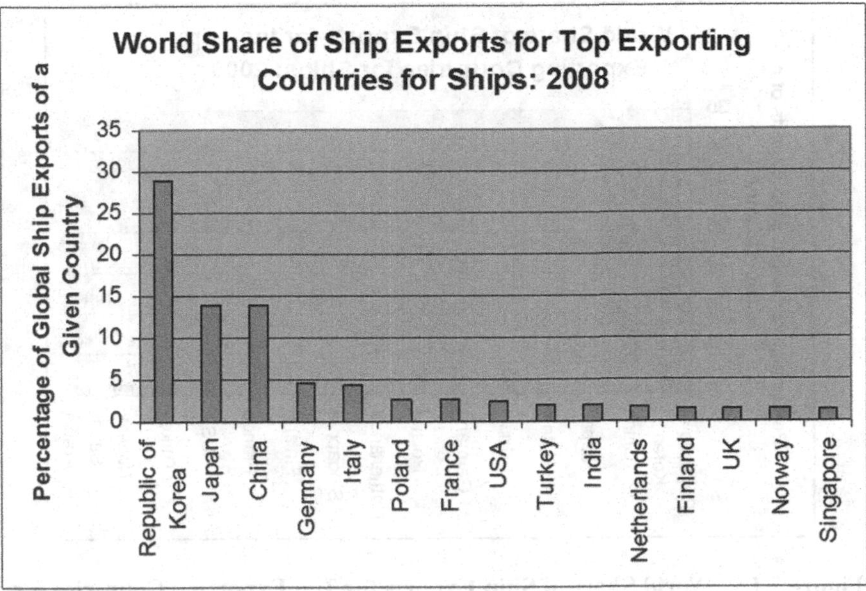

Figure 7.13 World Share of Ship Exports for Top Exporting Countries for Ships: 2008

Source: UN International Merchandise Trade Statistics, 2008, p. 793.

imports of aircraft, followed a similar growth pattern and expanded from $112.2 billion in 1998 to $120.7 billion in 2004, $161.7 billion in 2006 and $197.3 billion in 2008. Unlike imports and exports of ships, which dropped between 2011 and 2012, however, the imports and exports of aircraft continued to grow between 2011 and 2012, with imports reaching $187.9 billion and exports reaching $172.6 billion in 2012.

In 2008, about 58% of the global export market for ships was led by the Republic of Korea, Japan, and China. Indeed, the Republic of Korea exported almost 30% of ships, followed by Japan and China, each of which exported almost 14% of the ships. Germany and Italy each exported about 4.5% of the ships, while the US only exported about 2.3% of the ships in the global market.

By 2009, about 66% of the global export market for ships was led by the Republic of Korea, Japan, and China, largely due to growth in the Chinese shipping industrial base. The Republic of Korea exported 30% of ships (as was the case in 2008), followed by China, which exported about 20% of ships (an increase from 2008), and by Japan, which exported about 15.8% of the ships. Italy exported about 4% of the ships (similar to 2008), while India, Germany and Poland exported a smaller share at 2.7%, 2.2%, and 2.1%, respectively. The US exported only 1.5% of ships in 2009 – a decline from 2008, as was also the case with Germany.

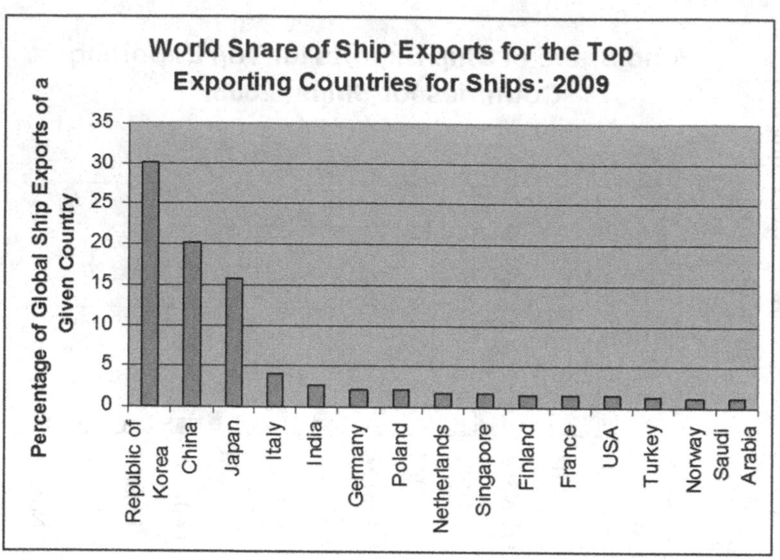

Figure 7.14 World Share of Ship Exports for Top Exporting Countries for Ships: 2009

Source: UN International Merchandise Trade Statistics, 2009, p. 793.

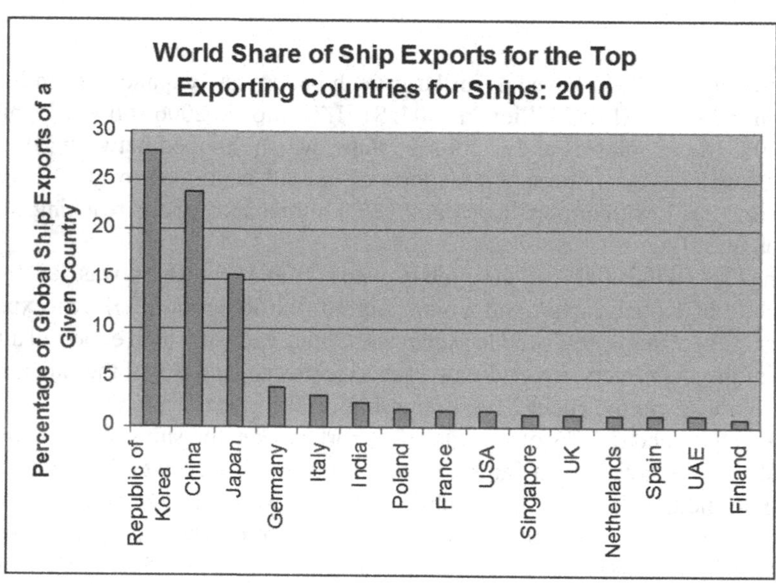

Figure 7.15 World Share of Ship Exports for Top Exporting Countries for Ships: 2010

Source: UN International Merchandise Trade Statistics, 2010, p. 793.

By 2010, about 67% of the global market for exporting ships was led by the Republic of Korea, Japan, and China (as was the case in 2009). Indeed, the Republic of Korea exported about 28% of the ships, followed by China, which exported about 24% of ships (an increase from 2008 and 2009, indicating growth in the Chinese shipping industrial base), and by Japan, which exported about 15.4% of the ships. Relative to 2009, Germany increased its shipments to 3.9% of the global market, while Italy exported a lower amount of shipments at 3.2% and India exported about 2.5% of the ships. The US exported only 1.6% of its ships in 2010, which was similar to 2009 and a decline from 2008.

By 2011, about 65% of the global market for exporting ships was led by the shipyards in the Republic of Korea, Japan, and China (as had been the case in 2009 and 2010). Indeed, the Republic of Korea exported about 28% of the ships, followed by China, which exported about 23% of ships (similar to 2010), and by Japan, which exported about 14% of the ships. India, Singapore, and Poland increased their share of global exports of ships to 3.8%, 3%, and 2.7%, respectively. Italy and Germany exported 2.6% and 2.1%, respectively, of the market, which was a lower share than in previous years. The US exported only 1.4% of ships in 2011, which was slightly lower than in 2010, similar to 2009 and a decline from 2008.

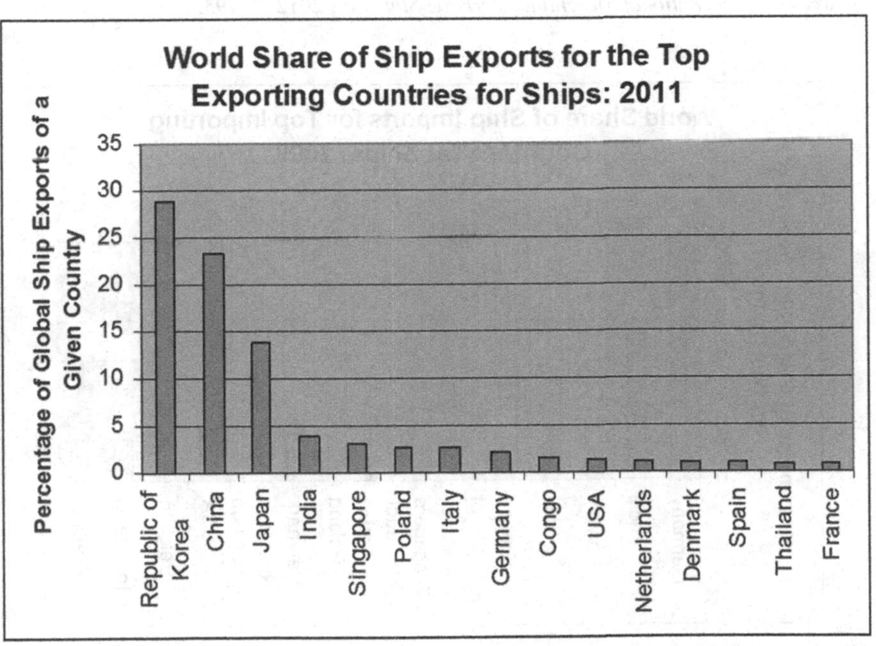

Figure 7.16 World Share of Ship Exports for Top Exporting Countries for Ships: 2011

Source: UN International Merchandise Trade Statistics, 2011, p. 793.

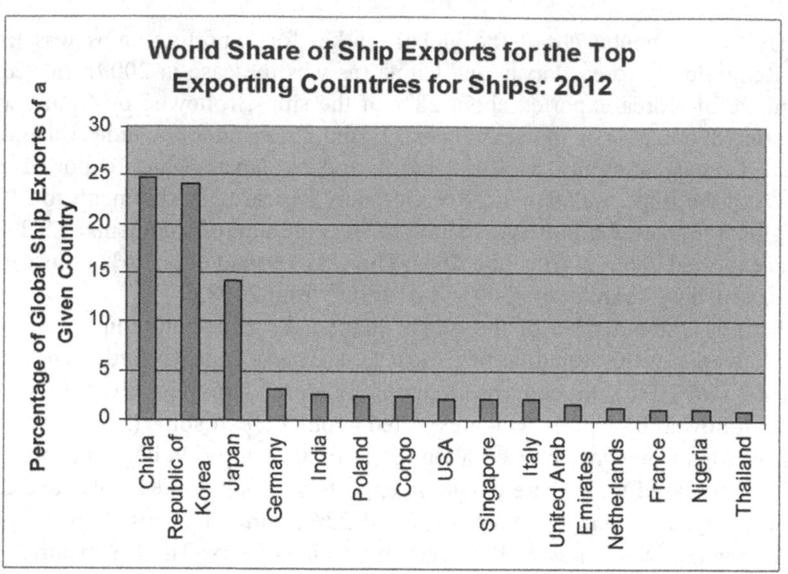

Figure 7.17 World Share of Ship Exports for Top Exporting Countries for Ships: 2012

Source: UN International Merchandise Trade Statistics, 2012, p. 793.

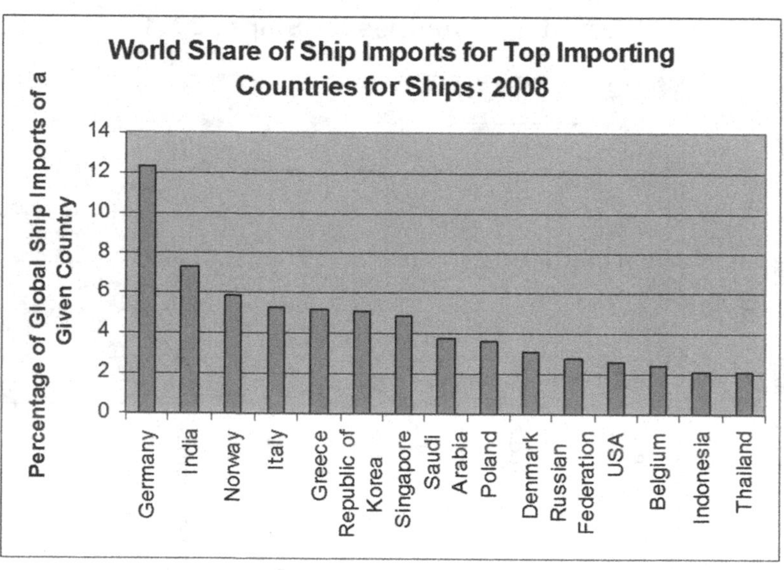

Figure 7.18 World Share of Ship Imports for Top Importing Countries for Ships: 2008

Source: UN International Merchandise Trade Statistics, 2008, p. 793.

By 2012, about 62.8% of the global market for exporting ships was led by the shipyards in the Republic of Korea, Japan, and China (as had been the case in 2009, 2010, and 2011). Unlike the global market in 2011, China was the leading exporter at 24.7% of the market since the Republic of Korea declined from exporting 28% of the market in 2011 to 24% in 2012. Japan remained constant at 14%. Germany increased its share of the market from 2.1% in 2011 to 3.2% in 2012, as did the Congo. India, Poland, Singapore, and Italy declined in their market shares between 2011 and 2012: India fell from 3.8% in 2011 to 2.6% in 2012; Poland fell from 2.7% to 2.5%; Singapore declined from 3% to 2.1%; Italy declined from 2.6% to 2.1%. The US, however, increased its share to 2.2%, which was an improvement from its market share of 1.4% in 2011.

The next set of figures provides a perspective on the development of the global market for importing ships. The market for ship imports was more fragmented than the moderately concentrated market for ship exports.

In 2008, the top six countries jointly comprised over 40% of the global market for importing ships, despite their smaller individual shares: Germany (12.3%), India (7.3%), Norway (5.9%), Italy (5.3%), Greece (5.2%), and the Republic of Korea (5.1%). The global market for exporting ships is much more highly concentrated than the global market for importing ships in that three countries – the

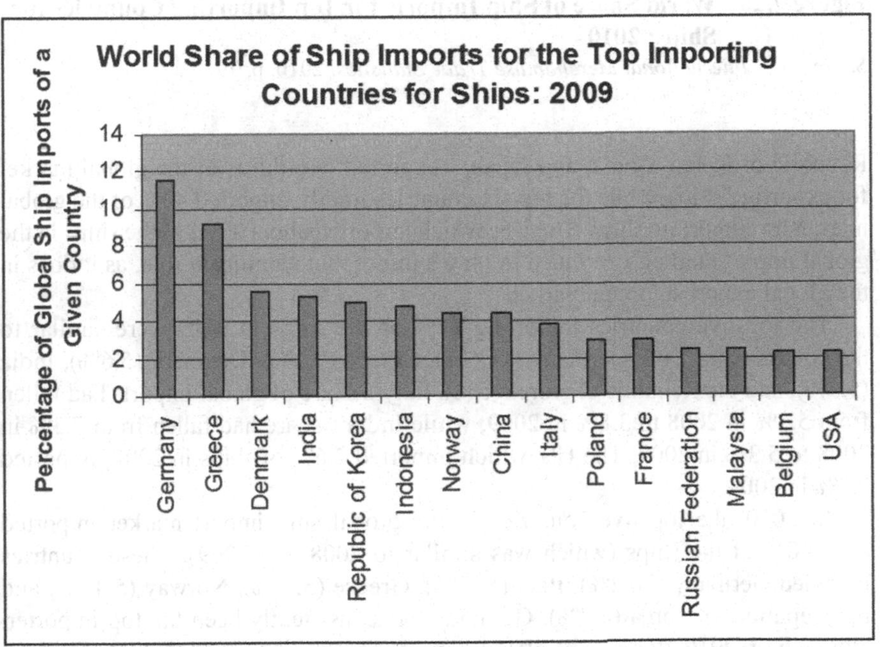

Figure 7.19 World Share of Ship Imports for Top Importing Countries for Ships: 2009

Source: UN International Merchandise Trade Statistics, 2009, p. 793.

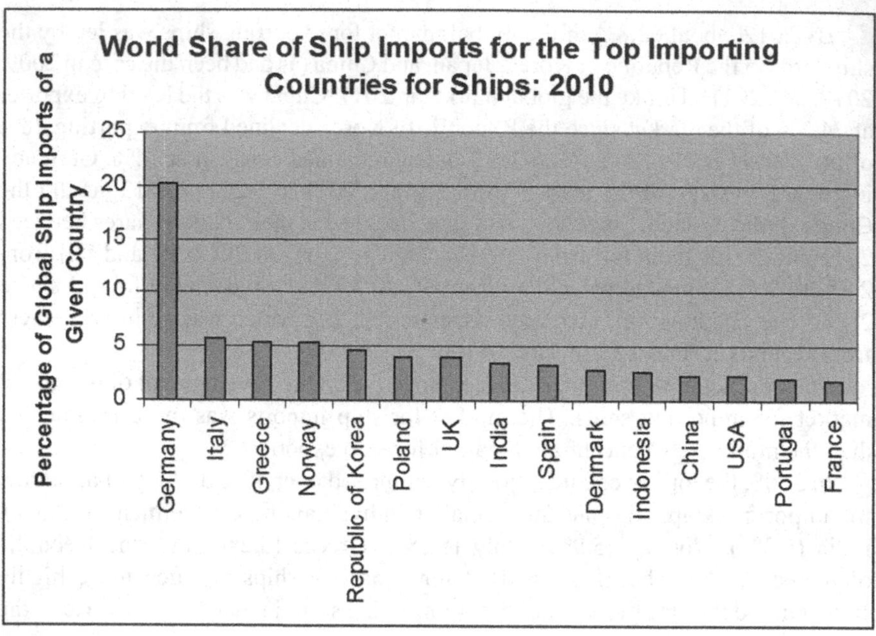

Figure 7.20 World Share of Ship Imports for Top Importing Countries for Ships: 2010

Source: UN International Merchandise Trade Statistics, 2010, p. 793.

Republic of Korea, China, and Japan – exported two-thirds of the global market for exporting ships, while the top six countries jointly imported 40% of the global market for importing ships. The US, which imported about 2.6% of the ships in the global import market, continued to play a minor, but significant role, as it does in the global export shipping market.

The top five countries importing 37% of the ships in 2009 were similar to the countries in 2008: Germany (11.6%), Greece (9.2%), Denmark (5.6%), India (5.3%), and the Republic of Korea (5%). Italy's share of global imports had fallen from 5.3% in 2008 to 3.8% in 2009, while India's share had fallen from 7.3% in 2008 to 5.3% in 2009. The US, which imported 2.6% of ships in 2008, imported 2.3% in 2009.

In 2010, the top five countries in the global ship import market imported over 40% of the ships (which was similar to 2008 and 2009). These countries included Germany (20.2%), Italy (5.7 %), Greece (5.4 %), Norway (5.4 %), and the Republic of Korea (4.7%). Germany has consistently been the top importer, although, in 2010, its share of global imports of ships increased to 20.2 %, which is higher than the 12% range in the prior two years. Italy's share of global imports had risen from 2009 levels, when it imported 3.8% of the global ship market, back to the 5.3% share in 2008. India's share of imports in the global shipping market continued to fall, from 7.3% in 2008 to 5.3% in 2009 to 3.6% in 2010. The US, as in 2009, imported 2.3% of ships in the global shipping import market. France, which had not been one of the top ship importers in 2008, imported 2.9% of the market in 2009, then dropped to 1.9% of the market in 2010.

In 2011, the top six countries imported over 37% of the ships, which was similar to the prior year. These countries included: Germany (13.3%), Norway (6.4%), the Russian Federation (5%), India (4.6%), Poland (4.3%), and Italy (4.2%). Germany, which had imported around 12% of the global shipping market in 2008 and 2009, continued at this level in 2011, falling from its 20.2% share in 2010. The Russian Federation's 5% share of imports of ships in 2011 reflected an increase from the 2.4% and 2.8% shares in 2009 and 2008, and a significant increase from its share in 2010, when the Russian Federation was not one of the top global importers. Norway increased its share to 6.4% in 2011, which was higher than 5.9% in 2008, 4.4% in 2009, and 5.4% in 2010. Similarly, India, like Norway, experienced a reduced share of shipping imports between 2008 and 2009 (declining from 7.3% to 5.3%). India's share continued to decline to 3.6% in 2010, then rose to 4.6% in 2011, which, although at a lower level than the share of shipping imports in 2008 and 2009, showed a significant increase from the prior year. Poland similarly experienced a decline from 2008 to 2009 (3.6% to 2.9%), then experienced growth

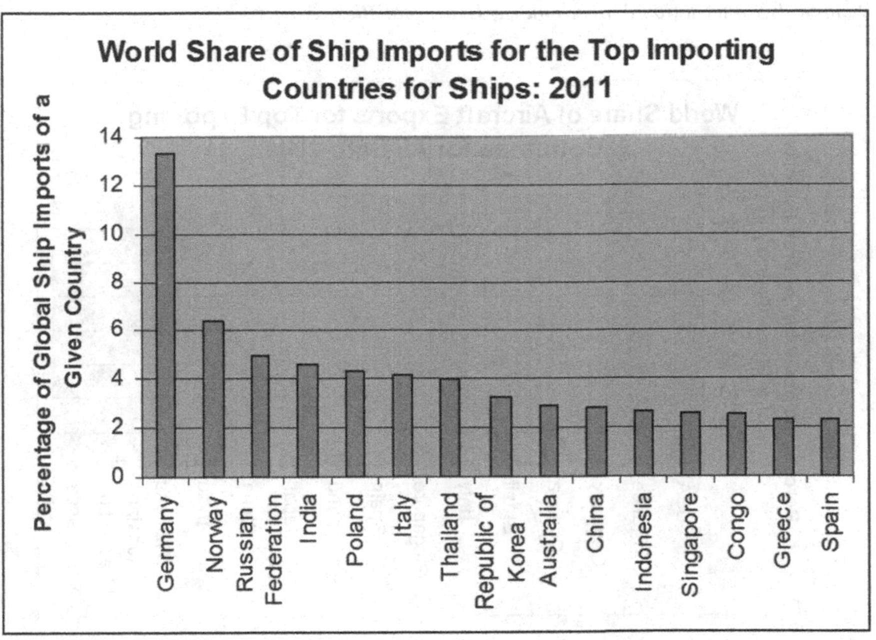

Figure 7.21 World Share of Ship Imports for Top Importing Countries for Ships: 2011

Source: UN International Merchandise Trade Statistics, 2011, p. 793.

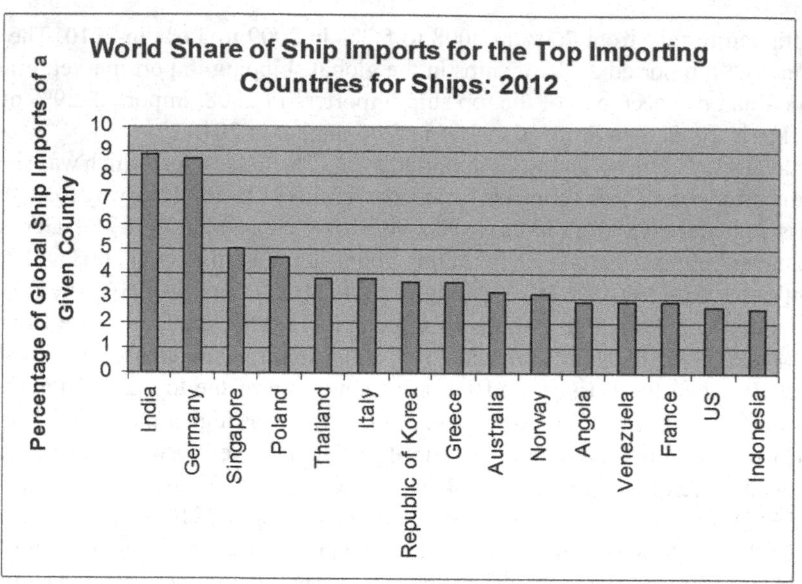

Figure 7.22 World Share of Ship Imports for Top Importing Countries for Ships: 2012

Source: UN International Merchandise Trade Statistics, 2012, p. 793.

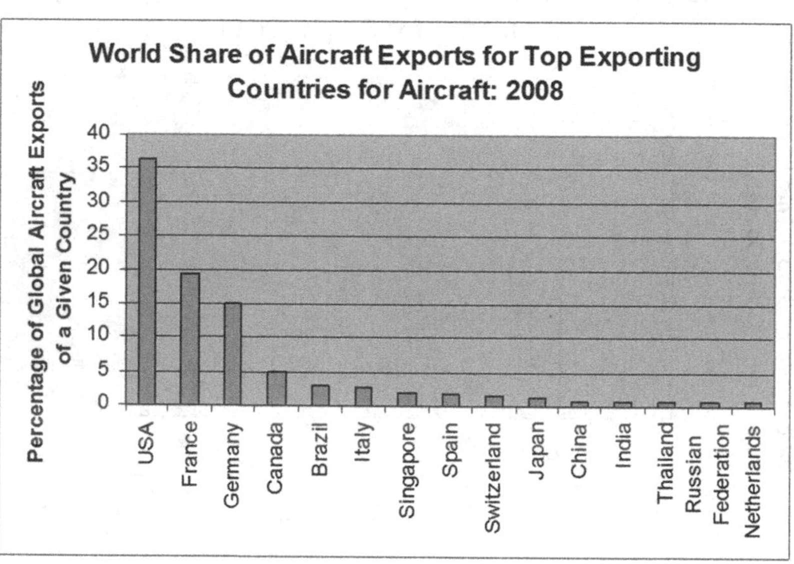

Figure 7.23 World Share of Aircraft Exports for Top Exporting Countries for Aircraft: 2008

Source: UN International Merchandise Trade Statistics, 2008, p. 792.

in 2010 and 2011 at 4% and 4.3%, respectively. Unlike the prior years, the US was not one of the top countries importing ships.

In 2012, the top six countries imported almost 35% of the ships, which was similar to 2010 and 2011. These countries included: India (8.9%), Germany (8.7%), Singapore (5.0%), Poland (4.7%), Thailand (3.8%), and Italy (3.8%). Germany declined from 20.2% in 2010 to 13.3% in 2011 and 8.7% in 2012. India, however, almost doubled its market share in 2012 relative to 2011, exceeding its levels from 2008 and continuing to reverse the decline that began in 2009, while Singapore also rose from 2.6% in 2011 to 5% in 2012. The Russian Federation dropped off the list as one of the top 15 importers, while Norway similarly declined from 6.4% in 2011 to 3.2% in 2012. Unlike 2011 when the US was not one of the top countries importing ships, the US, with a 2.7% market share, was ranked 14th of the top 15 importing countries for ships in 2012.

The next set of figures examines the global markets for imports and exports of aircraft. The market for aircraft exports, like the market for exports of ships, is fairly concentrated.

The US was the leading exporter of aircraft in 2008 and exported 36.3% of the global aircraft market. The top three countries – the US, France, and Germany – exported 70% of the market; indeed, the US and France alone exported

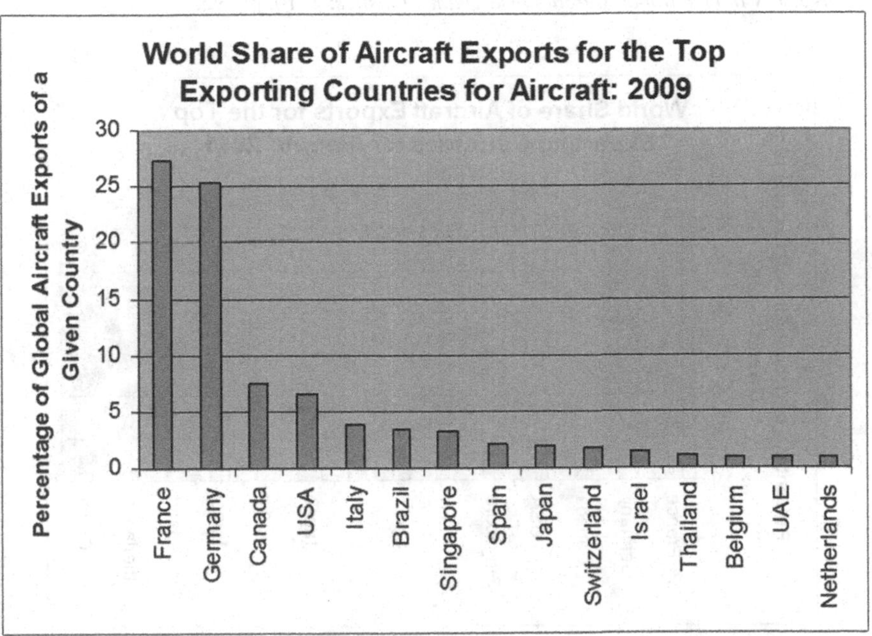

Figure 7.24 World Share of Aircraft Exports for Top Exporting Countries for Aircraft: 2009

Source: UN International Merchandise Trade Statistics, 2009, p. 792.

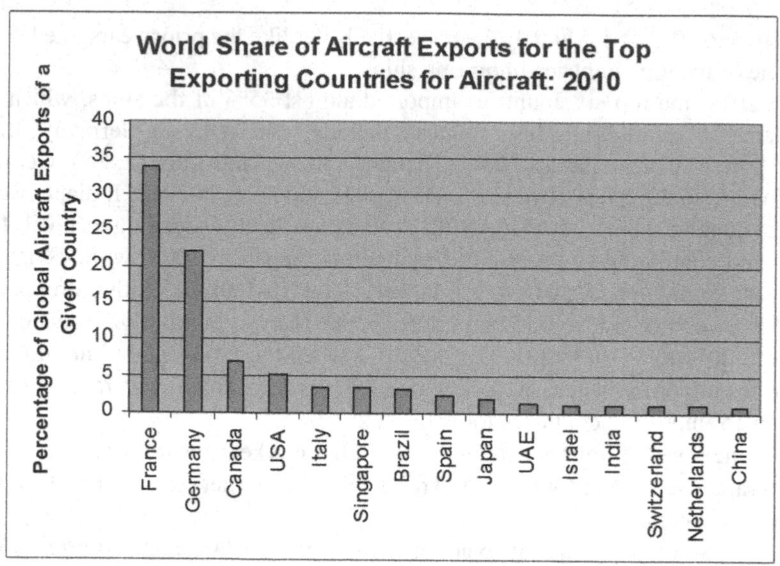

Figure 7.25 World Share of Aircraft Exports for Top Exporting Countries for Aircraft: 2010

Source: UN International Merchandise Trade Statistics, 2010, p. 792.

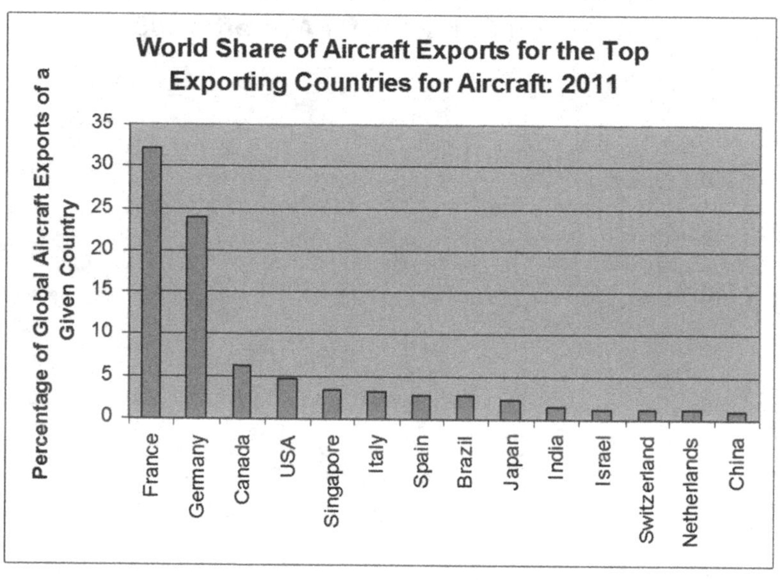

Figure 7.26 World Share of Aircraft Exports for Top Exporting Countries for Aircraft: 2011

Source: UN International Merchandise Trade Statistics, 2011, p. 792.

55.6% of the aircraft market, partially due to the role of Boeing and EADS/Airbus in the commercial aircraft market. Countries such as Canada (4.8%) and Brazil (3%) played a role in the exports of their business jets – Bombardier and Embraer.

In 2009, the top two countries – France and Germany – exported 52.6% of aircraft. They were followed by Canada (7.5%) and the US (6.5%). Partially due to the financial crisis, the US fell from exporting 36% of aircraft in 2008 to exporting 6.5% in 2009. France and Germany, on the other hand, increased their share of the aircraft export market; France's share of the aircraft export market increased from 19.3% in 2008 to 27.3% in 2009, while Germany's share increased from 15% in 2008 to 25.3% in 2009.

In 2010, the top two countries – France and Germany – exported 55.7% of aircraft. They were followed by Canada (7%) and the US (5.3%). The share of the top four exporting countries was similar to 2009, although the US had fallen from 6.5% in 2009.

In 2011, the top two countries – France and Germany – exported 56% of aircraft in the global market. They were followed by Canada (6.3%) and the US (4.7%). The share of the top four exporting countries was similar to 2009 and 2010, although the US continued to fall in its share of exports of aircraft from 36% in 2008 to 5.3% in 2010 to 4.7% in 2011.

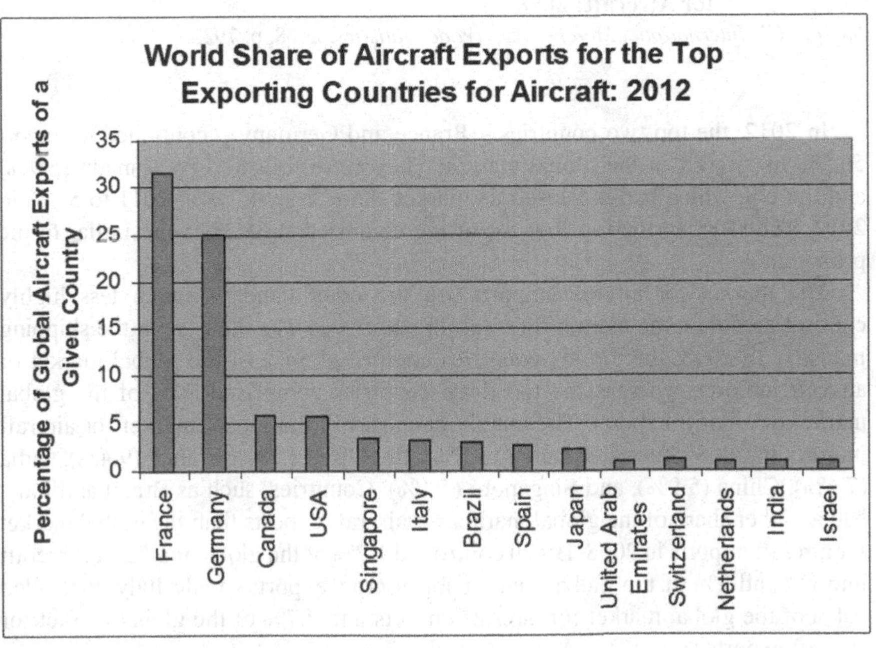

Figure 7.27 World Share of Aircraft Exports for Top Exporting Countries for Aircraft: 2012

Source: UN International Merchandise Trade Statistics, 2012, p. 792.

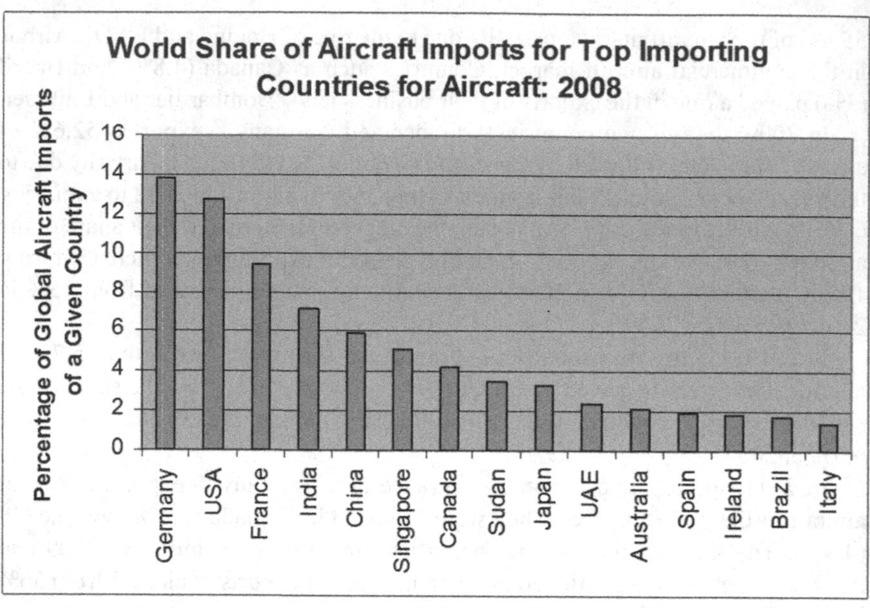

Figure 7.28 World Share of Aircraft Imports for Top Importing Countries for Aircraft: 2008

Source: UN International Merchandise Trade Statistics, 2008, p. 792.

In 2012, the top two countries – France and Germany – continued to export 56.5% of aircraft in the global market. They were followed by Canada (5.9%) and the US, which had increased its market share from 4.7% in 2011 to 5.8% in 2012. The share of the top four exporting countries at 68.3% was similar to the prior year.

The market for aircraft imports, on the other hand, is much less highly concentrated than the market for aircraft exports, as was the case in the shipping industry. In 2008, the top six countries comprised 54% of the global market of aircraft imports, whereas the top three countries comprised 70% of the global market of aircraft exports. The top six countries in the global market for aircraft imports in 2008 were: Germany (13.8%), the US (12.7%), France (9.4%), India (7.1%), China (5.9%), and Singapore (5.1%). Countries, such as Brazil and Italy, had a higher share of the global market for aircraft exports than the global market for aircraft imports in 2008. Brazil controlled 1.7% of the global market for aircraft imports and 3% of the global market for aircraft exports, while Italy controlled 1.4% of the global market for aircraft imports and 2.7% of the global market for aircraft exports.

In 2009, the top five countries comprised 52% of the global market for aircraft imports, which suggests similar concentration levels to 2008. These countries were Germany (16.8%), the US (12.4%), France (10.8%), China (7%), and Singapore

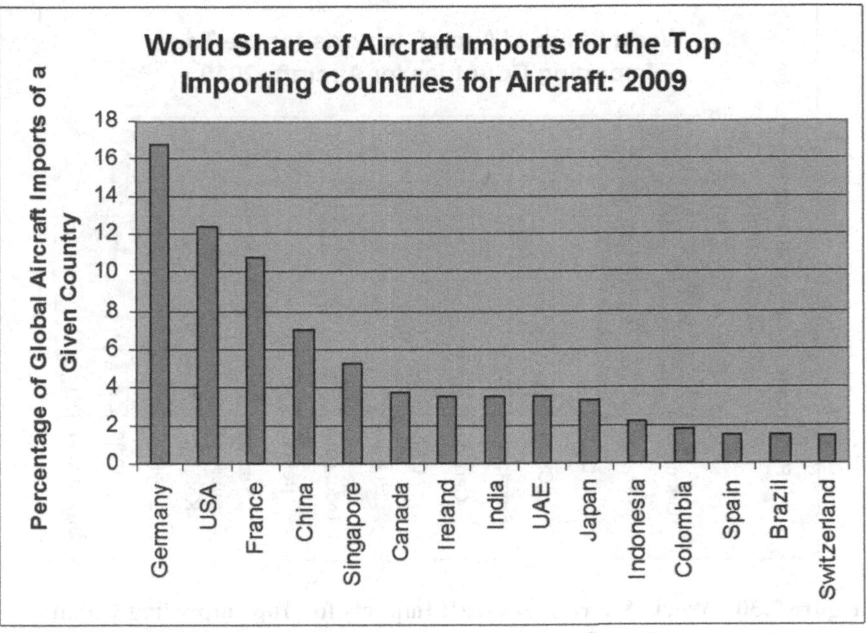

Figure 7.29 World Share of Aircraft Imports for Top Importing Countries for Aircraft: 2009

Source: UN International Merchandise Trade Statistics, 2009, p. 792.

(5.2%). The share of imports purchased by the top three countries (France, Germany, and the US) in 2008 – 35.9% – had increased to 40% by 2009, largely due to slight increases in imports by Germany and France. India, which comprised 7.1% of the global import market in 2008, had fallen to 3.5% of the global import market in 2009. As was the case in prior years, the market for aircraft exports was more concentrated in 2009 than the market for aircraft imports, such that the top two countries – France and Germany – exported 52.6% of aircraft in 2009.

The market for aircraft imports for the top three countries became somewhat more concentrated in 2010 at 44.7%, relative to 2008 (35.9%) and 2009 (40%). The top three countries included: Germany (16.7%), France (15.7%), and the US (12.3%), with France continuing to import a higher share of the global aircraft market, expanding from 9.4% in 2008 to 15.7% in 2010. China, the fourth most significant global importer of aircraft, imported about 8% of aircraft. which was a greater share than in 2008 (5.9%) and in 2009 (7%). India continued to play a smaller role in the global imports of aircraft, importing 1.7% of the market by 2010, which is a significantly smaller share than the 7.1% share in 2008 and the 3.5% in 2009. As was the case in prior years, the market for aircraft exports was more concentrated in 2010 than the market for aircraft imports, such that the top two countries – France and Germany – exported 55.7% of aircraft in 2010.

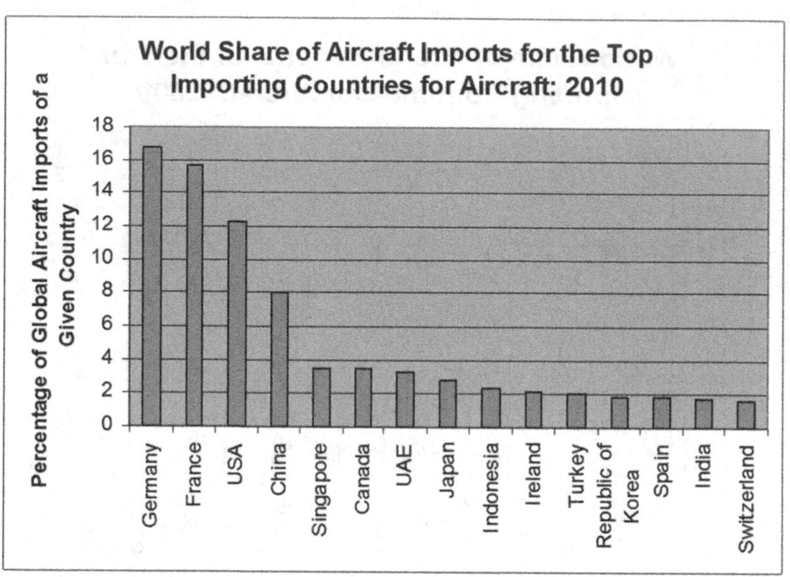

Figure 7.30 World Share of Aircraft Imports for Top Importing Countries for Aircraft: 2010

Source: UN International Merchandise Trade Statistics, 2010, p. 792.

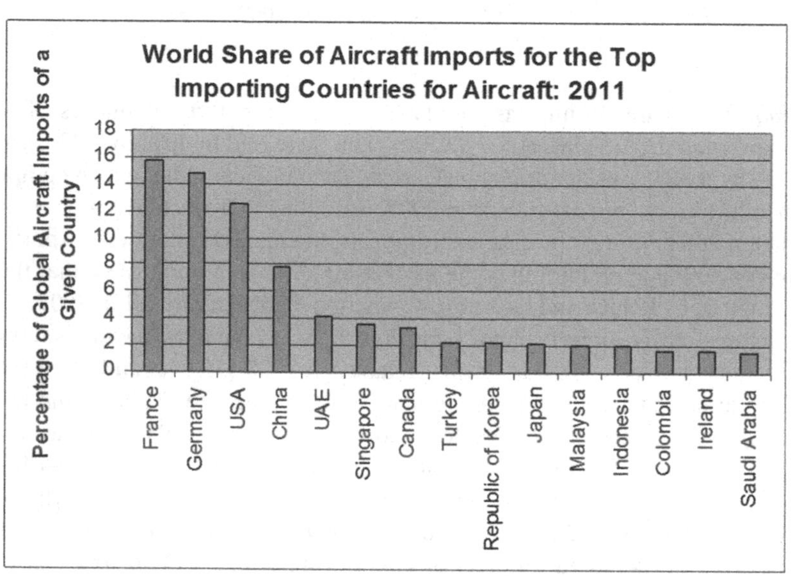

Figure 7.31 World Share of Aircraft Imports for Top Importing Countries for Aircraft: 2011

Source: UN International Merchandise Trade Statistics, 2011, p. 792.

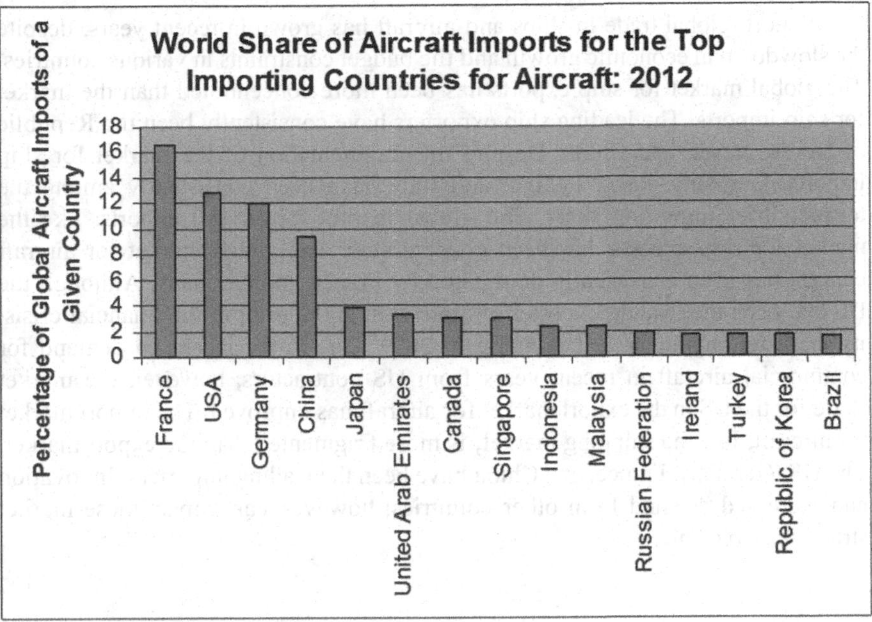

Figure 7.32 World Share of Aircraft Imports for Top Importing Countries for Aircraft: 2012

Source: UN International Merchandise Trade Statistics, 2012, p. 792.

The market for aircraft imports for the top three countries in 2011 remained similarly concentrated at 43% relative to the market in 2010. The top three importing countries included: France (15.7%), Germany (14.8%), and the US (12.5%). China, the fourth most significant global importer of aircraft, imported about 7.8%. Germany, relative to the prior year, dropped in its share of aircraft imports from 16.7% to 14.8%. India continued to play a smaller role in the global imports of aircraft and was not one of the top importing countries, as had been the case in prior years. Saudi Arabia, however, gained a higher share of aircraft imports at 1.5%, joining the end of the list of top importing countries, unlike prior years. As was the case in previous years, the market for aircraft exports was more concentrated in 2011 than the market for aircraft imports, such that the top two countries – France and Germany – exported 56% of aircraft in 2011.

The market for aircraft imports for the top three countries in 2012 continued to remain concentrated at 41.6%. The top three importing countries were: France (16.6%), the US (12.9%), and Germany (12.1%). Germany continued its decline in its share of imports in the global aircraft market, dropping its share of aircraft imports from 16.7% to 14.8% to 12.1%, as was also the case in the global shipping market. China continued to increase its share of imports, reaching 9.4% in 2012.

In short, global trade in ships and aircraft has grown in recent years, despite the slowdown in economic growth and the budget constraints in various countries. The global market for ship exports has been more concentrated than the market for ship imports. The leading ship exporters have consistently been the Republic of Korea, Japan, and China. Despite the fragmentation of the market for ship imports, Germany, India, Poland, and Italy have been consistently among the top countries importing ships. The global market for aircraft exports, like the market for ship exports, has been concentrated. The global market for aircraft exports has been consistently dominated by France and Germany. Although the US had been the leading exporter of aircraft in 2008, prior to the financial crisis, its share fell significantly beginning in 2009. Due to the increased demand for commercial aircraft in recent years from US contractors, however, the market share for the US in the export market for aircraft has improved. The import market for aircraft, like the shipping market, is more fragmented than the export market. The US, Germany, France, and China have been the leading importers. Innovation and increased demand from other countries, however, can impact these market structures over time.

Strategies for Countries in Preserving Their Defense Industrial Base

There are a variety of potential opportunities for the defense industrial base under tighter federal and defense budgets. The governments of various countries can make policy decisions which will assist their defense contractors. Examples of these strategies include:

a. countries re-selling used defense equipment to other financially constrained nations, which may provide opportunities for maintenance, operations, and upgrades from their original defense manufacturers;
b. governments ordering upgrades of existing equipment due to the financial constraints in buying newer equipment, which would help to preserve manufacturing jobs; and
c. governments deciding to use existing types of equipment to replace older types of equipment, which may increase order books for existing types of equipment, which can help defense contractors bolster production of existing products.

Countries facing budgetary constraints have been pursuing strategies involving selling used defense equipment to other countries, which are also facing financial constraints, but which need the equipment. One example is Italy, in which defense spending was projected to drop 28% for 2012, with $3.7 billion eliminated from defense budgets through 2014. As of June 2012, Italy had 82 ships and six submarines and was assessing the likelihood of reducing maintenance costs and obtaining funds by donating or selling about one-third of its naval fleet. It considered selling or donating 28 vessels, some of which would have otherwise been retired, over the next few years with the Philippines as a potential buyer and donating used Italian army vehicles to Peru or Colombia. One of the benefits of providing former Italian ships to foreign navies is that it could provide business for Italian defense contractors to provide maintenance or to upgrade.

A second strategy for countries facing budgetary constraints has been the purchase of used equipment from other countries to save money and to meet equipment needs. For example, in 2012, the Libyan government expressed interest in acquiring US Army used Chinook helicopters since it was unclear what proportion of its CH-47C helicopters, acquired in the 1970s, were in usable condition.[1] Similarly, Romania bought 21 second-hand F-16 A/B jet fighters from Portugal to replace its fleet of 49 MiG-31 Lancer jet fighters built by the Soviet Union.[2]

A third strategy for countries facing budgetary constraints has been to upgrade existing equipment, rather than to buy new equipment, which still preserves the defense industrial base through some degree of manufacturing, operations and maintenance. For example, in the US, as of 2012, General Dynamics, under the "Stryker Exchange" program, planned to upgrade 47 Strykers with a double V-hull configuration, to provide them with an improved engine using parts from older Strykers, and to provide them with greater protection. This strategy was designed partially to assist in preserving jobs in Anniston, AL and would enable the US Army to obtain durable combat vehicles at a lower cost rather than purchasing new vehicles. The strategy would involve shipping new hulls from the Lima, OH plant and reusing portions from the flat-bottomed Strykers in the conversion program which could reduce the cost from $2.4 million for a new double V-hull Stryker to $1.6 million for the refitted Strykers.[3] A second example is in Australia in 2012, when the Royal Australian Air Force planned to convert 12 F/A-18F Super Hornets to the EA-18G Growler configuration which would increase its capabilities for air combat. The Growler additions would help to ensure that no fighter capability gap would exist between the arrival of the F-35As and the retirement of the F/A-18A/B fleet. At the time, Australia was committed to buy 14 F-35As, although it had only ordered two aircraft and had deferred the remaining 12 for up to two years. It had an overall requirement of up to 100 F-35As.[4]

A fourth strategy for countries facing budgetary constraints has been altering the use of various types of existing equipment to replace other types of aging equipment. For example, in the US in 2012, the Navy was considering the V-22 Osprey as the next-generation carrier onboard delivery (COD) aircraft. The V-22s had already operated on carriers on an as-needed basis and their operations

1 Weisgerber, 2012.
2 Adamowski, 2012; Mehta, 2012.
3 McLeary, 2012.
4 Pittaway, 2012.

would not lead to any modifications or changes to the carriers. As a result, the V-22s could replace the C-2, which delivered supplies and personnel to aircraft carriers. Although the first C-2s flew in 1964, the second batch of planes was ordered in 1984, so the C-2s were aging. Unlike the C-2, the V-22 Osprey did not require catapults, runways, etc. and could land on a variety of platforms. Nevertheless, replacement of one type of equipment with another type can also have disadvantages, since the Osprey could only fly about 1,150 miles without refueling, while C-2s could fly 1500 miles.[5]

Finally, a fifth strategy for countries facing budgetary constraints would be to emphasize continued demand for products which could have other uses within their federal governments beyond defense. In many cases, these products would also have civilian uses in the private sector. Examples of these types of products include UAVs, cybersecurity products, trucks, etc.

Case Study: European Defense Industrial Base and Fiscal Constraints

As described in the prior chapters, the US defense industrial base has faced many challenges in terms of fiscal austerity and shifting defense priorities. This section discusses the challenges and opportunities facing the European defense industrial base, which are similar to the US in terms of budget constraints, the risk of attenuating skillsets with the closure of manufacturing facilities, and the focus of European defense contractors on restructuring and overseas sales.

European military spending overall declined by 0.7% between 2012 and 2013, while East European military spending increased by 5.3% and Western and Central European spending declined by 2.4%. Over the period from 2004 to 2013, European military spending overall increased by 7.6%, while East European military spending increased by 112% and Western and Central European spending declined by 6.5%. Between 2012 and 2013, the European countries with the greatest increases in military spending were the Ukraine (16%), Belarus (15%), Latvia (9.3%), and Switzerland (9%). The European countries with the greatest declines in military spending were Spain (-13%), Albania (-13%), Hungary (-12%), and the Netherlands (-8.3%).[6]

Figure 7.33 reflects the impact of fiscal austerity in Europe on defense spending and compares military expenditures in 2008 to military expenditures in 2013. France and the UK continued to have the highest military expenditures of the European nations. While some countries, such as Germany and Poland, have increased their military expenditures, many other countries have declined in their military expenditures, such as France, the UK, and Italy.

Under the minimum defense investment obligations of the NATO alliance, NATO members must invest at least 2% of GDP in defense. The UK (2.3% of

5 "Osprey Could Become Part of Carrier Mission," 2012.
6 Perlo-Freeman and Solmirano, 2014.

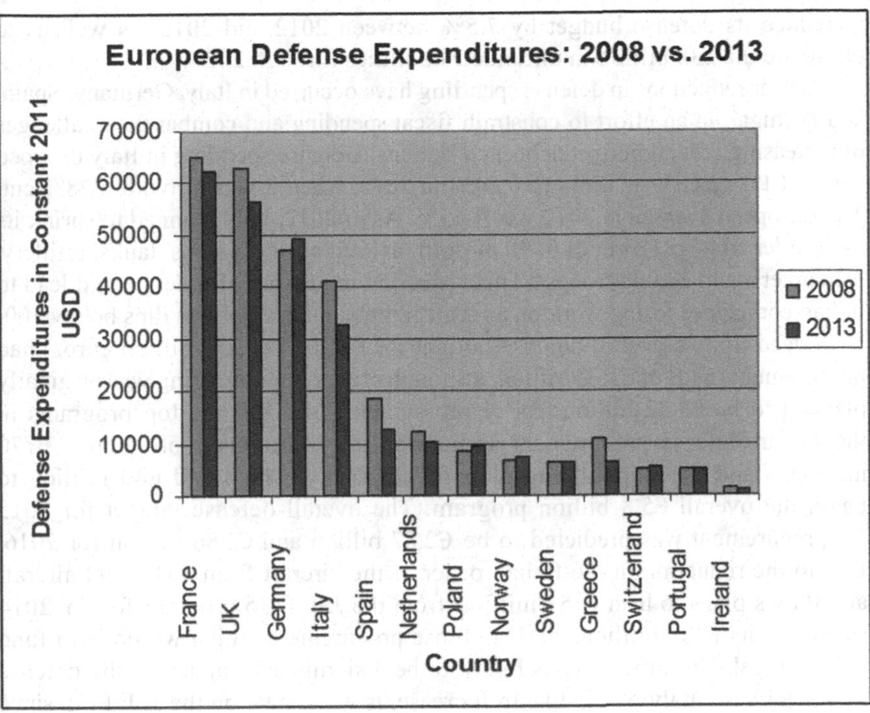

Figure 7.33 European Military Expenditures: 2008 vs. 2013
Source: SIPRI Military Expenditures Database.

GDP), France (2.2% of GDP), and Greece (2.4% of GDP) were among the only EU-NATO members that had met that standard by 2013. France planned to reduce its defense expenditures by €3.6 billion through 2013. In 2013, France announced that it would halve its purchases over the next six years of Dassault Rafale jet fighter planes from 11 planes a year over six years to a total of 26 jets during the next six years.[7] Similarly, Greece, which spent 3.2% of its GDP on military spending in 2010 has been cutting its spending sharply. Indeed, since 2009, the money in the armaments budget has largely been spent on maintenance. As of 2012/early 2013, Greece had cut €325 million in defense spending and was unlikely to launch any major arms programs until 2014 at the earliest. The cuts in defense spending limited Greece's capacity to purchase an air defense version of the FREMM frigate from DCNS (the French company), as well as to reduce the plans to upgrade 20 Mirage 2000 fighter jets. The UK similarly has been reducing its defense spending. As of 2012, it had planned

7 Rahir, 2013.

to reduce its defense budget by 7.5% between 2012 and 2015, as well as to eliminate the £38 billion in unfunded liabilities through 2020.[8]

Similar reductions in defense spending have occurred in Italy, Germany, Spain, and Portugal, in an effort to constrain fiscal spending and combat the challenges of increasing federal budget debts and deficits. Defense spending in Italy dropped from 1.01% of GDP in 2004 to 0.84% in 2012, when it was hit with a 28% cut. The European average in 2010 was 1.61%. As of 2012, Italy planned to shrink its F-35 order from 131 aircraft to 90 aircraft, as well as to cut ships, tanks, artillery, helicopters, and 33,000 troops.[9] This reduction in number of orders could lead to Italian companies losing work on aircraft if the number of orders dips below 100.

Indeed, Italy's overall defense budget for FY2014 (13.59 billion euros) had procurement funds of €2.73 billion, although procurement funding was originally planned to be €3.22 billion. As of the summer of 2014, the top programs in the Italian defense procurement arena were the Eurofighter program – €770 million – and the shipbuilding plan for the Navy, which had €40 million to begin the overall €5.8 billion program. The overall defense budget for 2015 for procurement was predicted to be €2.87 billion and €2.86 billion for 2016. Due to the reduction in the Italian order of the aircraft from 131 to 90 aircraft and Italy's plans to trim €153 million from the JSF F-35 program for FY 2014 as part of its €400 million cut in defense procurement which would help fund the tax breaks in Italy, there is likely to be a significant impact on the defense industrial base. Italy would like to increase its workshare on the JSF F-35 since its final assembly line at Cameri Air Base in northern Italy has faced a reduction in production. As of the summer of 2014, there was some possibility that some Dutch F-35 orders might be assembled at the plant.[10] Moreover, as a result of Italy's reduction in defense budgets, the Italian Navy, as of the summer of 2014, opened its shipyards to civilian construction and maintenance for builders of luxury yachts, but would not take cash payments, although the Navy could request fuel, materials, or work on the infrastructure of the yard. For example, the La Spezia Yard recently launched a yacht from the firm VSY.[11]

Similarly, Germany also has experienced a reduction in defense budgets. The FY 2014 German defense budget passed at €32.44 billion in June, 2014, which was slightly lower, by €800 million, than the FY 2013 budget of €33.36 billion.[12] Germany, which currently spends 1.3% of its GDP on defense, plans to increase the share of defense spending, although it expects it to be less than the NATO spending guidelines of 2%.[13] As of mid-March, 2013, Germany had reduced its purchases of NH90 helicopters from 122 to 82 and its purchases

8 *Defense News* Staff, 2012.
9 *Defense News* Staff, 2012; Kingston and Tran, 2012.
10 Kingston, 2014a.
11 Kingston, 2014b.
12 "German Defense Budget Passes," 2014b.
13 "German Defense Budget to Rise," 2014a.

of Tiger support helicopters from 80 to 57.[14] It was also likely to reduce the Bundeswehr's personnel from 225,000 military and 76,000 civilian posts to 185,000 soldiers and 55,000 civilians. The concern of the military downsizing has been that the German Army will lose its capabilities to succeed in high-intensity scenarios, such as Iraq, but will have the necessary capabilities for low or medium intensity scenarios.

Portugal reduced military procurement in 2012 by 40% and reduced military personnel by 10%. The Air Force, as of 2012, was having difficulty funding flying hours and 2–3 of the Navy's 10 corvettes were experiencing such significant maintenance problems that they were not operational partially because the age of some of the corvettes exceeded 40 years. Spain, as of 2012, required an outlay of €600 billion for ongoing programs, but only €400 million had been budgeted in 2012 for acquisitions. Concerns arose that the fiscal constraints could lead to cuts or delays in programs, such as the A400M and submarines, for which deliveries had not begun, as well as delays in payments for deliveries.[15]

At the 2014 Farnborough Airshow, the UK announced that it would spend £1.1 billion in cyber, special forces, airborne surveillance, a joint UCAV study with France, updating Typhoon capabilities, etc. Nevertheless, while current UK defense spending was above the NATO target of 2% for defense spending (£33 billion), the UK Defense Secretary did not commit to retaining UK defense spending at 2% at the Airshow. Indeed, the defense budget may be reduced in the 2016 spending review to reduce the UK's debt.[16]

Despite this, the UK's first new aircraft carrier in three decades, the Queen Elizabeth, was officially named on July 4, 2014. The Queen Elizabeth, which will be completed in 2017, will be the largest warship ever built in Europe, and will operate F-35s, Army Chinook and Apache attack helicopters, and Navy helicopters, such as the Wildcat and Merlin. The Prince of Wales, the second carrier, will start sea trials in January, 2019. Both of the aircraft carriers will be built by Aircraft Alliance, which consists of Thales, BAE, and Babcock Marine. The program is expected to cost $10.5 billion (£6.2 billion, which has grown from £3.8 million). The first warship will become fully operational in 2020, which will reinvigorate the carrier strike capability. The Royal Navy, which lost its carrier strike capability in the 2010 Strategic Defense and Security Review, also experienced budget cuts for destroyer and frigate fleets, and lost the Harrier jet, as well as lighter aircraft carriers built in the 1980s.[17]

Nevertheless, despite the funding of the aircraft carriers, Britain's maritime capabilities are being restructured, due to budget limitations, and the closure of the BAE systems shipbuilding yard in Portsmouth in southern England is one example. The shipbuilding factory is scheduled to close when it completes its

14 Muller, 2013.
15 Kingston and Tran, 2012.
16 Chuter, 2014b.
17 Chuter, 2014a.

work on the Queen Elizabeth and the Prince of Wales, which, for the company, has led to the largest increase in shipbuilding since World War II.[18] The Royal Navy has planned to purchase three offshore patrol boats to provide work for the company's remaining naval yards in Scotland. The Portsmouth shipbuilding yard has been around for over 500 years and closing the shipbuilding yards could lead to the loss of 940 jobs in Portsmouth and an additional 835 jobs across BAE's two naval shipyards in Glasgow, the Rosyth site (also in Scotland) and Filton (in southwest England). The closure of Portsmouth shipbuilding facilities could result in BAE having no naval surface shipbuilding capabilities in England. BAE plans to move some of its shipbuilding operations to Scotland to build Type 26 frigates in Glasgow.[19]

BAE's maintenance and support facilities at Portsmouth will still be open for the Type 45 destroyers and two 65,000-ton carriers entering service by 2020, which will maintain 3,200 BAE positions. English naval shipbuilding capabilities will be centered on BAE's nuclear submarine yard at Barrow and the small Babcock-owned facility at Appledore.[20]

The number of European defense contractors is likely to shrink in the coming years due to reductions in defense budgets. For example, defense cutbacks, especially from Spain, Germany, Italy, the UK, and France, are likely to reduce the number of contractors in land systems. Nevertheless, with the concerns about fiscal austerity and scarce defense resources, European countries have attempted to engage in joint efforts. For example, Britain and Norway will collaborate for the support and training of the F-35 joint strike fighter, while Lithuania, Latvia, and Estonia are discussing collaboration on joint arms purchases. Similarly, in mid-2013, the European Union announced that it would offer a $132.7 million incentive to European firms and governments to launch a joint UAV program.

As of 2013, Britain and France had planned to jointly fund contracts with BAE Systems and Dassault to set specifications for a demonstrator unmanned combat aerial vehicle and to study the technical risks of a new medium-altitude long endurance (MALE) UAV, as well as to collaborate in the UCAV area.[21] Moreover, in the missile industry, France and Britain planned to launch through MBDA an assessment of an upgrade of the Storm Shadow/Scalp cruise missile and the feasibility of an anti-surface tactical weapon and to sign a contract for development and production of an anti-ship guided weapon. Unfortunately, however, it is unclear whether the joint efforts will succeed or move sufficiently rapidly and efficiently. For example, bilateral talks between the UK and France on a MALE program stalled.[22] The UK and France agreed in July 2014 to sign a MOU on studying a combat drone (unmanned combat aerial system) which would

18 Bennhold, 2013, p. B2.
19 Chuter, 2013a; Bennhold, 2013, p. B2.
20 Chuter, 2013a; Bennhold, 2013, p. B2
21 Chuter and Tran, 2013b.
22 Chuter and Tran, 2013a.

involve Dassault Aviation, and BAE studying the platform, Thales and Selex focusing on sensors and electronic systems, and Rolls Royce and Snecma focusing on the engine. This study would help in achieving the goal of replacing the Rafale and Typhoon fourth-generation fighters beginning in 2035.[23]

Another example of collaboration is the Eurofighter consortium, owned by BAE, EADS and Finmeccanica, which has been successful with some of its products and may achieve some profitability through overseas sales. For example, beginning in October 2013, the Eurofighter Typhoon was to be upgraded and would be delivered by the end of 2015. The Typhoon remains a possibility if its price becomes more competitive for South Korea, since it rejected the Boeing F-15 Silent Eagle which had emerged as the remaining bidder when Eurofighter and the Lockheed Martin F-35A had prices which weren't competitive. Other countries, such as Saudi Arabia, Oman, and Austria, have purchased the Typhoon and, as of the fall of 2013, potential additional sales for the Typhoon included: Saudi Arabia, Bahrain, Kuwait, Qatar, the UAE, Malaysia and South Korea.[24]

Although partially due to fiscal austerity, the reorganization of EADS also has the potential for greater efficiency and is an example of the restructuring of the European defense industrial base. In early December, 2013, EADS announced that it would cut 5,800 positions – 4,500 permanent positions and 1,300 temporary positions – in its space and military divisions, due to the current and expected reductions in the European defense budgets. This comprised 5% of its total workforce of 133,000. In 2014, Airbus, which had been growing rapidly relative to the other divisions, attached its name to the Airbus Defense and Space division and the Airbus Eurocopter. The greatest share of employees who were to be laid off were in the military and electronics division, Cassidian, in Germany. EADS also planned to consolidate 12 of its current plants in Spain, Germany, France, and Britain in an effort to streamline EADS' facilities, which have been spread throughout Europe, due to the merger of aerospace firms in 2000 in Spain, France, and Germany to create EADS. Damage to the military divisions occurred when, in late 2012, German Chancellor Angela Merkel opposed the merger between BAE Systems and EADS, which, as a result, failed. Merkel's concern was that there would be reduced influence of Germany if the merger happened. The hope for the merger had been that it might create an opportunity for EADS to expand its strength in the space and military divisions.[25]

Despite the shrinkage in the defense budgets for the UK and many of the continental European nations, some of the Nordic countries are, however, expanding their defense budgets. Sweden is one example. In September, 2013, the Swedish Defense Ministry announced that its expenditures on defense during 2014–2017 would expand by $220 million, which would result in an increase of $60 million a year. This was an increase from the 2013 defense budget of $6.2

23 Tran and Chuter, 2014, p. 6.
24 Chuter, 2013b.
25 Clark, 2013.

billion.[26] The Swedish government was concerned that the military was not sufficiently prepared. Indeed, it was noted in the spring of 2012 that the military could "defend Sweden for less than a week."[27]

The AFC's Budget Perspective 2013 report noted that the Swedish defense budget made up 1.35% of Sweden's GDP in 2009 and that if, by 2016, the lack of investment in the military continued, defense expenditures could fall to below 1.1% of GDP. As of 2012, Sweden's defense budget as a share of GDP was lower than any of the Baltic or Nordic states at 1.2% of GDP. The AFC Budget Perspective concluded that "the military's capacity to defend Sweden against all land, sea and air threats will continue to weaken unless defense funding is scaled up to a level that ensures adequate manpower, trained forces and modern equipment exist to repel possible attacks against the country's territorial sovereignty" and also suggested greater cross-border alliances and military operational cooperation between Nordic states.[28]

Another example of a Nordic country that has expanded its defense budget, despite slower global economic growth, is Norway. Its 2014 budget was $7.2 billion, which was an increase from the FY 2013 budget of $7.06 billion. Indeed, its 2014 budget was among the 10 largest European defense budgets for 2014 and had the highest defense spending on a per capita basis relative to the other NATO countries in Europe. The 2014 budget continued to focus on modernizing Norway's defense systems, including purchasing F-35s. Indeed, the acquisition portion of the 2014 budget comprised 25% of the total government budget at $1.83 billion, which was an increase from the FY 2013 $1.46 billion.[29] The Army's core front-line units were to experience an increase of $87 million to their operational budget. Moreover, Norway's reduction of its military presence in Afghanistan was to be finalized in 2014 and the reduction of forces would enable $57 million in 2014 to be redistributed to the Air Force and the Navy.[30]

Overall, the tightening fiscal situation for Europe suggests that the European defense market is likely to weaken and may experience attenuation in skillsets, particularly due to reduced internal sales. Indeed, Stefan Zoller, head of defense and security at EADS, noted that, "European markets will decline or be stable at best ... Strategically, we have to go where the money is and the money is around the globe." The company will have to "generate growth to maintain our industrial base at home." European defense contractors, such as EADS, BAE, and Finnmeccanica, are targeting markets beyond Europe. EADS is targeting markets such as Brazil, the Middle East, and India. As of 2010, EADS planned to increase the share of its employees who are located outside Europe from 5%

26 O'Dwyer, 2013b.
27 O'Dwyer, 2013b.
28 O'Dwyer, 2013b.
29 O'Dwyer, 2013a.
30 O'Dwyer, 2013a.

in 2009 to 20% in 2020.[31] Finmeccanica (Italy), which has established a strong presence in the US and in the UK (where it owns BAE's former avionics group and Agusta Westland's helicopter group) is a second example. Finnmeccanica is targeting India, the Middle East, Turkey, Algeria, Brazil, and Libya, as well as Russia.[32] BAE is a third example; it only obtains 20% of its defense revenues from the UK. The investors' perception regarding the budget difficulties in European countries, as well as slower economic growth, may lead to increased weakness in the euro. The weaker euro could promote European exports of defense products. The traditionally growing markets which would have been good sources of exports for European defense contractors are, however, also facing economic difficulties.

In short, while some of the Nordic countries are expanding their defense budgets, many of the European countries (the UK, Greece, France, etc.) are contracting their budgets. Although alliances between European countries in developing equipment have been slow, they still provide opportunities. As in the case of the US, closure of key European defense manufacturing facilities can lead to a loss of skillsets. Expanding overseas sales and corporate re-organization may be helpful strategies for European defense contractors in preserving their industrial base.

The Impact of the Ukrainian Crisis on the Defense Industrial Base

The ongoing Ukrainian crisis is likely to have a significant impact on the defense industrial base and is an example of how, although fiscal constraints may lead to reductions in the defense industrial base for European countries, perceived military threats in the region may lead to an increase in defense spending. The crisis may counter the tendencies of many European countries to reduce their defense budgets due to concerns about economic growth or budget deficits. Moreover, greater collaboration among various countries may lead to sharing of equipment, which could reduce defense procurement costs. Sanctions imposed on Russia may limit the ability of Russia to import or export defense equipment from other countries, which could impact both Russia's military and its defense industrial base, as well as impact the military of other countries which are dependent on importing products from Russia.

The Ukrainian crisis has resulted in the Baltic countries increasing their defense budgets, since Latvia, Lithuania, and Estonia have significant ethnic Russian populations. For example, Lithuania plans to double its defense spending to $800 million by 2020. Certain procurement projects, such as the expansion of additional radar capabilities, have been prioritized. The long-range radar acquisition program will cost $80 million by its completion in 2019 and is part of the broader strategy to link air surveillance with NATO's Integrated Air Defense

31 Pfeifer and Clark, 2010.
32 Pfeifer, 2010.

System. NATO has committed to establish a more permanent, stronger force presence in the Baltic Sea, led by the US. In fact, the Baltic states have encouraged the US to revisit the European Anti-Missile Shield System (EAMSS). NATO has encouraged Lithuania and Latvia to increase defense spending to 2% of GDP, and to engage in an expansion of their military capabilities, as Estonia did in recent years.[33] The Ukrainian crisis has further stimulated the UAV efforts of Eastern European countries. Indeed, Estonia is also upgrading its military drone fleet and plans to buy RQ-4 Global Hawk UAVs made by Northrop Grumman as part of a joint NATO procurement effort. It has also set aside $8 million for 2015–2016 to purchase military equipment.[34]

The Ukrainian crisis has led countries, such as Poland and the Czech Republic, to expand their defense budgets. The Czech Republic plans to increases its defense spending from 1% of GDP to 1.4% of GDP. Poland plans to increase its defense spending and to expand from 1.95% of GDP to 2% of GDP. Programs to expand the Polish Air Force include the purchase of 70 new helicopters, several hundred drones, and fifth-generation fighters.[35] On March 5, 2014, the Polish Prime Minister noted that the government was reassessing its plans under a $42.7 billion program to modernize the military by 2022. In late April, 2014, Poland announced that in 2016, Poland would acquire its first UAVs. Moreover, the Polish Navy is building an unmanned fleet and, in November, 2013, acquired two Gavia underwater vehicles which were constructed by Teledyne Gavia of Iceland.[36]

Russia's annexation of the Crimean Peninsula has resulted in a greater focus on naval and air force equipment for the Black Sea region.[37] Russia has been upgrading its long-range aircraft in recent years. While they originally had a strategic plan to do so in the 1980s, the plans collapsed with the end of the Cold War. In 2007, the upgrading of long-range aviation missions by the Russian Air Force resumed.[38] Indeed, the increase in air defense capability in the Crimea by Russia is leading to other East European countries expanding their capabilities. Poland had purchased 70 helicopters which would be delivered after 2015, as well as helicopter purchases to replace Mi-24s, which will be delivered after 2020.[39]

Although Russia has been expanding its purchases of military equipment in recent years, its involvement in the Ukrainian crisis may lead to a reduction in its ability to purchase equipment. As of the spring of 2014, Russia planned to spend $564 billion (20 trillion rubles) by 2020 on defense procurement.[40] On July 22, 2014, the 28 EU countries agreed to increase sanctions against Russia; these included

33 O'Dwyer, 2014a.
34 Adamowski, 2014b.
35 Adamowski, 2014c.
36 Adamowski, 2014b.
37 Adamowski, Jaroslaw, 2014d.
38 Weisgerber, 2014.
39 Adamowski, 2014d.
40 Adamowski, 2014a.

banning new arms sales. The delivery of a $1.7 billion Mistral-class amphibious warship to Russia by France was to be exempted from this ban, although countries, such as the UK, considered this to be unacceptable.[41] In early September, 2014, France announced that they would not deliver the Mistral warship due to the role of Russia in eastern Ukraine. The contract, from January 2011, provided Russia with four Mistral warships, of which two were to be built by DCNS and the other two were to be built in Russia, These ships can carry 16 military helicopter and landing craft, as well as 160 troops.[42] Moreover, the development of the Russian defense industry may be limited by the Western sanctions on the role of Russian firms in arms trade and Western sanctions could reduce defense exports. For example, the Russian Aircraft Corporation MiG experienced a 68% increase in revenues in 2013 relative to 2012, which was due to the expansion of foreign sales, such as the delivery to India of 10 modernized MiG-29 fighters.[43]

Finally, the Ukrainian crisis has stimulated the tendencies of Nordic countries to strengthen regional partnerships. Finland, which has historically focused on self-reliance and non-alignment, signed an MOU on April 22 with NATO in which it would invest in a NATO-based military force and assist in maintenance of NATO equipment, as well as provide facilities for maintenance and refueling of land forces, although the MOU does not require Finland to join NATO. Sweden and Finland would be able to deploy troops jointly and to participate in UN, EU, or NATO-led operations.[44]

Conclusion

The global defense market continues to exhibit significant growth, despite the challenges faced by various countries and their defense contractors. The US significantly leads the other countries in military expenditures, however defense spending as a share of GDP is higher in many other countries. Indeed, the importance of defense spending for the economy in Saudi Arabia – as measured by military expenditures as a share of GDP – is almost twice that of the US. The countries with the highest military expenditures have, in many cases, experienced burgeoning debt and slow GDP growth, although economic growth has experienced a recovery from the downswing in 2009. Due to flat or declining GDP growth and rising government debt as a share of GDP, contractors in the leading export countries in defense equipment are likely to increasingly depend on foreign sales (rather than domestic sales). Nevertheless, many of the top importing countries, such as China, India, and Saudi Arabia, experienced recovery in GDP growth from the downswing in 2009, but, between 2011 and 2013, experienced declining GDP

41 Tran, 2014.
42 Lamothe, 2014.
43 Adamowski, 2014a.
44 O'Dwyer, 2014b.

growth. Similarly, although some of these top importing countries experienced an improvement in debt as a share of GDP, others experienced a decrease. As a result, defense contractors face challenges both domestically and overseas from their customers in producing particular quantities and types of equipment and in innovating to develop new products.

One example is the global trade market in ships and aircraft. The export market for ships and aircraft is more highly concentrated than the import market for ships and aircraft. Consequently, defense contractors in the export market for this equipment are dependent on demand from a greater diversity of overseas countries on the import side. Economic growth, shifting defense priorities, federal (as well as defense) budgets, and government debt can impact the demand of importers. Nevertheless, the greater diversity of importing countries helps to hedge the risk of defense contractors in ships and aircraft on the more concentrated export side.

Despite fiscal constraints and slow economic growth, however, the governments of various countries can initiate strategies to assist their defense contractors in the global defense industrial base. One strategy is to ensure continuation of operations and maintenance functions on equipment to support the defense contractors by selling the used equipment to overseas countries, rather than retiring the equipment. This leads to demand from the acquiring countries over the years for continued maintenance from the defense contractors. A second strategy is to order greater quantities of existing equipment to support manufacturing jobs, which otherwise would have been hurt by the reduced demand for new equipment due to fiscal constraints. Moreover, increased orders of existing equipment can be placed partially to serve the functions of another type/model of equipment which could then be retired. A third strategy would be to stimulate demand for products that can have both civilian and defense uses, both domestically and overseas.

As the case study of Europe showed, many European countries (UK, Greece, France, etc.) are experiencing reduced defense spending, similar to the US, although some of the Nordic countries are expanding their defense budgets. Moreover, the Ukrainian crisis is an example of military threats which can result in Baltic and East European countries increasing their defense spending, despite concerns about slow economic growth or fiscal constraints. As a result, many of the European countries are forming alliances to consolidate knowledge and skillsets, despite the closure of some of their key facilities. As the importing and exporting countries of the global defense industrial base face challenges domestically, their defense contractors can be hurt by similar challenges facing overseas countries, which can damage their international sales. This can lead to a self-reinforcing mechanism which can lead to atrophy of skillsets for defense contractors and a reduced incentive for innovation. Consequently, alliances between defense contractors globally can pool knowledge, while diversification of equipment by defense contractors to both civilian and defense markets can help to reinforce and strengthen the global defense industrial base. The similarity of challenges in terms of the needs for equipment, as well as fiscal constraints, across importing and exporting countries make sustainability and continued profitability of the global defense industrial base key for maintaining global peace and security.

References

Adamowski, Jaroslaw, 2012. "Romania to Buy F-16 Fighters from Portugal." *Defense News*, August 29.

——. 2014a. "Ukraine Crisis Could Hinder Russian Market Expansion." *Defense News*, April 4.

——. 2014b. "Ukraine Crisis Speeds West European UAV Efforts." *Defense News*, May 12.

——. 2014c. "Russian, NATO Arms Race Takes Shape." *Defense News*, June 9, 2014.

——. 2014d. "Russia Bolsters Crimean Air Defense, E. Europe Eyes Countermeasures." *Defense News*, July 14.

Bennhold, Katrin, 2013. "Job Cuts, and No More Shipbuilding, at Historic Home of Britain's Warships." *New York Times*, November 7, p. B2.

Chuter, Andrew, 2013a. "BAE To Close Portsmouth Naval Yard as UK Restructures Shipbuilding." *Defense News*, November 6.

——. 2013b. "Eurofighter Announces Typhoon Upgrades." *Defense News*, October 30.

——. 2014a. "UK Carrier Capability Retunes." *Defense News*, June 30.

——. 2014b. "Plotting a New Course," *Defense News*, July 21.

Chuter, Andrew and Tran, Pierre, 2013a. "Anglo-French UCAV Development Takes a Step Forward." *Defense News*, December 6.

——. 2013b. "BAE, Dassault Submit Plan to Jointly Develop UCAV." *Defense News*, September 13.

Clark, Nicola, 2013. "Parent of Airbus to Cut 5400 Jobs as Europe's Military Budgets Shrink." *New York Times*, December 10, p. B4.

Defense News Staff, 2012. "Europe's Biggest Powers Balance Cuts, Capabilities." *Defense News*, October 24.

"German Defense Budget to Rise," 2014a. *Defense News*, June 23.

"German Defense Budget Passes," 2014b. *Defense News*, June 30.

Kingston, Tom, 2014a. "Italy Seeks Bigger JSF Workshare." *Defense News*, June 30.

——. 2014b. "Italian Navy Opening Yards to Civilian Work." *Defense News*, August 4.

Kingston, Tom and Tran, Pierre, 2012. "Eurosatory Preview: An Austere Landscape." *Defense News*, June 4.

Lamothe, Dan, 2014. "France Backs Off Sending Mistral Warship to Russia in $1.7 Billion Deal." *Washington Post*, September 3.

McLeary, Paul, 2012. "General Dynamics Hopes Hull 'Exchange' Program Keeps Stryker Line Open." September 6.

Mehta, Aaron, 2014. "Market Still Flat, But Signs of Hope." *Defense News*, July 14.

Muller, Albrecht, 2013. "German Military Reduces its Future Helicopter Fleet." *Defense News*, March 15.

O'Dwyer, Gerald, 2013a. "Departing Norwegian Government Boosts Defense Spending for 2014." *Defense News*, October 14.

——. 2013b. "Sweden Plans Defense Spending Boost." *Defense News*, October 15, 2013.

——. 2014a. "Baltics to Hike Budgets Pursue Permanent NATO Troop Presence." *Defense News*, April 28.

——. 2014b. "Finland Builds Multiple Defense Partnership with NATO, Sweden." *Defense News*, May 12.

"Osprey Could Become Part of Carrier Mission." 2012. *Navy Times*, August 19.

Perlo-Freeman, Sam and Carina Solmirano, 2014. *SIPRI Fact Sheet: Trends in World Military Expenditure, 2013*. April.

Pfeifer, Sylvia, 2010. "Finmeccanica: Group is Keen to Target Emerging Markets." *Financial Times*, July 19.

Pfeifer, Sylvia and Clark, Pilita, 2010. "EADS Eyes International Markets to Fuel Growth." *Financial Times*, July 19.

Pittaway, Nigel, 2012. "Australia Details Plan to Convert F/A-18's to Growlers." *Defense News*, September 3.

Rahir, Patrick, 2013. "France to Cut Rafale Order: Betting on Exports." *Defense News*, August 2.

Tran, Pierre, 2014. "Europe Struggles to Unify its Response to Russia." *Defense News*, July 28.

Tran, Pierre and Andrew Chuter, 2014. "France, UK to Sign Pact Kicking Off Combat Drone Study." *Defense News*, July 14.

Weisgerber, Marcus, 2012. "Boeing: Libya Has Interest in Used Army Chinooks." *Defense News*, August 31.

——. 2014. "Interceptions Rise as Russia Boosts Air Power." *Defense News*, June 30.

Chapter 8
Global Arms Deliveries

The global integration of the defense industrial bases across countries is achieved partially through global trade and global arms deliveries. These enable countries to deal with emerging defense threats with the appropriate equipment, as well as to have a wider choice of affordability by expanding beyond their domestic defense sector. Moreover, the contracts with foreign defense companies can assist countries in developing their own defense industrial bases through exchanging knowledge and collaborating in expanding skillsets. This chapter examines the global arms deliveries in the worldwide defense market in detail. It discusses the global arms deliveries by supplier country, as well as evaluates the arms deliveries to developing nations in a variety of regions – Asia, the Middle East, Latin America, and Africa. Moreover, it explores the development of domestic industrial bases and shifting defense priorities for many of the countries in those regions. Finally, it assesses the ways in which the demand for equipment can impact the global defense industrial base through the demand of developing countries for various categories of equipment from US defense contractors, as well as contractors from other countries.

Global Arms Deliveries by Supplier Country

The global arms deliveries by supplier country during 2004–2011, as represented in millions of US dollars, are shown in Figure 8.1. The US was the leading supplier during the period and, like many of the countries, experienced a decline in global arms deliveries in 2010 due to the economic crisis and then recovered. Russia generally ranked as the second most significant supplier in global arms deliveries between 2004 and 2011. France tied with Russia in 2004 as the second leading supplier of global arms deliveries, but dropped from $5,600 (in millions) in 2004 to $2,700 in 2005 and remained largely in the range between $1,300 and $2,500 in the successive years. The UK rose from $3,200 in 2004 in arms deliveries to $4,900 in 2006, dropped to $2,200 in 2007, and then recovered, reaching $3,000 in 2011. Italy expanded its arms deliveries, especially between 2009 ($800) and 2011 ($1,700). Germany expanded from $2,000 in 2004 to $3,700 in 2008 before dropping in the successive years, reaching $1,600 in 2011.

The aggregate value of global arms deliveries during the period from 2004 to 2011 is summarized in Figure 8.2. The US delivered the highest value of global arms deliveries, followed by Russia and the UK.

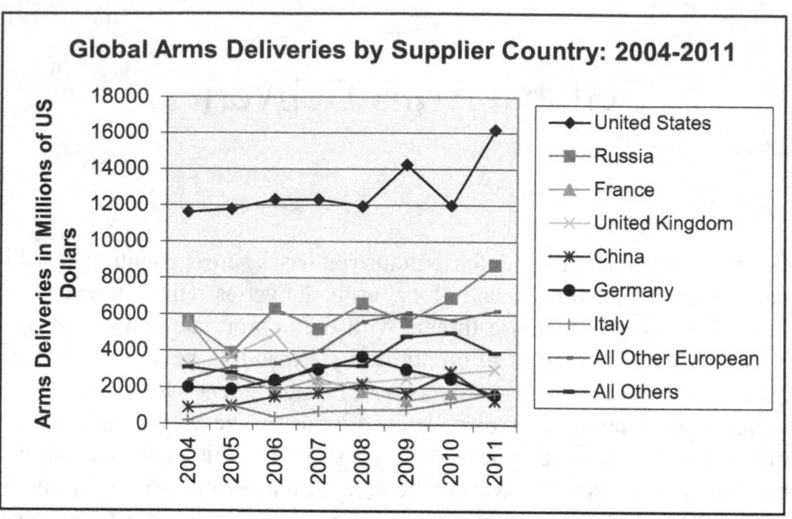

Figure 8.1 Global Arms Deliveries by Supplier Country: 2004–2011

Source: Grimmett, Richard E. and Kerr, Paul K., *Congressional Research Service Report (R42678): Conventional Arms Transfers to Developing Nations: 2004–2011*, Washington DC, August 24, 2012.

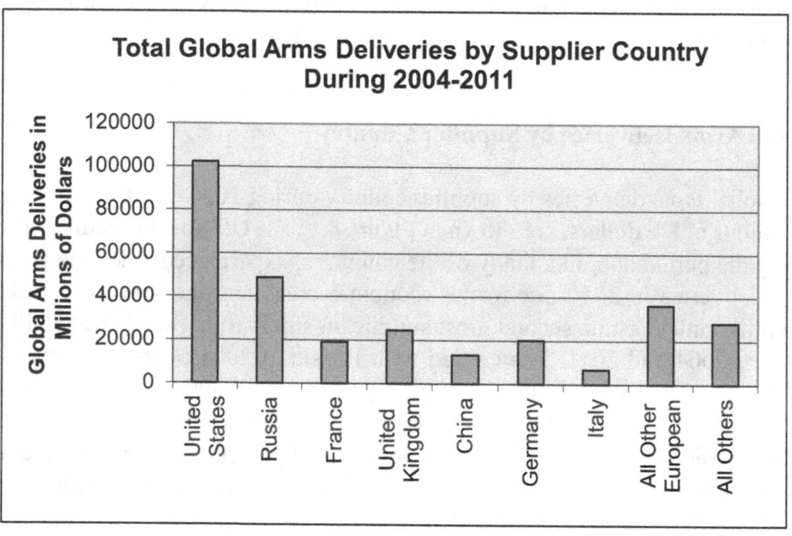

Figure 8.2 Total Global Arms Deliveries by Supplier Country During 2004–2011

Source: Grimmett, Richard E. and Kerr, Paul K., *Congressional Research Service Report (R42678): Conventional Arms Transfers to Developing Nations: 2004–2011*, Washington DC, August 24, 2012.

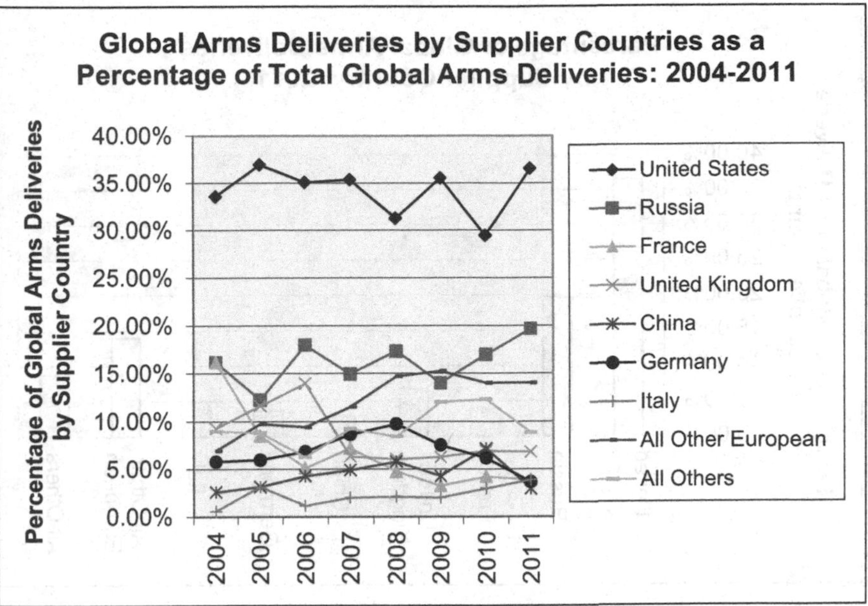

Figure 8.3 Global Arms Deliveries by Supplier Countries as a Percentage of Total Global Arms Deliveries: 2004–2011

Source: Grimmett, Richard E. and Kerr, Paul K., *Congressional Research Service Report (R42678): Conventional Arms Transfers to Developing Nations: 2004–2011*, Washington DC, August 24, 2012.

The global arms deliveries by each supplier country as a percentage of total global arms deliveries are depicted in Figure 8.3. The US was the leading supplier of global arms deliveries as a percentage of total global arms deliveries between 2004 and 2011, followed by Russia. The US varied from delivering 29.44% of global arms deliveries (2010) to 36.92% of global arms deliveries (2005). Russia varied from delivering between 12.24% (2005) and 19.66% (2011) of global arms deliveries. France significantly declined from delivering 16.18% of arms deliveries in 2004 to 3.84% of arms deliveries by 2011.

Figure 8.4 provides a yearly snapshot and shows the percentage of global arms delivered by supplier countries in 2011. Over half (56.17%) of the global arms deliveries were delivered by the US (36.51%) and Russia (19.66%). France, Germany, Italy, the UK, and other European countries delivered 32.09% of the global arms deliveries in 2011. Consequently, the US and Russia have consistently been the leading suppliers in global arms deliveries. Countries such as France and Germany have declined in their deliveries in recent years, while Italy has expanded.

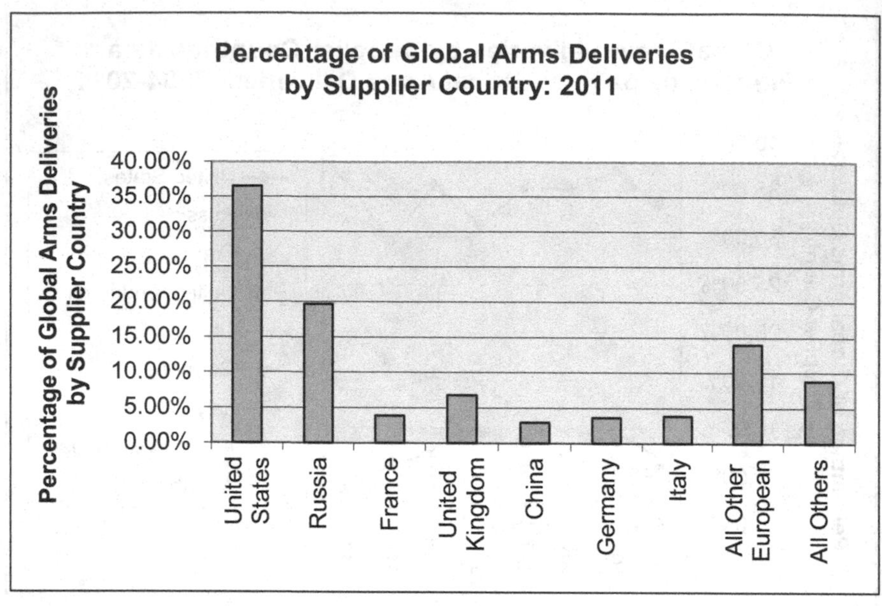

Figure 8.4 Percentage of Global Arms Deliveries by Supplier Country: 2011

Source: Grimmett, Richard E. and Kerr, Paul K., *Congressional Research Service Report (R42678): Conventional Arms Transfers to Developing Nations: 2004–2011*, Washington DC, August 24, 2012.

Global Arms Deliveries to Various Regions

Figure 8.5 focuses on the arms deliveries to developing nations by supplier between 2004 and 2011, as represented in millions of US dollars. The US and Russia have been the leading suppliers (as was the case in the global market in the earlier figures), and France experienced the most significant decline as a supplier. The US delivered $7,385 of arms deliveries in 2004 and $10,522 of arms deliveries by 2011. France's arms deliveries to developing nations significantly fell from $5,200 in 2004 to $700 by 2011. Similarly, the UK and Germany also declined, such that the UK's arms deliveries to developing nations fell from $2,400 in 2004 to $1,500 by 2011, while Germany's deliveries declined from $800 in 2004 to $400 in 2011, after increasing to as high as $1,300 in 2008 and $1,100 in 2009. Italy's arms deliveries increased from $100 in 2004 to $1,100 in 2011. China also experienced growth in delivering arms to developing nations, growing from $900 in 2004 to a peak of $2,900 in 2010, before declining in 2011.

Data on the aggregate value of global arms deliveries to developing nations during the period from 2004 to 2011 indicates that the US has delivered the highest value of global arms deliveries, followed by Russia and the UK (Figure 8.6). This

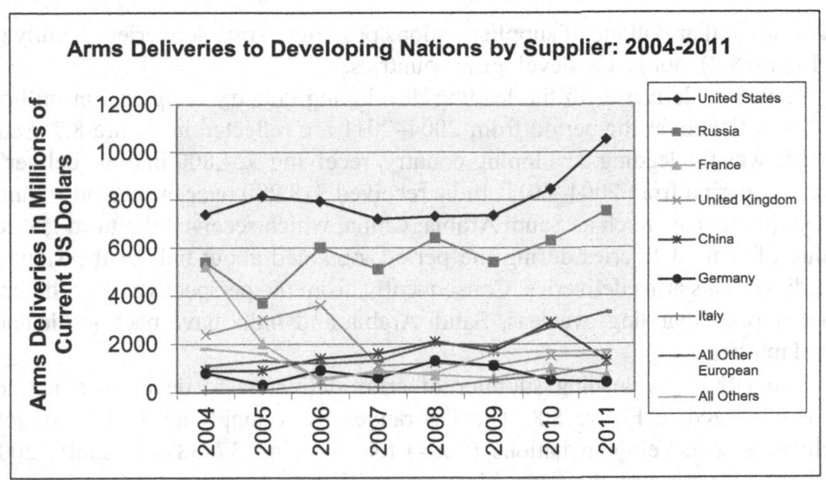

Figure 8.5 Arms Deliveries to Developing Nations by Supplier: 2004–2011

Source: Grimmett, Richard E. and Kerr, Paul K., *Congressional Research Service Report (R42678): Conventional Arms Transfers to Developing Nations: 2004–2011*, Washington DC, August 24, 2012.

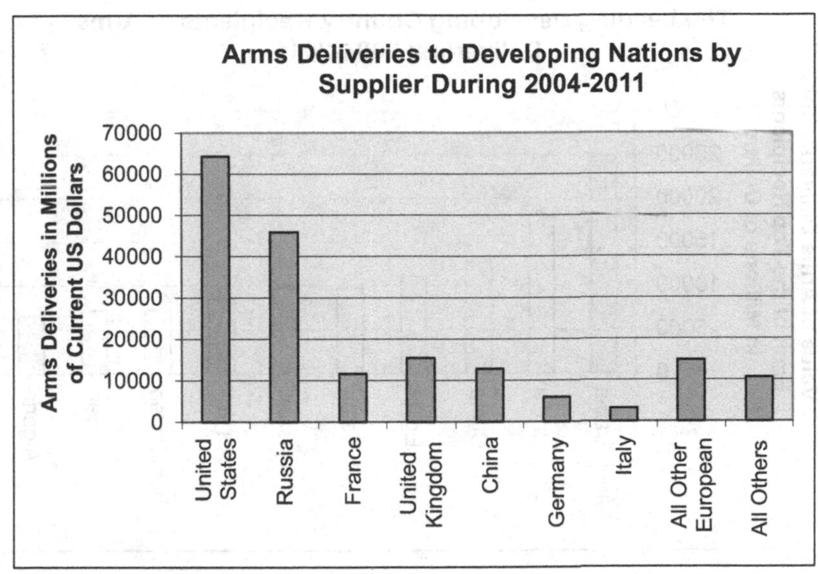

Figure 8.6 Arms Deliveries to Developing Nations by Supplier During 2004–2011

Source: Grimmett, Richard E. and Kerr, Paul K., *Congressional Research Service Report (R42678): Conventional Arms Transfers to Developing Nations: 2004–2011*, Washington DC, August 24, 2012.

has very similar ranking of supplier nations providing arms deliveries globally (as in Figure 8.2), not just to developing countries.

The arms deliveries to the leading developing country recipients in millions of US dollars over the period from 2004–2011 are reflected in Figure 8.7. Saudi Arabia was the leading developing country, receiving $24,800 in arms deliveries over the period from 2004–2011. India received $18,200 over the period – almost three-quarters as much as Saudi Arabia. China, which received the third greatest value of arms deliveries during the period, obtained about half of the value of Saudi Arabia's arms deliveries. Consequently, from the perspective of US defense contractors expanding overseas, Saudi Arabia and India have been particularly good markets.

The US is the leading supplier of arms deliveries to developing nations, as emphasized in Figure 8.8. The US ranges from supplying 29.44% of total deliveries to developing nations (2004) to supplying 37–38% in 2005, 2007, and 2011. Russia was the second leading supplier to developing nations, ranging from 21.53% in 2004, peaking at 29.09% in 2008, and reaching 26.76% in 2011. France's share of arms deliveries to developing nations fell from 20.73% in 2004 to 9.5% in 2005, and declined further, such that it reached 2.5% in 2011. The UK

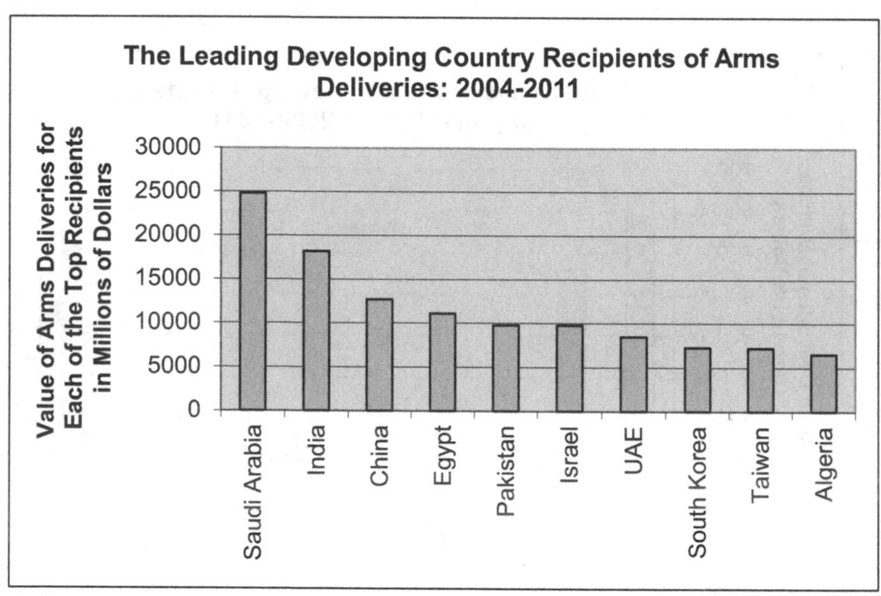

Figure 8.7 The Leading Developing Country Recipients of Arms Deliveries: 2004–2011

Source: Grimmett, Richard E. and Kerr, Paul K., *Congressional Research Service Report (R42678): Conventional Arms Transfers to Developing Nations: 2004–2011*, Washington DC, August 24, 2012.

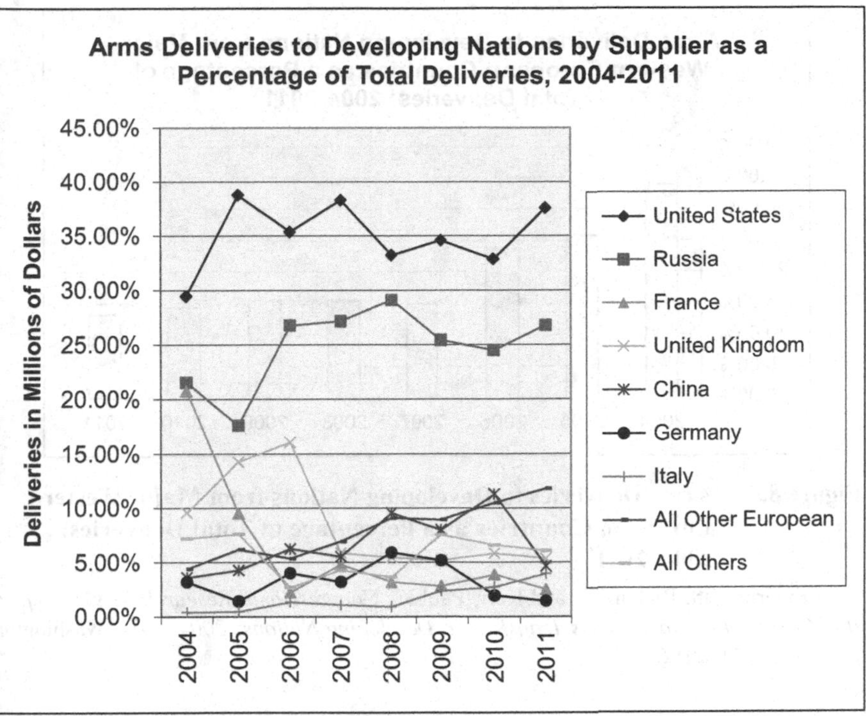

Figure 8.8 Arms Deliveries to Developing Nations by Supplier as a Percentage of Total Deliveries: 2004–2011

Source: Grimmett, Richard E. and Kerr, Paul K., *Congressional Research Service Report (R42678): Conventional Arms Transfers to Developing Nations: 2004–2011*, Washington DC, August 24, 2012.

rose from 9.57% in 2004 to 16.05% in 2006, and then declined from 2007 onwards at 5–6%. The patterns of leading arms suppliers to developing countries is similar to the pattern of leading arms suppliers to the entire global market in that the US and Russia have been the leading suppliers, and France's role as a supplier has significantly declined.

The share of arms deliveries to developing nations as a percentage of total deliveries from major western European countries (including France, Italy, Germany, and the UK) fell from 33.88% in 2004 to 13.20% in 2011. This significant decline, reflected in Figure 8.9, is partially due to the reduction in France's arms deliveries.

The quantity of weapons delivered to developing countries by supplier countries on a country-by-country basis between 2004 and 2007 can be seen in Figure 8.10. Overall, Russia was the leading supplier, followed by the US during this period.

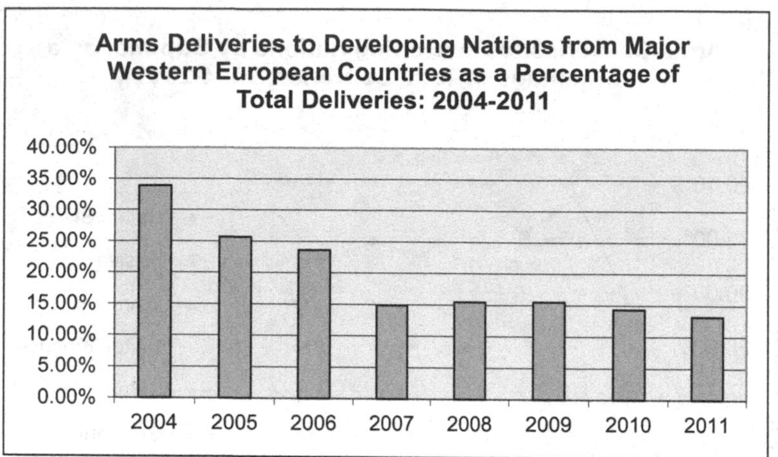

Figure 8.9 Arms Deliveries to Developing Nations from Major Western European Countries as a Percentage of Total Deliveries: 2004–2011

Source: Grimmett, Richard E. and Kerr, Paul K., *Congressional Research Service Report (R42678): Conventional Arms Transfers to Developing Nations: 2004–2011*, Washington DC, August 24, 2012.

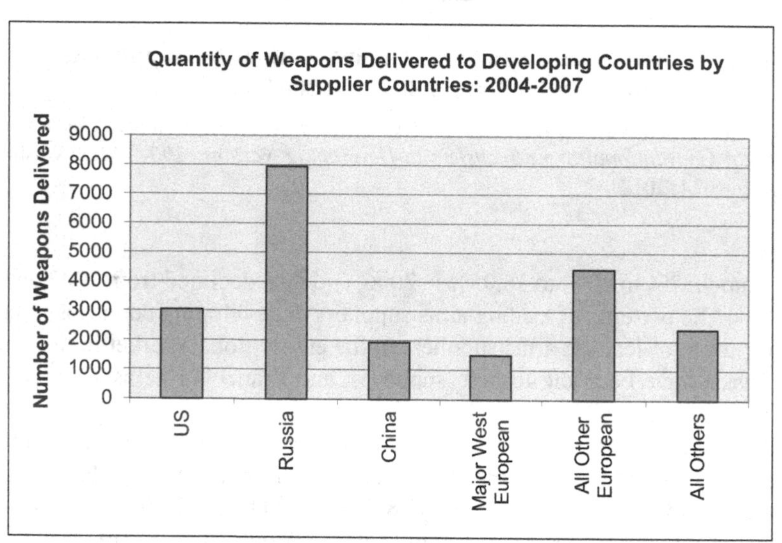

Figure 8.10 Quantity of Weapons Delivered to Developing Countries by Supplier Countries: 2004–2007

Source: Grimmett, Richard E. and Kerr, Paul K., *Congressional Research Service Report (R42678): Conventional Arms Transfers to Developing Nations: 2004–2011*, Washington DC, August 24, 2012.

Surface-to-air missiles (led by Russia) and armored cars (led by all other European nations, excluding major Western European nations) were the dominant markets. The US led the delivery of tanks and self-propelled guns (672 tanks and self-propelled guns were delivered), followed by Russia (300), China (160), and the major Western European countries (160). Similarly, the US, as a single country, led the delivery of armored cars (726 armored cars were delivered), followed by Russia (480), China (460), and the major Western European countries (260). Surface-to-air missiles delivered to developing nations were led by Russia (6,340), followed by the US (910), major Western European countries (650), and China (530). China led the delivery of artillery (450), followed by the US (240), while Russia and the major Western European countries delivered 20 and 10, respectively, in the artillery market. The major Western European countries (17) were the primary suppliers of major surface combatants. China (56) and the major Western European nations (57) were primary suppliers of minor surface combatants. Russia (8) and the major Western European countries (5) were primary deliverers of submarines. The major Western European countries also were significant suppliers of guided missile boats (7). The leading country providing delivery of supersonic combat aircraft was Russia (180), followed by

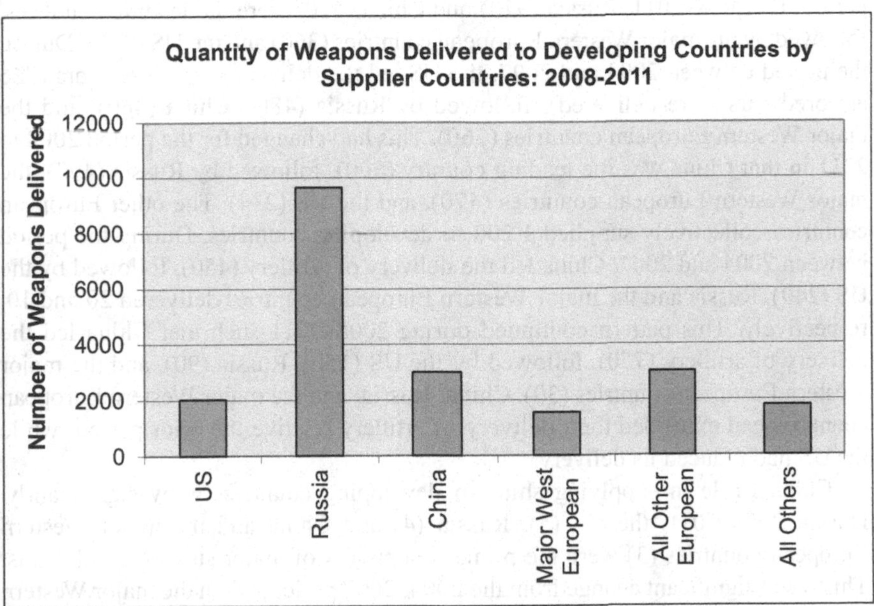

Figure 8.11 Quantity of Weapons Delivered to Developing Countries by Supplier Countries: 2008–2011

Source: Grimmett, Richard E. and Kerr, Paul K., *Congressional Research Service Report (R42678): Conventional Arms Transfers to Developing Nations: 2004–2011*, Washington DC, August 24, 2012.

the US (104) and the major Western European countries (70). The major Western European countries were the largest suppliers of subsonic combat aircraft (20), followed by China (10). Helicopters delivered by suppliers to developing nations were led by Russia (200), followed by major Western European countries (80) and the US (73). Anti-ship missiles delivered to developing nations were led by Russia (360), the US (262), major Western European countries (150), and China (120),

Figure 8.11 shows the changes in the quantity of weapons delivered to developing countries by supplier countries between 2008 and 2011, relative to the previous chart (Figure 8.10), which focused on the period between 2004 and 2007. During the 2008–2011 period, Russia was the leading supplier, followed by China and the US. During 2004–2007, the US was the second largest supplier and China was the third largest supplier, so China's quantity of weapons sold to developing countries grew faster than the US' weapons sales to developing countries. Russia's surface to air missiles continued to be the dominant market in the 2008–2011 period, as it was in the 2004–2007 period. Between 2004 and 2007, the US had led the delivery of tanks and self-propelled guns (672 tanks and self-propelled guns were delivered), followed by Russia (300), China (160), and the major Western European countries (160). The period between 2008 and 2011, however, had exhibited significant growth for the supplier nations relative to the previous period. By 2008–2011, Russia (570) and China (510) were the leading countries, followed by the major Western European countries (360) and the US (348). During the period between 2004 and 2007, the US led the delivery of armored cars (726 armored cars were delivered), followed by Russia (480), China (460), and the major Western European countries (260). This had changed for the period 2008 to 2011 in that China was the leading country (590), followed by Russia (490), the major Western European countries (470), and the US (234). The other European countries collectively supplied 1,200 to developing countries. During the period between 2004 and 2007, China led the delivery of artillery (450), followed by the US (240). Russia and the major Western European countries delivered 20 and 10, respectively. This pattern continued during 2008–2011 such that China led the delivery of artillery (770), followed by the US (150), Russia (90), and the major Western European countries (30). China, Russia, and the major Western European countries had expanded their delivery of artillery relative the prior period, while the US had reduced its delivery.

China's role in supplying ships to developing countries grew significantly. During 2008–2011, the US (5), Russia (4) and China and the major Western European countries (3) were the primary suppliers of major surface combatants. This was a significant change from the 2004–2007 period, when the major Western European countries (17) were the primary suppliers of major surface combatants. During 2008–2011, China was the primary supplier of minor surface combatants (108), followed by the major Western European countries (57). This shows significant growth for China in supplying minor surface combatants relative to the 2004–2007 period, when China (56) and the major Western European nations (57) were primary suppliers of minor surface combatants. Russia (8) and the major

Western European countries (5) were primary deliverers of submarines during 2004–2007. During 2008–2011, the market for external delivery of submarines did not grow significantly; Russia (2) contracted, while the major Western European countries (4) were the primary suppliers. The already small guided missile boat deliveries had also shrunk in 2008–2011 relative to 2004–2007.

China and Russia's roles in supplying aircraft to developing countries grew significantly, while the US had a declining role. The leading country providing delivery of supersonic combat aircraft in 2008–2011 was Russia, as had been the case in 2004–2007. The US had decreased its deliveries of supersonic combat aircraft (53 during 2008–2011 vs. 104 during 2004–2007), as had the major Western European countries (50 during 2008–2011 vs. 70 during 2004–2007). China's role as a supplier had increased to 30 during 2008–2011 relative to 20 during 2004–2007. The major Western European countries were the largest suppliers of subsonic combat aircraft, increasing from 20 during 2004–2007 to 50 during 2008–2011, followed by China, which had increased from 10 during 2004–2007 to 20 during 2008–2011. Helicopters delivered by suppliers to developing nations continued to be led by Russia (which had expanded from 200 during 2004–2007 to 270 during 2008–2011), followed by major Western European countries (which had expanded from 80 during 2004–2007 to 110 during 2008–2011) and the US (which had fallen from 73 during 2004–2007 to 57 during 2008–2011).

The dominant suppliers in delivering missiles to developing countries exhibited some changes across the periods. Surface-to-air missiles delivered to developing nations were led by Russia (which had expanded from 6,340 during 2004–2007 to 7,750 during 2008–2011), followed by the US (which had only slightly expanded from 910 during 2004–2007 to 944 during 2008–2011). China had expanded into third place, with 780 deliveries during 2008–2011, which indicated growth relative to the 530 deliveries during 2004–2007. Major Western European countries had declined to fourth place by contracting from 650 deliveries during 2004–2007 to 290 deliveries during 2008–2011. Anti-ship missiles delivered to developing nations shrunk in 2008–2011 relative to 2004–2007. Russia continued to be the leading supplier, although it delivered 220 anti-ship missiles during 2008–2011, relative to 360 during 2004–2007. The US was the second largest deliverer, although it contracted from 262 missiles to 176. Similarly, major Western European countries contracted from 120 during 2004–2007 to 60 during 2008–2011, while major Western European countries contracted from 150 to 60.

Global Arms Deliveries to Various Regions

Asia

This section explores the trends in global arms deliveries in Asia, as well as the shifting defense priorities and the fiscal constraints facing these nations. The Asian markets have exhibited significant growth in defense spending in recent years, partially due to concerns about the need to rebuild and expand their defense forces. The top five defense spenders in Asia – China, India, Japan, South Korea, and Taiwan – captured 87% of Asian defense spending in 2011. These five countries spent a combined $224 billion in 2011, which was twice the amount spent by those same five countries in 2000.[1] Between 2004 and 2013, spending of East Asian countries increased by 74%, Southeast Asian countries increased by 47%, and Central and South Asian countries increased by 49%. Between 2012 and 2013, spending of East Asian countries increased by 4.7%, Southeast Asian countries increased by 5%, and Central and South Asian countries increased by 1.2%. The 3.6% increase in overall Asian defense spending between 2012 and 2013 was largely driven by China's 7.4% increase.[2]

Tensions have increased in the Asia-Pacific region, which may ultimately impact the global defense spending patterns in the coming years. As a result of concerns for China's ambitions to expand its control in the East China Sea, Asian countries are increasing their military equipment. Former Secretary of Defense Robert M. Gates noted "in January 2011 that he believed the long-term goal of the Chinese was to push the US to 'the second island chain,' farther out in the Pacific, keeping American air and naval assets ever farther from the region around China's coast."[3] In an effort to defend lanes for shipping in the East China Sea, South Korea is building a base for 20 warships.[4]

As of mid-December, 2013, Japan announced that it would increase defense spending on new equipment, at $240 billion over the next five years to purchase ships, drones, and fighter planes, including other equipment for amphibious warfare, such as the V-22 Osprey. Nearly half of Japan's ground forces would be reconfigured for rapid deployment and a unit similar to the Marine Corps would protect some of Japan's southwest islands.[5] Moreover, the Japanese intended to build a new base for the army, to be completed in 2016, in the disputed Senkaku Islands (or the Diaoyu Islands in China). The deployment by the Japanese of radar planes and F-15s at Okinawa is part of their strategy, as is the possibility of purchasing US drones.[6] Japan is installing a radar base for the Japanese Ground Self-Defense Forces on Yonaguni Island to suggest that it will resist China, particularly regarding the dispute over the Senkaku Islands. Indeed, Chinese ships have repeatedly approached these waters since Japan nationalized them in September, 2012. On April 24, 2014, President Obama, on his state visit to Japan, noted that the US would defend the Senkaku Islands against any attack under Article 5 of the US–Japan security treaty and stated that Japanese territory included the Senkakus. Japan's MoD is also attempting to provide greater surveillance and

1 Minnick, 2012.
2 Perlo-Freeman and Solmirano, 2014.
3 Sanger, 2013.
4 Spitzer, 2013.
5 Spitzer, 2013.
6 Spitzer, 2013.

warning in the southwestern area of the sea by reorganizing its airborne early warning group.[7]

In December, 2013, China created a new air defense zone in the East China Sea, requiring commercial airlines to file flight plans. The FAA and the State Department advised commercial airlines to follow China's notices.[8] Moreover, China "has begun asserting ownership of thousands of shipwrecks within a vast U-shaped area that covers almost all of the South China Sea," which, they say, will reduce the theft of antiquities.[9]

China is constructing a seaport and an airstrip in the South China Sea at the Fiery Cross Reef, which continues China's active expansion efforts as it establishes claims within its 9-Dash Line, "which outlines China's purported 'indisputable sovereignty' of the South China Sea." The airbase would enable China to have shorter-range tactical aircraft with significant weapons to operate in the airspace over the South China Sea, without having to send them from the mainland. As a result, there will be greater demand for US Navy forces in the area due to the presence of the PLA Navy, the PLA Air Force, and the Chinese Army, which also operates the DF-21D anti-ship ballistic missiles. Moreover, the result of the construction of the Fiery Cross airstrip and seaport may be greater fortification by the Philippines, Malaysia, and Vietnam with reefs, etc. in the area.[10]

As China stakes greater claims in the South China Sea, Australia, Japan, Singapore, and the Philippines have emphasized the need for regional stability. The US has been attempting to solidify its presence in the area since some of the busiest trade routes run through the Strait of Malacca and the South China Sea. The US has troops at long-standing bases in Japan, South Korea, and Guam, as well as on the island of Diego Garcia in the Indian Ocean. Moreover, as of 2013, it had established agreements to base Marines in northern Australia and to station warships in Singapore, while the Navy had undertaken negotiations with the Philippines to possibly operate Navy ships from the Philippines, and to deploy troops on a rotational basis or to stage more frequent exercises.

As of 2013, there were also possibilities of bolstering military partnerships with Vietnam and Thailand. In these cases, US forces would effectively be guests at foreign bases, which would enable a "light footprint" to be maintained rather than the large, expensive bases of the Cold War era. During 2012, DoD worked on the development of an "afloat forward staging base," which is a platform that could be configured to carry and refuel helicopters, drones, and patrol boats, as well as to place a small strike force ashore or to launch commando operations without requesting access to a land base in another country. Simultaneously, the Navy has been adjusting its troop and equipment capacity in the wake of defense pressures in the Pacific. Some sources suggest that the majority of cruisers, littoral

7 Kallender-Umezu, 2014.
8 Hayashi and Pasztor, 2013.
9 Page, 2013.
10 Minnick, 2014b.

combat ships, submarines, destroyers, and six aircraft carriers will be deployed to the Pacific. By 2020, the Navy reportedly plans to have shifted its forces from 50% in the Pacific to 60% in the Pacific, and decreased them from 50% in the Atlantic to 40% in the Atlantic.

Malaysia, Brunei, the Philippines, Vietnam and Taiwan have overlapping claims regarding key sea routes which are important for global shipping traffic. This has resulted in confrontations between China and countries, such as the Philippines and Vietnam, although Malaysia has had fewer tense confrontations with China, which is the top export market for Malaysia.[11]

As a result, other countries, such as Vietnam and Malaysia, expanded their defense equipment. Vietnam has been expanding their surface and submarine fleet with new purchases from Russia and potentially the Netherlands, while Malaysia was planning to have a multi-role combat aircraft (MRCA) competition for a $2–$3 billion contract to replace MiG-29 aircraft.

As of mid-December, 2013, DoD planned to provide $40 million in maritime aid to the Philippines, as well as to have more American military forces in the country, in order to deal with the growing ambitions of China in the East China Sea and the South China Sea and the concern among neighboring countries. It is likely that the Philippines would not want American forces to appear to be permanently located there, especially since the US closed four small bases and two large bases in the Philippines during 1991–1992.[12]

Similarly, the main ally for Taiwan has been the US, which sent two aircraft carrier battle groups in 1996 to the area near Taiwan after an incident in which China launched missiles in that area. About 1600 Chinese ballistic missiles are said to currently target Taiwan. Nevertheless, Taiwan plans to reduce its defense spending, as well as its military manpower,[13] although it continues to receive military equipment from orders with foreign defense contractors. Moreover, Taiwan will need to replace its ageing aircraft, since 55 Mirage 2000s and 50 F-5s will be retired over the next decade, leaving 144 F-16 A/Bs and 126 upgraded indigenous defense fighters for Taiwan. While upgrading F-16s with US assistance has been important for Taiwan, the acquisition of 66 F-16 C/D fighters has been stalled since 2006.[14] As of September 2013, Taiwan received the first of 12 anti-submarine aircraft from the US, which, in 2007, had agreed to provide Taiwan with the refurbished P-3C Orion patrol aircraft for $1.96 billion, which would replace the older S-2T anti-submarine aircraft.[15] In November, 2013, Taiwan received the first six of 30 Apache advanced attack helicopters from the US, which were a portion of a 2008 $6.5 billion arms deal.[16]

11 Agence France-Presse, 2013a.
12 Bradsher, 2013.
13 Agence France-Presse, 2013g.
14 Minnick, 2014a.
15 Agence France-Presse, 2013e.
16 Agence France-Presse, 2013e.

South Korea has been rethinking its naval capitalization efforts due to island territorial disputes with Japan and China's increased naval activity in the region. Indeed, over the past five years, South Korea has shifted funds from the Army to the Air Force and the Navy. Japan has also been facing challenges from China and North Korea. Its purchase of 42 F-35s for $7 billion in late December 2011 has been partially due to Japan's concerns about regional instability (North Korea and China). Finally, during 2012 and 2013, Japan, Taiwan, and South Korea bolstered their ballistic missile defenses with new Patriot Advanced Capability systems, long-range early warning radars, and improved command and control systems.[17]

In September 2013, South Korea made the decision not to purchase Boeing's $7.7 billion jet fighter, which suggested that it would re-open the competition for this defense contract. This contract is the largest contract that South Korea has had in its history. The goal of the contract had been for South Korea to replace its older fleet of F-5s and F-4s and purchase 60 advanced combat fighters, which resulted in bids from EADS, Lockheed Martin, and Boeing. Although EADS' Eurofighter was turned down despite the $2 billion investment to provide South Korea with the opportunity to develop its own advanced jet and to construct Eurofighters, EADS was positive, as of the fall of 2013, that it might have a viable opportunity in the re-opened competition.[18]

China's defense industrial base has expanded, which is key for its own military strategy, as well as the increasing demand of other countries for Chinese equipment. Indeed, about 72% of Sudan's purchases of small arms, ammunition, and artillery came from China. Significant purchasers of Chinese weapons include: Bangladesh, Pakistan, Lebanon, Qatar, Jordan, and Egypt. Iran has also been a significant buyer of arms since the 1980s during its war in Iraq.[19]

Evidence suggests that China has increased its international small arms and light weapons sales to developing nations where, in some cases, terrorist groups utilize Chinese-made small arms. Examples of small arms include machine guns, rifles, and pistols, while examples of light weapons include mortars, missiles, and rockets. Chinese small arms are only allowed to be exported by state-owned companies; due to their low prices, many of the poorer countries in South America, the Middle East, and Africa are the purchasers.[20]

China has recently done well in competition, more due to pricing and less due to experience and innovation. One example of this was the selection by Turkey of China's proposal in the fall of 2013 to build its missile defense system. This was surprising in that Turkey hosts a US-made radar as part of a larger NATO missile shield.[21] Traditionally, the US-made antimissile systems have been popular with countries in the Middle East and the Pacific Rim partially because the sales

17 Minnick, 2012.
18 Agence France-Presse, 2013b.
19 "Chinese Small Arms Sales May Fan Conflicts, Equip Terrorists," 2013.
20 "Chinese Small Arms Sales May Fan Conflicts, Equip Terrorists," 2013.
21 Erwin, 2013.

are supported by the Pentagon and the US firms continue to support the missile defense systems for 20–30 years even if they are no longer in the US inventory.[22] China's missile defense systems lack the lengthy history of the US' missile defense system, including the history of the defense contractors supporting the missile defense systems in the future decades. As a result, technical capabilities sometimes do not compete successfully against lower prices.[23]

Nevertheless, in this case, pricing was a significant variable in Turkey's choice of missiles. China's state-controlled Precision Machinery Export-Import Corporation would produce the HQ9 long-range surface-to-air missile which, at $4 billion, would be $1 billion cheaper than the products produced by Russia, the US, and the French–Italian alliance bid. The US bid involved Lockheed Martin and Raytheon teaming, such that Raytheon's Patriot air defense batteries would fire Lockheed Martin's Patriot PAC-3 interceptor missiles.[24]

The selection of the Chinese equipment led to issues being raised by the US and NATO allies regarding the potential undermining of the interoperability of NATO's missile shield, as well as concerns about maintaining security. The Chinese "HQ9 would not be able to operate with Patriot and with Raytheon's AN/TPY-2 radar, which currently is based in Turkey."[25]

Taiwan's 2013 National Defence Report suggested that China's annual spending on defense had average growth in double-digit growth rates over the past two decades. The Chinese military has manpower of 2.27 million individuals, of which over half includes the army (1.25 million). Many of China's new weapons, such as strategic bombers, stealth fighters, early warning aircraft, air defense and ballistic missiles, and conventional and nuclear-powered submarines, are purchased from Russia.[26] China's first carrier was fielded in 2012 as the Liaoning, as was its first carrier-based fighter, the J-15 Flying Shark. The Liaoning carrier was previously the Varyag, which was a Kuznetsov carrier and which was refurbished. The third carrier is likely to be bigger than the first carrier since the first carrier can only carry 20 fighter jets; moreover, more operational carriers will be needed if China plans to project power to a greater degree into the East China Sea and the South China Sea.[27]

As a result of the growth in domestic and international demand for China's defense products, China is expanding areas of its defense industrial base. For example, China plans to establish a UAV industrial base in Beijing's southern Daxing District. The base, which will be the first of its kind, will represent the entire supply chain for UAVs made in China. The estimated output for the UAV industrial base is projected to be $1.6 billion by 2015, $4.8 billion by 2020, and

22 Erwin, 2013.
23 Erwin, 2013.
24 Erwin, 2013.
25 Erwin, 2013.
26 Agence France-Presse, 2013g.
27 Minnick, 2014c.

$16.1 billion by 2025.[28] China is developing UAVs which are similar to US UAVs, such as the RQ-170 Sentinel (Lockheed Martin), the Predator (General Atomics), and the Global Hawk (Northrop Grumman). Moreover, the Pterodactyl UAV made by Chengdu Aircraft in China is similar to the US Predator/Reaper UAVs.[29]

Indonesia is another country which is expanding its defense industrial base. The Indonesian defense budget for purchasing new equipment during the 2010–2014 period was 57 trillion rupiahs (US $5 billion), which was about one-third of the overall defense budget over that period (156 trillion rupiahs).[30] Indonesia has pursued two mechanisms for procurement – imports and domestic development. Apart from PT PAL, Indonesia also has PT Pindad, a state-owned arms producer, and PT Dirgantara Indonesia (PT DI), which produces military aircraft. A number of aircraft for the Indonesian Air Force have come through the cooperation with PT DI, such as the Bell 412 helicopter, Bolcow 105 and Cassa 212.[31]

Nevertheless, despite the development of its industrial base, Indonesia continues to purchase equipment made in the US, Russia, etc. in order to expand its military capabilities. Indeed, as of the fall of 2013, Indonesia planned to add to its Air Force two squadrons of F-16s, 16 Russian-made Sukhoi fighters, and a squadron of South Korean-built T-50 Golden Eagle trainer jets.[32] As of the fall of 2013, the US announced that it intended to provide the Indonesian Army with a fleet of AH-64E Apache attack helicopters for $500 million, such that the first two Apaches would arrive by 2014 and the last Apache will be delivered to Indonesia by 2019. Another type of aircraft which Indonesia is likely to receive are 24 used F-16 Block 25 fighter aircraft, which, per the 2011 agreement, will cost Indonesia US $700 million. Under this arrangement, Indonesia also planned to have the US upgrade the fighter jets to Block 52, which would include supplying 18 air-to-ground missiles and 36 captive air training missiles.[33]

Examples of improved equipment for the Indonesian Navy, as of the fall of 2013, included the planned receipt in 2014 of light patrol vessels, amphibious tanks and rockets, as well as two South Korean-made Chang Bongo-class submarines. Moreover, a joint project between Indonesia and South Korea through a technology transfer agreement would produce a similar type of submarine. Nevertheless, due to the importance of guarding the Malacca Strait, which is important for trade for Indonesia, the country requires between 18–24 submarines, rather than only three.[34]

The defense markets for military equipment of other countries have faced pressure to expand due to defense priorities. Nevertheless, the defense budgets may become more constrained, due to pressures in other areas of the federal budget, if

28 Minnick, 2013.
29 Minnick, 2014d.
30 Siboro, 2013.
31 Siboro, 2013.
32 Siboro, 2013.
33 Siboro, 2013.
34 Siboro, 2013.

economic growth slows. For example, as of the summer of 2014, the new Indian government did not propose to increase defense spending significantly more than the previous administration. On July 10, the defense budget was proposed to Parliament for FY 2014–2015 at $38.16 billion, which was slightly higher than the proposal of the outgoing government of $37.3 billion. The significant defense programs included the $12 billion Medium Multi-Role Combat Aircraft (MMRCA), which some expected could be as high as $20 billion; $1.5 billion for eight mine countermeasures ships from South Korea; $1.2 billion for six Airbus 330 tankers; $1.1 billion for 22 Boeing Apache helicopters; and $1 billion for 197 light utility helicopters. While the government has increased FDI to 49% (from 26%), foreign investors would not be able to attain a majority 51% stake in companies.[35]

India has been expanding its Tri-Command at the Andaman and Nicobar Islands partially to bolster defenses against China and partially due to its key role in India's littoral warfare strategy. Indeed, India has planned to spend $100 billion to upgrade its forces over the next 7–10 years. The need to modernize the Air Force, Navy, and land forces has partially been due to concerns about internal threats of terrorism and perceived threats from China and Pakistan.[36] By 2027, the Indian Navy plans to expand the number of its warships from 127 to 150, as well as to double the size of its airfleet from 220 aircraft/helicopters to 500. Nevertheless, despite the demand for military equipment due to the evolution of defense priorities, India also has to balance this with the greater need for federal spending to stimulate economic growth.

The need to modernize equipment has resulted in India engaging in defense purchases from overseas countries. India's competition for a $12 billion contract for 126 fighter jets – the largest fighter jet competition since the 1990s – involved US, European, and Russian defense contractors, such as Saab, Dassault, Boeing, Eurofighter, and RAC MiG (Russia).[37] The Dassault Rafale triumphed over the Eurofighter Typhoon on the basis of cost, while, after the flight trials in August 2011, the MiG-35, the Gripen, the F/A-18, and the F-16 were dropped from the competition.[38] Although India selected France's Dassault Aviation for the acquisition of 126 Rafale fighter jets in January 2012, the deadlines for completion of the contract had not taken place, as of June, 2014.[39] The contract has continued to be delayed, both because Hindustan Aeronautics (HAL) has not finalized the cost of Rafale's made in India, as well as because HAL has been reluctant to provide written guarantees on the schedule for deliveries of Rafale's made in India, suggesting that Dassault Aviation should guarantee them.[40] India has also

35 Raghuvanshi, 2014d.
36 Raghuvanshi, 2012.
37 Lerner, 2010; Misquitta, 2009; Blumenthal, 2010; Raghuvanshi, 2014c.
38 Raghuvanshi, 2014c.
39 Agence France-Presse, 2013c.
40 Raghuvanshi, 2014c.

bought other overseas equipment; for example, it purchased six C-130 Js in 2007 for $962 billion, eight Boeing P-8I aircraft for $2.1 billion in 2009, etc.

Defense sales have developed over the years between India and Russia. For example, as of October 2013, the Russian Navy planned to finance the construction of an Akula-class vessel and then lease it to India for over US $1.2 billion, as well as deal with maintenance and overhauls. India and Russia have also considered developing work-sharing agreements. For example, in 2010, Russia's Sukhoi Desin Bureau and India's state-owned Hindustan Aeronautics Ltd. planned to jointly develop the FGFA, but, as of the fall of 2013, a final contract had still not been reached.[41]

Overseas contracts and alliances between developing countries and defense contractors in developed countries could help to expand the defense industrial base in the developing countries. For example, in procuring defense equipment overseas, India is also trying to develop its own defense industrial base. Although it makes aircraft, land vehicles, and small arms, it is trying to develop more in the high-tech equipment area of its domestic defense industrial base and hopes to reduce imports from 70% to 30% over the next decade as its own base develops. It engages in alliances with Western defense contractors to gain knowledge; examples include Larsen & Toubro, which teamed with Raytheon in February, 2010 to upgrade the Indian Army's Russian origin T-72 tanks, and Hindustan Aeronautics which is building the Hawk 132 Advanced Trainer Jet under license from BAE.[42] As of the summer of 2014, India was considering restarting its $10 billion infantry vehicle replacement program through funding the Future Infantry Combat Vehicle (FICV) program, which would involve replacing upgraded Russian combat vehicles with 3,000 vehicles. In 2009, when the program was conceived, Mahindra & Mahindra created a joint venture with BAE Systems. Restarting the program would include domestic firms, such as Ashok Leyland, Punj Lloyd, Force Motors, and Bharat Forge, while overseas firms, such as Rafale, Thales, Nexter, General Dynamics, Rosoboronexport, Doosan Group, and Krauss-Maffei Wegmann would also be included.[43] Another example of the development of the domestic defense industrial base involves the tender which was issued to local shipyards to construct shallow water anti-submarine warfare (ASW) vessels. The $2.25 billion program would replace Abhay-class corvettes, made in Russia and commissioned in 1989 and 1991. The tender was issued in June, 2014 to companies, including the state-owned Goa Shipyard, Garden Reach Shipbuilders and Engineers, Larsen & Toubro, ABG Shipyard, and Pipavav Defense and Offshore Engineering.[44]

A second avenue of expanding defense industrial bases in developing countries is through "offset agreements." The global value of "offset" agreements, in which western defense contractors who have won defense contracts agree to directly

41 Raghuvanshi, 2013a.
42 Misquitta, 2009; Krishna, 2010.
43 Raghuvanshi, 2014b.
44 Raghuvanshi, 2014a.

or indirectly help to develop the industrial base of the purchasing country, are worth between $75 billion and $100 billion and require a certain percentage of the contract value to remain in the purchasing country. These agreements began after the World War II and, since 1999, about 22 countries have developed formal "offset" legislation. Contractors consider "offsets" to be part of the cost for doing business overseas. In India, foreign companies winning import orders over $62 billion must use domestic suppliers for at least 30% of the order.[45]

As of late October 2013, India planned to spend over $2 billion over the subsequent five years in order to expand its UAV fleet, which has resulted in further development of the UAV defense industrial base. Many Indian companies are developing UAVs; examples include Idea Forge, Dynamatrics, Hi-tech Robotics, Ufcon, Omnipresent Technologies, Datapattern, Tata Advance Systems and state-owned Bharat Electronics. Moreover, the government's Defence Research and Development Organization has focused on research regarding the development of various types of UAVs. As of the fall of 2013, the Army and Air Force had an immediate requirement for more than 700 mini UAVs. The Army also needed 1,600 mini UAVs by 2017.[46]

Despite slowing economic growth, some countries are increasing their defense budgets. For example, Sri Lanka grew at 8% in the prior two years, but fell to 6.4% last year. Nevertheless, as of the fall of 2013, Sri Lanka planned to increase defense spending to a record 253 billion rupees ($1.95 billion), which (with the police department) accounts for almost 12% of the government's total estimated spending of 2.54 trillion rupees in 2014.[47]

The trends in suppliers of equipment and types of equipment to Asian countries are illustrated in Figure 8.12, which shows the arms deliveries, as represented in millions of dollars, to Asian countries by supplier countries between 2004–2007 and 2008–2011. Russia was the leader in supplying weapons to the Asian countries during both periods, followed by the US. During 2004–2007, China was the third most significant supplier ($2,500), followed by France ($2,000), Germany ($1,700), and the UK ($1,100). During 2008–11, China continued to be the third most significant supplier ($4,000), followed by Germany ($2,500), the UK ($1,700), France ($1,200), and Italy, which had expanded from $200 in deliveries during 2004–2007 to $1,000 in deliveries during 2008–2011. While Russia, the US, and France shrunk their deliveries to the Asian supplier countries between the two time periods, the UK, China, Germany, and Italy increased their deliveries, as their domestic defense industrial bases expanded. Of the major countries, China experienced the most significant increase in deliveries to Asian countries. The total deliveries to Asian countries from supplier countries experienced modest growth between 2004–2007 ($36,377) and 2008–2011 ($40,177).

45 Pfeiffer, 2010.

46 Vivek Raghuvanshi, 2013b.

47 Agence France-Presse, 2013d.

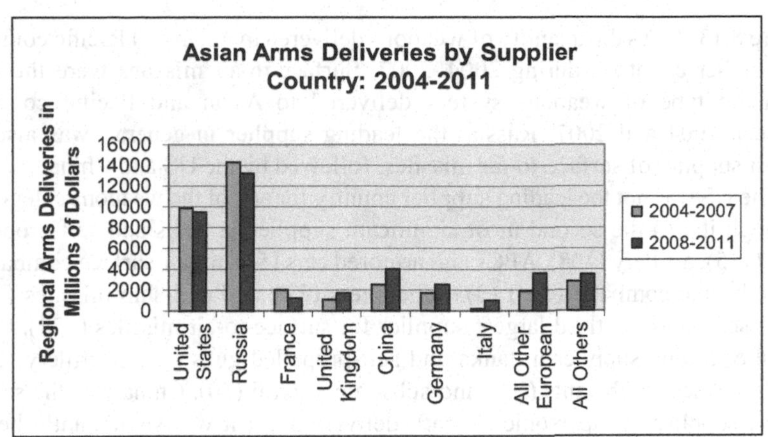

Figure 8.12 Asian Arms Deliveries by Supplier Country: 2004–2011

Source: Grimmett, Richard E. and Kerr, Paul K., *Congressional Research Service Report (R42678): Conventional Arms Transfers to Developing Nations: 2004–2011*, Washington DC, August 24, 2012.

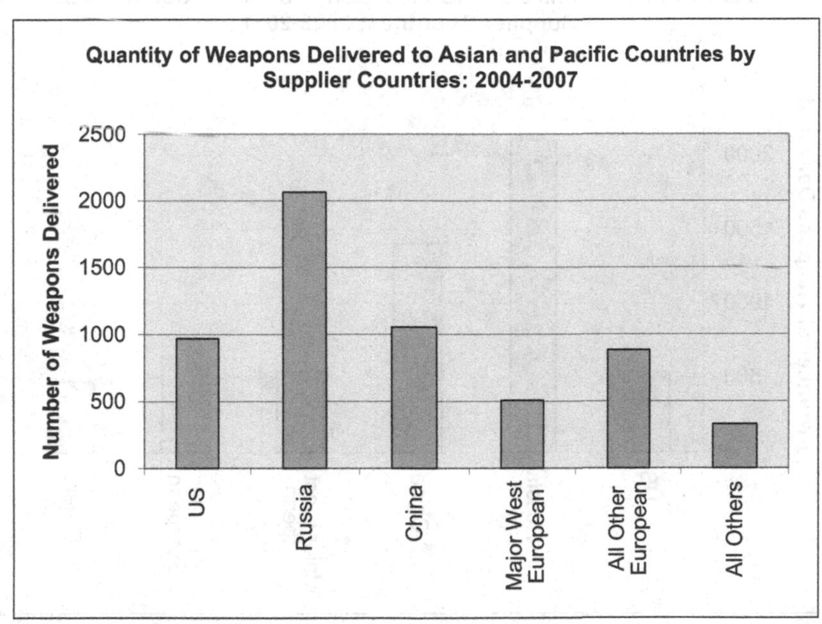

Figure 8.13 Quantity of Weapons Delivered to Asian and Pacific Countries by Supplier Countries: 2004–2007

Source: Grimmett, Richard E. and Kerr, Paul K., *Congressional Research Service Report (R42678): Conventional Arms Transfers to Developing Nations: 2004–2011*, Washington DC, August 24, 2012.

Figure 8.13 shows the quantity of weapons delivered to Asian and Pacific countries by supplier countries during 2004–2007. Surface-to-air missiles were the most dominant type of weapons system delivered to Asian and Pacific countries between 2004 and 2007. Russia, the leading supplier in general, was also the largest supplier of surface-to-air missiles, followed by the US and China.

The US was not the leading supplier country for any of the weapons categories, although it was the second most significant supplier in tanks and self-propelled guns (115), artillery (108), APCs and armored cars (54), minor surface combatants (6), subsonic combat aircraft (2), helicopters (22), and anti-ship missiles (175), and it served as the third largest supplier for surface-to-air missiles (474). China was the leading supplier in tanks and self-propelled guns (160), artillery (210), minor surface combatants (22), and subsonic aircraft (10). China was the second largest supplier in supersonic aircraft, delivering 10; it was significantly behind Russia at 110. It was also the second largest supplier in surface-to-air missiles (530). China was the third largest supplier in APCs and armored cars (50), closely behind the US at 54 and substantively behind Russia, the leading supplier, at 220.

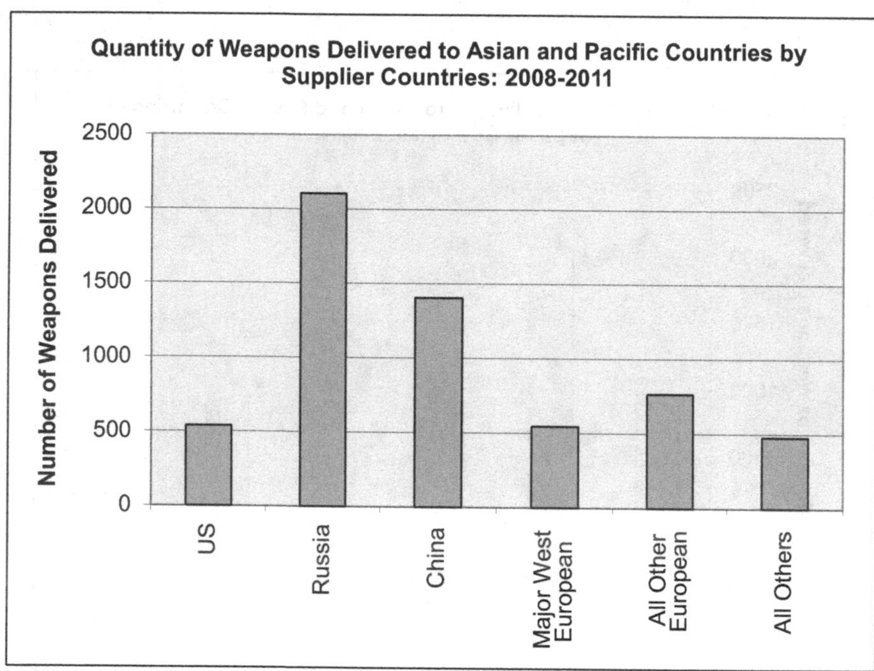

Figure 8.14 Quantity of Weapons Delivered to Asian and Pacific Countries by Supplier Countries: 2008–2011

Source: Grimmett, Richard E. and Kerr, Paul K., *Congressional Research Service Report (R42678): Conventional Arms Transfers to Developing Nations: 2004–2011*, Washington DC, August 24, 2012.

Russia was the leading supplier in APCs and armored cars at 220 (significantly ahead of the US at 54 and China at 50), major surface combatants at 3, supersonic combat aircraft at 110 (distantly followed by China at 10), submarines at 8, helicopters at 90 (followed by the US at 22), and surface-to-air missiles at 1180 (significantly higher than China at 530 or the US at 474). Russia was the third largest supplier in tanks and self-propelled guns (40), which was much less than China at 160 or the US at 115; artillery at 20, which was much less than China at 210 or the US at 108; minor surface combatants at three (significantly behind the leading supplier, China, at 22); and anti-ship missiles at 360 (followed by the US at 175 and China at 40).

Collectively, the major Western European nations, in the APC and armored cars category, were significant suppliers and delivered 120, which was less than Russia (220), but much more than the US and China (54 and 50, respectively). The countries, collectively, were also significant suppliers of surface-to-air missiles at 240, following Russia (1180), China (530), and the US (474).

The quantity of weapons delivered to Asian countries by supplier nations during 2008–2011 are shown in Figure 8.14. As in the prior period, surface-to-air missiles were the largest equipment category, with Russia being the leading supplier to Asian countries. Also, as in the prior period, Russia was the largest supplier of overall weapons systems equipment to Asian countries. China grew rapidly in this period relative to the prior one, and continued in second place, significantly outdistancing the US in third place.

During 2004–2007, the US was not the leading supplier country for any of the weapons categories. During 2008–2011, the US was the leading supplier in anti-ship missiles (176), followed by Russia (110) and China (60). Although the US was the second largest supplier during 2004–2007 at 175 missiles, the largest supplier was Russia with 360 missiles during that period. Consequently, while the US remained stable in supplying anti-ship missiles to Asian countries, Russia reduced its role as a supplier. The US was the leader in supplying major surface combatants (5), followed by Russia (4) and China (3). During 2004–2007, the US supplied zero major surface combatants, and Russia supplied three. The US was a distant third in supplying APCs and armored cars (25), while Russia supplied 250 and China supplied 100. Relative to the 2004–2007 period, the US had shrunk from supplying 54 APCs and armored cars to supplying 25, although China had grown from supplying 50 APCs and armored cars to supplying 100, and Russia experienced modest growth in supplying 220 during 2004–2007 and supplying 250 during 2008–2011.

During 2008–2011, Russia was the leading supplier of tanks and self-propelled guns (360), followed by China at 260. Relative to 2004–2007, it had grown significantly since it supplied 40 during that period. Similarly, China also grew from supplying 160 tanks and self-propelled guns during 2004–2007 to supplying 260 during 2008–2011. The US, however, contracted from supplying 115 during that period to supplying zero during 2008–2011. Russia was also the leading supplier of APCs and armored cars, supplying 250 during 2008–2011, followed by China at 100 and the US at 25. Russia's supply of APCs and armored cars was similar to the size of the supply of weapons systems during the earlier period. Russia was the leading supplier of minor surface combatants, delivering six during 2008–2011, followed by China, which supplied two. This was also quite similar to the size of the supply of weapons systems from the respective supplier nations during the earlier period. Russia was the leading supplier of supersonic combat aircraft, supplying 140, with the US only supplying 19 and China supplying only 10.

Russia was the leading supplier of helicopters in 2008–2011, providing 110 of them, and distantly followed by China at 10 and the US at two. While Russia expanded from 90 helicopters during 2004–2007 to 110 in the 2008–2011 period, the US declined from 22 during 2004–2007 to two in the subsequent period. Russia was also, by far, the leading supplier of surface-to-air missiles; it supplied 1,080 missiles, followed by China at 760 and the US at 297. In the surface-to-air missile category, although Russia was leading at 1,080 during 2008–2011, it had contracted from the 1,180 missiles, which it supplied during 2004–2007. The US had also contracted from 474 missiles in 2004–7 to 297 in 2008–2011. China, however, had expanded from supplying 530 missiles to Asian countries during 2004–2007 to supplying 760 missiles during 2008–2011.

Collectively, the major Western European nations were significant suppliers in the tanks and self-propelled guns category, where they supplied 100, making them the third largest supplier (when compared with individual countries), following Russia at 360 and China at 260. During the 2004–2007 period, these nations collectively supplied none of these to Asian countries. The major Western European nations collectively, when compared to individual nations, were the leading supplier of minor surface combatants at 10, followed by Russia at six and China at two minor surface combatants. While the major Western European countries, when compared to the prior period, exhibited almost no growth in terms of the number of deliveries of minor surface combatants to Asian countries, China shrunk significantly from 22 in 2004–2007 down to two in 2008–2011. Western European countries continued to be the leading suppliers, collectively, of subsonic combat aircraft, supplying 50 to Asian countries during both 2004–2007 and 2008–2011. China, the second largest supplier of subsonic combat aircraft when compared collectively to west European countries, dropped from 20 deliveries during 2004–2007 to 10 deliveries during 2008–2011. In the surface-to-air missiles category, the Western European nations collectively delivered 290 missiles, which is less than Russia (1080), China (760), and the US (297). During the 2004–2007 period, the Western European nations collectively delivered 240 surface-to-air missiles (indicating some growth between 2004–2007 and 2008–2011), while Russia delivered 1,180 (indicating some shrinkage by the later period since it supplied 1,080 between 2008 and 2011), the US delivered 474 (indicating significant shrinkage over time since 297 were delivered during the 2008–2011 period), and China delivered 240.

Middle East/Near East

Military expenditures in the Middle East grew by 56% between 2004 and 2013, and grew 4% between 2012 and 2013.[48] Due to concerns about Iran, countries such as Kuwait, Saudi Arabia, and the UAE are estimated to spend $175 billion over the next five years on defense.[49] Saudi Arabia's defense spending of $46 billion in 2011 was more than the combined defense budgets of neighbors Bahrain, Egypt, Israel, Iraq, Jordan, Kuwait, Oman, Syria, and Lebanon.[50] Indeed, Saudi Arabia's defense spending increased by 14% between 2012 and 2013. Nevertheless, Iraq and Bahrain surpassed it in growth at 27% and 26%, respectively, although they had smaller budgets.[51]

The UAE, which spends 6% of its GDP on defense equipment, is second only to Saudi Arabia in defense spending in the Gulf region. The UAE has been expanding its defense due to concerns about Iran's nuclear and ballistic missile capabilities, as well as due to the desire to increase employment, to build technology skills, and to diversify its oil-based economy.

The Middle East has been a successful market for US defense contractors. For example, during the 18 months ending in November 2012, US aerospace companies earned almost $1 billion from Oman, which was primarily spent on purchasing a dozen new F-16 jets and on modernizing an older F-16 squadron. Part of Oman's rising defense spending has been due to available funds for defense from oil revenues, as well as the need to improve military capabilities to deter Iran. Another example is Saudi Arabia's agreement with Boeing to upgrade 68 F-15s to the F-15SA configuration for $4 billion in early November, 2012. It also plans to purchase 84 new F-15SA fighters to be delivered in 2015.[52] In February, 2014, the Royal Saudi Air Force contracted for 72 Eurofighter Typhoon jets. It is unclear, however, whether the recent growth in defense spending in the Middle East will continue at such a significant pace in that if political tensions continue to increase in certain countries, more significant portions of the national budgets may need to be used for non-defense purposes to lessen political instability.

The Middle East is the second largest market for munitions, following the US. The munitions market includes guided rockets, guided missiles, guided projectiles, hand-off missiles, and anti-armor weapons. The entire market is expected to grow to $5.3 billion by 2018, while the Middle East, is estimated to grow to $712.1 million in 2018, which is an increase from $350.9 million in 2013. Countries, such as Saudi Arabia, are likely to be significant purchasers. Indeed, Saudi Arabia started upgrading its Tornado fighter jets in 2011, with the second phase of the program extending until 2020. Significant acquisitions of Pavewa,

48 Perlo-Freeman and Solmirano, 2014.
49 Cameron, 2013.
50 Mehta and Fryer-Biggs, 2012.
51 Perlo-Freeman and Solmirano, 2014.
52 Mehta and Fryer-Biggs, 2012.

Brimstone, and Storm Shadow launched weapons are forecast to occur between 2014 and 2018. Finally, Saudi Arabia put in an order of $6.7 million in 2013 for JDAMs – Joint Direct Attack Munitions.[53]

It has been estimated that the global armored vehicles market will have grown to $28.62 billion by 2019 with the Middle East being a significant buyer. In the Middle East, the role of various countries, such as Qatar, Saudi Arabia, and the UAE, on peace initiative missions, in areas such as Afghanistan, is likely to lead to significant purchases of armored vehicles. The impact of interstate conflict and the importance of securing borders, has led (and will likely lead) to greater demand for armored vehicles. Indeed, Oshkosh has received orders for 10,000 MRAP All-Terrain Vehicles from the Middle East, including Saudi Arabia and the UAE, and the US. Indigenous armored vehicle industries are being developed in countries such as Algeria, Jordan and the UAE, which have exported their vehicles to Yemen, Somalia, and Libya.[54]

Several of the Middle Eastern countries are expanding their defense industrial bases. Israel is an example. Indeed, one of Israel's defense contractors, Israel Aerospace Industries Ltd, is developing a new warship, called the Multirole Super Dvora, which is being designed in the IAI Ramta plant in Beersheba. This plant has built over 120 ships, of which 90 of them have been purchased by a variety of countries, including: India, Chile, and Sri Lanka, as well as other countries.[55] Turkey is another example; as of the end of October, 2013, Turkey announced that it would contract with the Turkish Aerospace Industries (TAI), which would provide the first batch of indigenous UAVs – 10 drones – beginning in 2016.[56]

The development of the domestic defense industrial base is important for many of these countries. Indeed, the UAE requires that at least 50% of the contract's value has to be reinvested in the local economy by the recipient of the contract. Many "offsets," however, have not provided the purchasing country with what they had wanted: when Saudi Arabia, in 1985, purchased the $3.8 billion US Peace Shield program, the "offset" was targeted to create 75,000 jobs, yet, as of 2009, the four resulting joint venture companies only had 3,540 employees.

The comparison of Near Eastern arms deliveries in millions of dollars by supplier country between the 2004–2007 period and the 2008–2011 period is illustrated in Figure 8.15. The US was the leading supplier during 2004–2007 and 2008–2011 by a significant margin. The UK and France were the second and third leading suppliers, respectively, during 2004–2007, however they reduced their deliveries of weapons systems to the Near East during 2008–2011. Russia grew significantly in 2008–2011 relative to 2004–2007 and replaced the UK in the 2008–2011 period as the second leading supplier. While the US, Russia, China, and Italy grew during the 2008–2011 period

53 Mustafa, 2014b.

54 Mustafa, 2014a.

55 "Israel Aerospace Targets the Warship Market," 2013.

56 Burak, 2013.

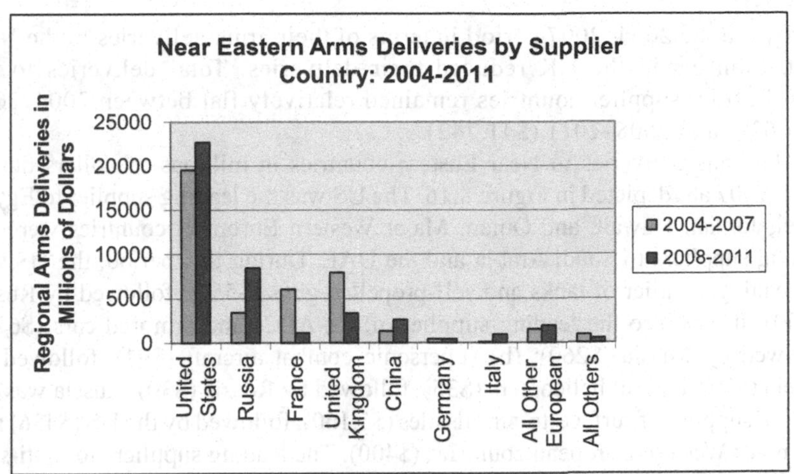

Figure 8.15 Near Eastern Arms Deliveries by Supplier Country: 2004–2011

Source: Grimmett, Richard E. and Kerr, Paul K., *Congressional Research Service Report (R42678): Conventional Arms Transfers to Developing Nations: 2004–2011*, Washington DC, August 24, 2012.

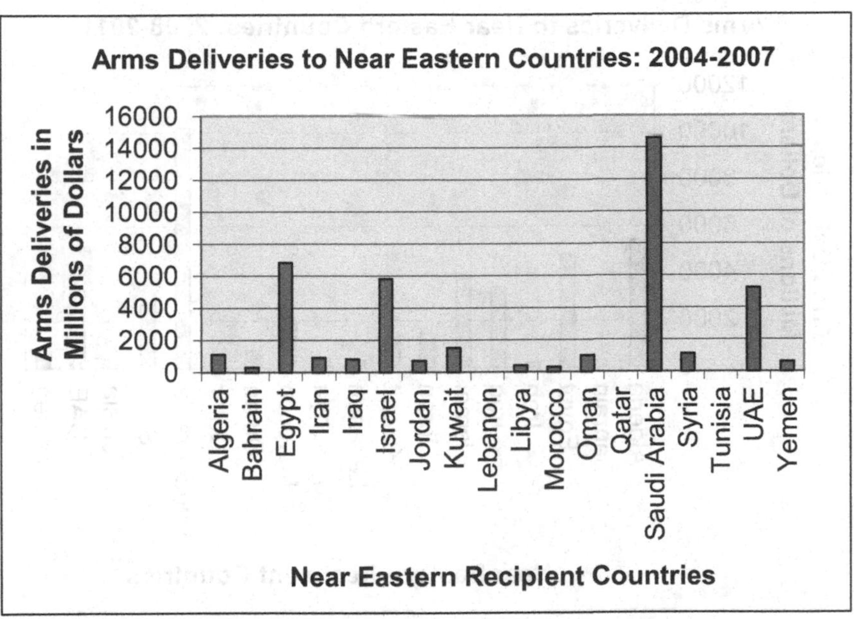

Figure 8.16 Arms Deliveries to Near Eastern Countries: 2004–2007

Source: Grimmett, Richard E. and Kerr, Paul K., *Congressional Research Service Report (R42678): Conventional Arms Transfers to Developing Nations: 2004–2011*, Washington DC, August 24, 2012.

relative to the 2004–2007 period in terms of their arms deliveries to the Near East, France and the UK reduced their deliveries. Total deliveries to the Near East by supplier countries remained relatively flat between 2004–2007 ($41,037) and 2008–2011 ($41,742).

The arms deliveries to Near Eastern countries in millions of dollars during 2004–2007 are depicted in Figure 8.16. The US was the leading supplier to Egypt, Israel, Jordan, Kuwait, and Oman. Major Western European countries were the leading suppliers to Saudi Arabia and the UAE. During this period, the US was the leading supplier of tanks and self-propelled guns ($557), followed by Russia ($260). It was also the leading supplier of: (a) APCs and armored cars ($672), followed by Russia ($260); (b) supersonic combat aircraft ($94), followed by Russia ($30); and (c) helicopters ($35), followed by Russia ($30). Russia was the leading supplier of surface-to-air missiles ($5,160), followed by the US ($436) and the major Western European countries ($400). The leading suppliers for anti-ship missiles were China ($80) and the US ($77), although major Western European countries collectively supplied $90.

The arms deliveries to Near Eastern countries during 2008–2011 are depicted in Figure 8.17. The US continued to be the leading supplier of arms deliveries

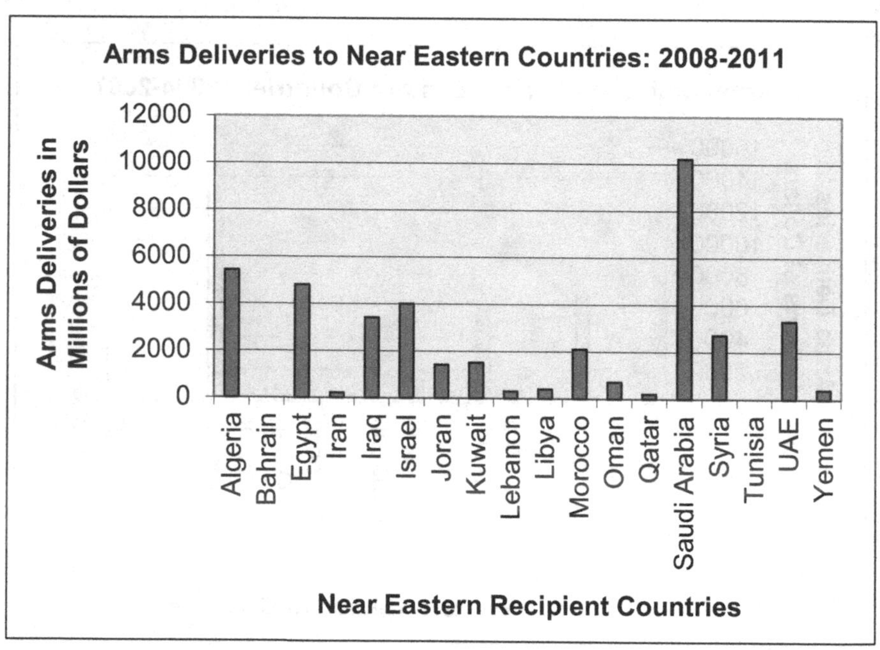

Figure 8.17 Arms Deliveries to Near Eastern Countries: 2008–2011

Source: Grimmett, Richard E. and Kerr, Paul K., *Congressional Research Service Report (R42678): Conventional Arms Transfers to Developing Nations: 2004–2011*, Washington DC, August 24, 2012.

to Near Eastern countries, such as Egypt, Iraq, Israel, Jordan, Kuwait, Morocco, and the UAE in a variety of weapons categories during 2008–2011. Russia was the leading supplier to Algeria and Syria. The US and the major Western European nations were the major suppliers to Saudi Arabia during 2004–2007 and 2008–2011. The US became a more significant supplier in 2008–2011 for the UAE than it was in 2004–2007, while major Western European countries became smaller suppliers.

The US was the dominant supplier in: (a) tanks and self-propelled guns (348), followed by China (60), and Russia (50); (b) APCs and armored cars (170), closely followed by China (160) and Russia (130); (c) supersonic combat aircraft (35), followed by Russia (30); and (d) helicopters (36), followed by Russia (30) and the major Western European nations collectively (50). Nevertheless, relative to 2004–2007, the quantities of equipment delivered in these categories by the US and other countries were lower.

China was the leading supplier of arms deliveries to Near Eastern countries in artillery (230), followed by the US (149), which shows significant growth since China delivered no artillery and the US delivered only 31 during 2004–2007.

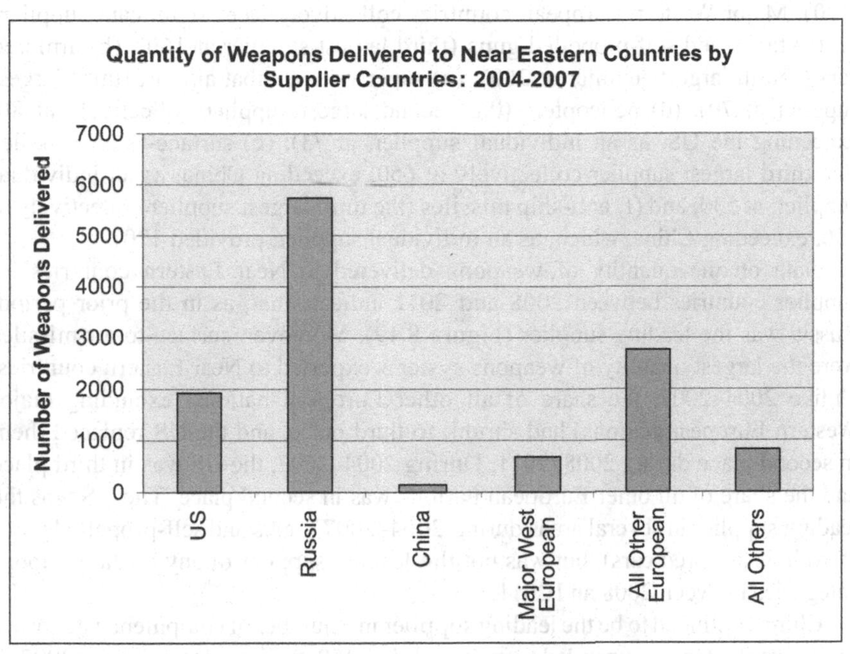

Figure 8.18 Quantity of Weapons Delivered to Near Eastern Countries by Supplier Countries: 2004–2007

Source: Grimmett, Richard E. and Kerr, Paul K., *Congressional Research Service Report (R42678): Conventional Arms Transfers to Developing Nations: 2004–2011*, Washington DC, August 24, 2012.

Russia was the leading supplier in several areas: (a) surface-to-air missiles (3,480), followed by the US (647); (b) surface-to-surface missiles (50): and (c) anti-ship missiles (110), followed by Western European countries collectively (50).

Figure 8.18 shows the quantity of weapons delivered to Near Eastern countries by supplier countries between 2004 and 2007. Russia was the most significant supplier during this period in terms of the quantity of equipment supplied, much of which was driven by its deliveries of surface-to-air missiles. Surface-to-air missiles were the largest quantity of weapons systems supplied overall to Near Eastern countries, while APCs and armored cars were the second largest quantity of weapons (all other European countries outside major Western European countries were the leading suppliers).

The US was the leading supplier of: (a) tanks and self-propelled guns (672), followed by Russia (300) and China (160); and (b) armored cars (726), followed by Russia (480) and China (460). China was the leading supplier of artillery (450), followed by the US (240). Russia was the leading supplier of: (a) supersonic combat aircraft (180), followed by the US (104); (b) helicopters (200), followed by the US (73); (c) surface-to-air missiles (6,340), followed by the US (910) and China (530); and (d) anti-ship missiles (360), followed by the US (262), and China (120). Major Western European countries collectively were significant suppliers in: (a) tanks and self-propelled guns (third largest supplier at 160); (b) armored cars (fourth largest supplier at 260); (c) supersonic combat aircraft (third largest supplier at 70); (d) helicopters (the second largest supplier collectively at 80, exceeding the US, as an individual supplier, at 73); (e) surface-to-air missiles (the third largest supplier collectively at 650, exceeding China, as an individual supplier, at 530; and (f) anti-ship missiles (the third largest supplier collectively at 150, exceeding China, which, as an individual supplier, provided 120).

Data on the quantity of weapons delivered to Near Eastern countries by supplier countries between 2008 and 2011 indicate that, as in the prior period, Russia was the leading supplier (Figure 8.19). Moreover, surface-to-air missiles were the largest quantity of weapons systems exported to Near Eastern countries. Unlike 2004–2007, the share of all other European nations (excluding major Western European nations) had shrunk to third place, and the US replaced them in second place during 2008–2011. During 2004–2007, the US was in third place and the share of all other European nations was in second place. The US was the leading supplier in several areas during 2004–2007 (tanks and self-propelled guns, as well as armored cars), but was not the leading supplier of any of the weapons categories between 2008 and 2011.

China continued to be the leading supplier in a number of equipment categories. For example, China expanded from supplying 450 artillery during 2004–2007 to supplying 770 during 2008–2011. It was followed by the US (150) and Russia (90). It also replaced the US as the leading supplier of APCs and armored cars, supplying 590 (higher than 460 during 2004–2007).

China continued to lead the suppliers for minor surface combatants, supplying 108 during 2004–2007, relative to the 56 during 2008–2011.

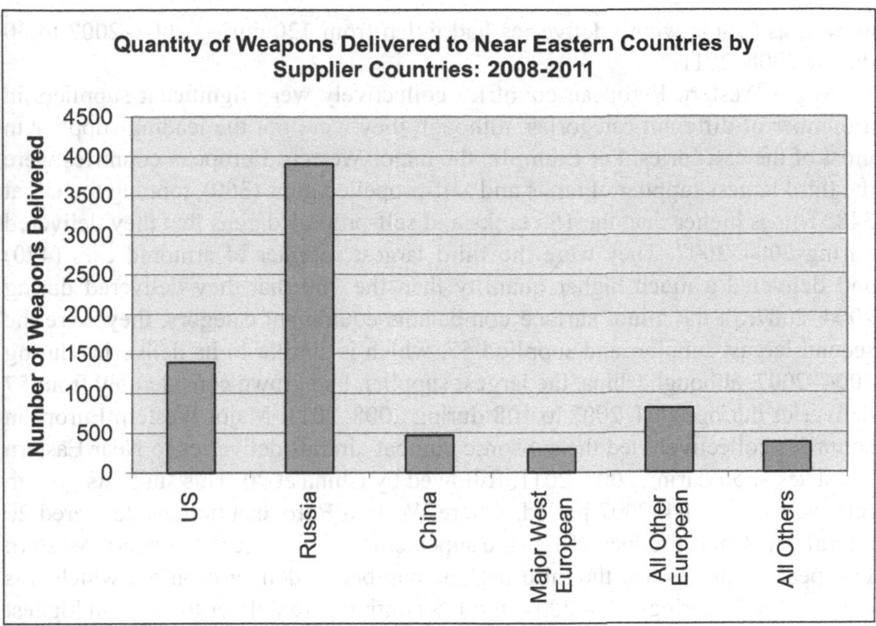

Figure 8.19 Quantity of Weapons Delivered to Near Eastern Countries by Supplier Countries: 2008–2011

Source: Grimmett, Richard E. and Kerr, Paul K., *Congressional Research Service Report (R42678): Conventional Arms Transfers to Developing Nations: 2004–2011*, Washington DC, August 24, 2012.

Russia also continued to be a leading supplier in a variety of equipment categories. For example, it continued to be the leading supplier of supersonic combat aircraft (180), followed by the US (which supplied 53 during 2008–2011 rather than 104 during 2004–2007) and China (which supplied 30 during 2008–2011 rather than the 20 in 2004–2007). It also continued to be the leading supplier of helicopters (270 during 2008–2011 compared to 200 during 2004–2007), followed by the US (which had declined from 73 in 2004–2007 to 53 in 2008–2011). In the surface-to-air missiles category, Russia supplied 7,750 missiles during 2008–2011, which was an increase from 6,340 during 2004–2007. It was followed by the US, which supplied 944 surface-to-air missiles during 2008–2011, which is similar to the 910 missiles that it supplied during 2008–2011. China also followed Russia in the surface-to-air missiles category; China delivered 530 missiles during 2004–2007 and 780 missiles during 2008–2011. Finally, Russia was the leading supplier in the quantity of anti-ship missiles delivered to Near Eastern countries, although the quantity that it delivered fell from 360 missiles in 2004–2007 to 220 during 2008–2011. Russia was followed by the US, whose deliveries of anti-ship missiles had also fallen from 262 during 2004–2007 to 176 during 2008–2011, as well as China, whose deliveries had fallen from 120 during 2004–2007 to 60 during 2008–2011.

Major Western European countries collectively were significant suppliers in a number of different categories, although they were not the leading supplier in most of the categories. For example, the major Western European countries were the third largest supplier of tanks and self-propelled guns (360), topping the US at 348. This is higher than the 160 tanks and self-propelled guns that they delivered during 2004–2007. They were the third largest supplier of armored cars (470) and delivered a much higher quantity than the 260 that they delivered during 2004–2007. In the minor surface combatants equipment category, they were the second largest supplier and supplied 57, which is similar to its deliveries during 2004–2007, although China, the largest supplier, had grown considerably from 57 deliveries during 2004–2007 to 108 during 2008–2011. Major Western European countries collectively led the subsonic combat aircraft deliveries to Near Eastern countries at 50 during 2008–2011, followed by China at 20. This suggests growth relative to the 2004–2007 period, where Western European nations delivered 20 aircraft and China produced 10. In the supersonic combat aircraft category, Western European countries had the third highest number of deliveries at 50, which was down from 70 during 2004–2007; the US continued to deliver the second highest number of deliveries at 53 during 2008–2011, down from 104 in 2004–2007. Major Western European countries were the second largest supplier of helicopters (110), although Russia was the leading supplier, delivering 270 helicopters during 2008–2011. The major Western European countries showed some growth since 2004–2007 in the helicopter category since they delivered 80 during that period, while Russia delivered 200. In the surface-to-air missiles category, the Western European nations were collectively the fourth largest supplier at 290, following China at 780. This was a significant decline from the prior 2004–2007 period when the Western European countries were the third largest suppliers at 650, with China supplying 530. Finally, in the anti-ship missiles category, the major West European countries were the third largest supplier collectively, tying with China. Nevertheless, the deliveries of anti-ship missiles to Near Eastern countries have fallen overall between 2004–2007 and 2008–2011. Although Russia continued to be the leader in the anti-ship missile category, it had fallen from delivering 360 missiles during 2004–2007 to 220 during 2008–2011. Similarly, the US fell from delivering 262 anti-ship missiles during 2004–2007 to delivering 176 anti-ship missiles during 2008–2011, while the Western European countries fell from delivering 150 missiles to delivering 60. China's deliveries of the anti-ship missiles also fell from 120 during the 2004–2007 period to 60 during the 2008–2011 period.

Latin America

Military spending in Latin America expanded by 61% between 2004 and 2013, and grew by 2.2% between 2012 and 2013. Military spending in South America grew by 58% between 2004 and 2013, but only grew 1.6% between 2012 and

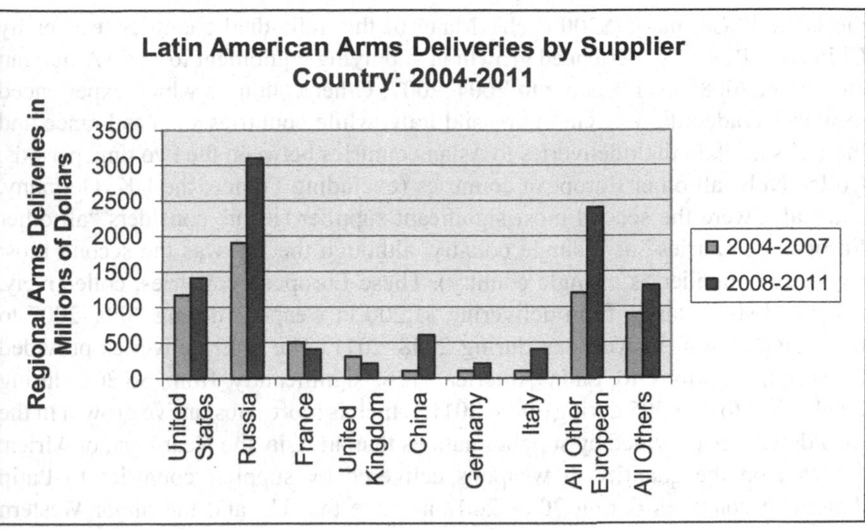

Figure 8.20 Latin American Arms Deliveries by Supplier Country: 2004–2011

Source: Grimmett, Richard E. and Kerr, Paul K., *Congressional Research Service Report (R42678): Conventional Arms Transfers to Developing Nations: 2004–2011*, Washington DC, August 24, 2012.

2013. This decline in South American military spending was largely due to the reduction in growth in military spending from Brazil, which traditionally had the most substantial military spending in the area and which declined by 3.9% between 2012 and 2013. Central American countries, such as Honduras, Nicaragua, and Guatemala, expanded their military spending significantly between 2012 and 2013.[57]

What have been the trends in arms deliveries for Latin America? Figure 8.20 shows the arms deliveries by supplier countries in millions of dollars to Latin America and compares the 2004–2007 period with the 2008–2011 period. Russia was the leading supplier to Latin America during both periods, exhibiting significant growth by delivering 1900 weapons during 2004–2007 relative to delivering $3,100 (in millions of dollars) in weapons during 2008–2011. The US was the second most significant supplier, but it showed much less growth than Russia, delivering $1,165 in weapons during 2004–2007 relative to $1,405 in weapons during 2008–2011. The other countries delivered fewer weapons. During 2004–2007, France was the third largest supplier ($500), followed by the UK ($300). China, Germany, and Italy each supplied $100 of weapons. During 2008–2011, China had become the third largest supplier ($600), followed by Italy and France ($400 each) and by

57 Perlo-Freeman and Solmirano, 2014.

the UK and Germany ($200 each). Many of the individual countries, especially China and Russia, experienced growth in supplying equipment to Latin American nations in 2008–2011 relative to 2004–2007. Other countries which experienced growth included the US, Germany, and Italy, while countries such as France and the UK shrunk in their deliveries to Asian countries between the two time periods. Collectively, all other European countries (excluding France, the UK, Germany, and Italy) were the second most significant supplier (if one considers "all other European countries" as a single country, although the US was the second most significant supplier as a single country). These European countries, collectively, expanded significantly from delivering $1,200 in weapons during 2004–2007 to delivering $2,400 in weapons during 2008–2011. The total deliveries provided by supplier nations to Latin America grew significantly from $6,265 during 2004–2007 to $10,005 during 2008–2011, which is more substantive growth in the total deliveries provided by supplier nations than in Asia, the Near East, or Africa.

Data on the quantity of weapons delivered by supplier countries to Latin American countries during 2004–2007 indicate that US and the major Western European countries (collectively) were the leaders in delivering weapons to

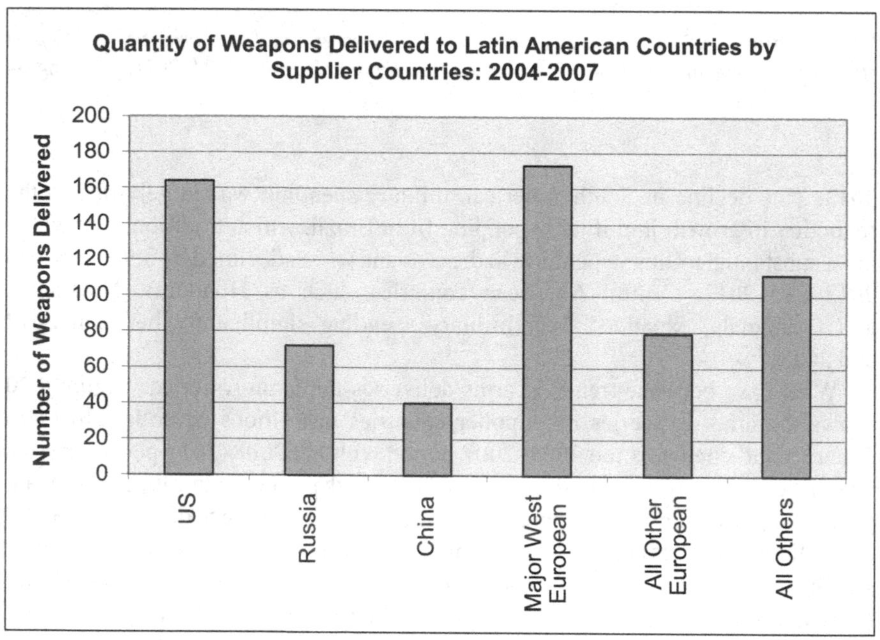

Figure 8.21 Quantity of Weapons Delivered to Latin American Countries by Supplier Countries: 2004–2007

Source: Grimmett, Richard E. and Kerr, Paul K., *Congressional Research Service Report (R42678): Conventional Arms Transfers to Developing Nations: 2004–2011*, Washington DC, August 24, 2012.

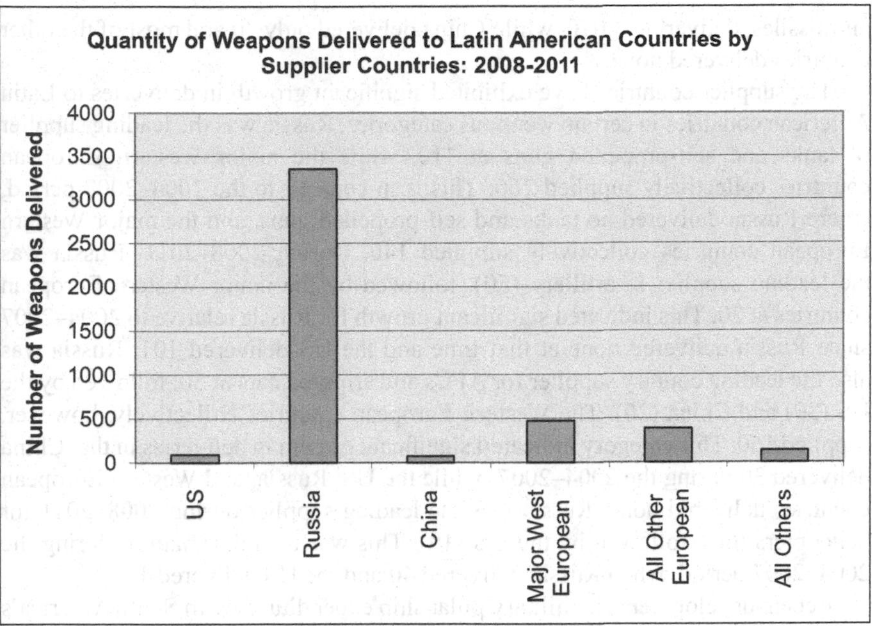

Figure 8.22 Quantity of Weapons Delivered to Latin American Countries by Supplier Countries: 2008–2011

Source: Grimmett, Richard E. and Kerr, Paul K., *Congressional Research Service Report (R42678): Conventional Arms Transfers to Developing Nations: 2004–2011*, Washington DC, August 24, 2012.

Latin America (Figure 8.21). Artillery (the US leading) and tanks (major Western European countries leading) were the most abundant systems delivered to Latin America. The US was also the primary supplier of minor surface combatants at nine, followed collectively by major Western European countries at four and Russia at two. The US was the primary supplier of anti-ship missiles at 10, with few other countries providing missiles. Russia was the leading supplier of supersonic combat aircraft at 20, followed by the US and the Western European countries supplying 10 each. Russia was the leading supplier of helicopters at 40, followed by the US at 16 and, collectively, major Western European countries at 10. China was one of the few suppliers (and therefore the leading supplier) of APCs and armored cars at 30.

Data on the quantity of weapons delivered by supplier countries to Latin American countries during 2008–2011 indicate that Russia had become the major supplier to Latin American countries (Figure 8.22). This was largely due to the demand for surface-to-air missiles during this period. Surface-to-air missiles were not in significant demand through imports by Latin American countries in the earlier period. During 2008–2011, Russia was the leading supplier in surface-to-air missiles, delivering 3,070, while China delivered only 20 and most of the other countries delivered none.

The supplier countries have exhibited significant growth in deliveries to Latin American countries in certain weapons categories. Russia was the leading supplier of tanks and self-propelled guns at 110, while the major Western European countries collectively supplied 260. This is in contrast to the 2004–2007 period, where Russia delivered no tanks and self-propelled guns, and the major Western European countries collectively supplied 140. During 2008–2011, Russia was the leading supplier in artillery (50), followed by the major Western European countries at 20. This indicated significant growth for Russia relative to 2004–2007 since Russia delivered none at that time and the US delivered 101. Russia was also the leading country supplier for APCs and armored cars at 50, followed by the US (39) and China (20). The Western European countries collectively, however, supplied 160. This category indicated significant growth in deliveries in that China delivered 30 during the 2004–2007, while the US, Russia, and Western European countries delivered none. Russia was the leading supplier during 2008–2011 for helicopters (60), followed by the US (19). This was a similar pattern during the 2004–2007 period, when Russia delivered 40 and the US delivered 16.

Recent developments in military polar ship expenditures with South America's growing interest in Antarctica may lead to additional purchases from overseas countries. A number of South American countries are expanding their assets in Antarctica, purchasing key types of ships for the area, and opening bases. As of August, 2014, Columbia, which was developing a permanent base and which held its first Antarctic expedition in 2013, also planned to procure a polar ship. Argentina has six bases in Antarctica, in addition to owning an icebreaker built in Finland in 1978 which has been in an extended repair program since 2009. It is currently developing plans for a second icebreaker. The Chilean Navy has a 6,500-ton vessel, and is involved in building an 8,000-ton icebreaker, in addition to having four bases in Antarctica. Brazil began expanding in Antarctica in 1982 when it bought a polar ship, which was subsequently joined by two additional ships. The Peruvian Navy plans to replace its existing icebreaker, built in 1980, with a new icebreaker ship. The Uruguayan Navy has a 1,700-ton logistic ship, made in 1976 by Poland and which was bought in the 1990s from East Germany. Ecuador, which has one base in Antarctica, is considering purchasing a polar ship.[58]

Africa

Military spending in African countries, which collectively grew 81% between 2004 and 2013, expanded by 8.3% between 2012 and 2013. Ghana increased its military spending by 129% between 2012 and 2013, although its military spending was only 0.6% of GDP. Algeria's military spending reached $10.4 billion in 2013, which was the first time that a nation in Africa exceeded $10 billion in military

58 Higuera, 2014.

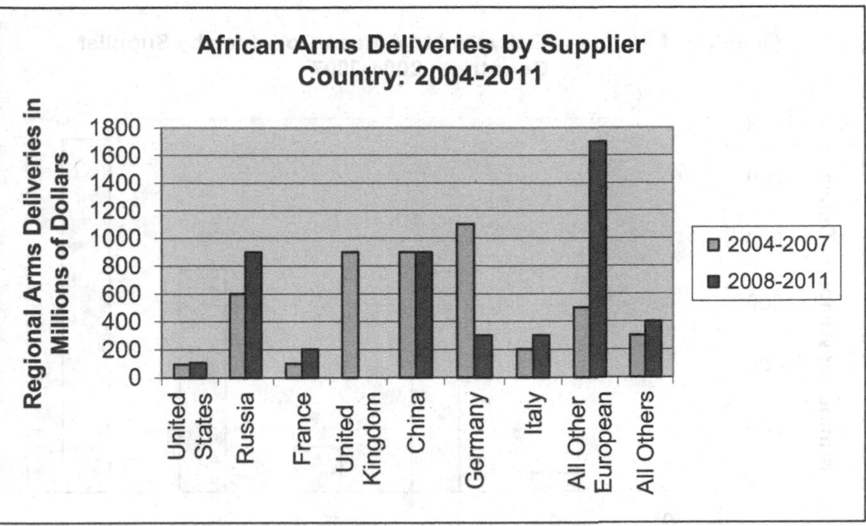

Figure 8.23 African Arms Deliveries by Supplier Country: 2004–2011

Source: Grimmett, Richard E. and Kerr, Paul K., *Congressional Research Service Report (R42678): Conventional Arms Transfers to Developing Nations: 2004–2011*, Washington DC, August 24, 2012.

spending. Angola similarly grew 36% between 2012 and 2013, with its military spending exceeding that of South Africa. Moreover, Angola and Algeria had the highest share of military expenditures relative to GDP among the African countries at 4.8% of GDP.[59]

Figure 8.23 illustrates the trends in arms deliveries to Africa and compares the arms deliveries in millions of dollars by supplier countries to Africa between 2004–2007 and 2008–2011. The arms deliveries showed significant growth or contraction for some countries. For example, Germany's arms deliveries exhibited a significant reduction, declining from $1,100 during 2004–2007 to $300 during 2008–2011. Russia and China tied for the leaders in deliveries during 2008–2011 at $900 each; China did not experience much growth, while Russia increased from $600 in deliveries (as the third largest supplier in 2004–2007) to $900 in deliveries (as the second larger supplier in 2008–2011). The UK, which tied with China during 2004–2007 as the second most substantive supplier at $900 in deliveries, shrunk significantly to no deliveries by 2008–2011. Italy, France, and the US remained less significant suppliers than Russia, Germany, and China, but maintained some modest growth: Italy expanded from $200 in deliveries in 2004–2007 to $300 in deliveries in 2008–2011; France expanded from $100 to $200 in deliveries across the time periods, and the US expanded from $93 to $105 in deliveries. The other

59 Perlo-Freeman and Solmirano, 2014.

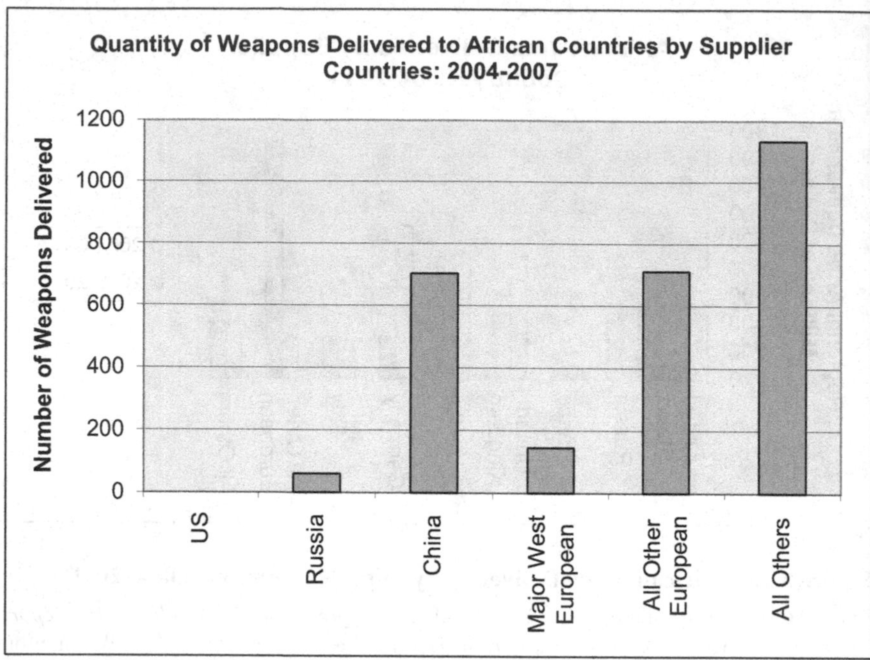

Figure 8.24 Quantity of Weapons Delivered to African Countries by Supplier Countries: 2004–2007

Source: Grimmett, Richard E. and Kerr, Paul K., *Congressional Research Service Report (R42678): Conventional Arms Transfers to Developing Nations: 2004–2011*, Washington DC, August 24, 2012.

European nations, excluding France, Germany and Italy, experienced significant growth from $500 in deliveries during 2004–2007 to $1,700 in deliveries during 2008–2011. Nevertheless, the total of deliveries from supplier nations to Africa experienced only modest growth, from $4,693 in deliveries during 2004–2007 to $4,805 deliveries during 2008–2011.

Data on the quantity of weapons delivered to African countries by supplier countries during 2004–2007 indicate that the US delivered no weapons to African countries (Figure 8.24).

During this period, China was the leading country which supplied weapons systems to African countries. African countries had the most significant demand for artillery, which was largely supplied by China and other countries (excluding European countries). African countries had the second greatest demand for APCs and armored cars, for which China was the leading supplier.

Russia was the leading supplier in supersonic combat aircraft, delivering 20, followed by China, which delivered 10. Russia was also the leading supplier in helicopters, delivering 40, while major Western European countries collectively

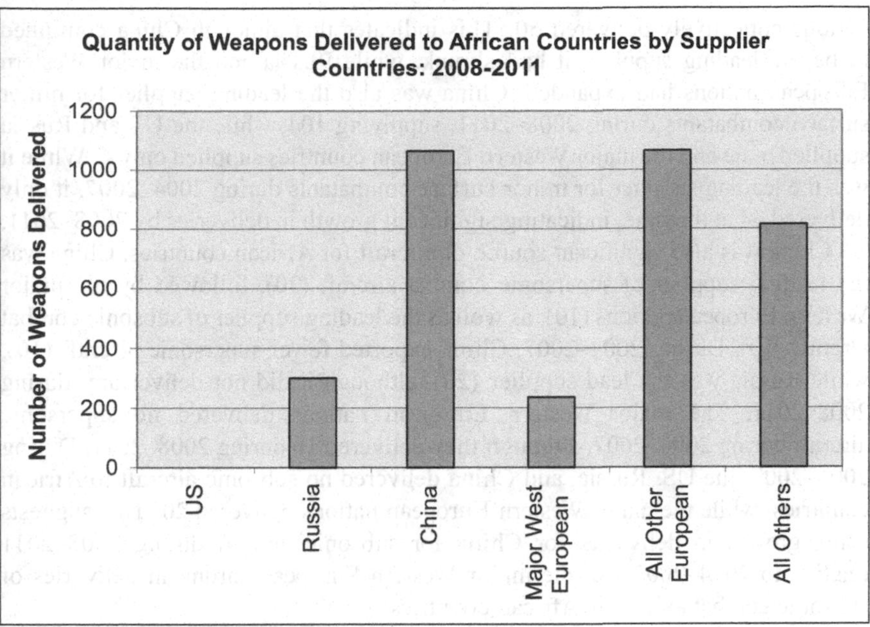

Figure 8.25 Quantity of Weapons Delivered to African Countries by Supplier Countries: 2008–2011

Source: Grimmett, Richard E. and Kerr, Paul K., *Congressional Research Service Report (R42678): Conventional Arms Transfers to Developing Nations: 2004–2011*, Washington DC, August 24, 2012.

delivered 40. In the minor surface combatants category, China was the leading supplier (34), while the major Western European countries delivered nine.

Figure 8.25 shows the quantity of weapons delivered by supplier countries to African countries during 2008–2011. As was the case during the 2004–2007 period, the US was not a supplier, whereas China was the leading supplier in a number of categories.

China has played a significant role in providing African countries with military equipment, such as artillery, tanks, armored cars, and ships. In the tanks and self-propelled guns category, China delivered 190, followed by Russia (50). This shows significant growth in deliveries during 2004–2007, since China and Russia delivered no tanks/self-propelled guns at that time. China was the leading supplier of artillery during 2008–2011, supplying 410 of them, while the US, Russia, and the major Western European nations delivered none. During 2004–2007, China delivered 230, while the others delivered none. China was the leading supplier in APCs and armored cars during 2008–2011, delivering 310, followed by Russia (60), while Western European nations collectively delivered 180. During 2004–2007, however, China delivered 380, while the major western European nations collectively delivered 50. This indicates that although China continued to be the leading supplier, it had shrunk, while Russia and the major Western European nations had expanded. China was also the leading supplier for minor surface combatants during 2008–2011, supplying 104, while the US and Russia supplied none and the major Western European countries supplied only 4. While it was the leading supplier for minor surface combatants during 2004–2007, it only delivered 34 at the time, indicating significant growth in deliveries by 2008–2011.

China was also significant source of aircraft for African countries. China was the leading supplier of supersonic combat aircraft (20), followed by the major Western European nations (10), as well as the leading supplier of subsonic combat aircraft (10). During 2004–2007, China exported fewer supersonic aircraft (10), while Russia was the lead supplier (20), although it did not deliver any during 2008–2011. The major Western European nations delivered no supersonic aircraft during 2004–2007, although they delivered 10 during 2008–2011. During 2004–2007, the US, Russia, and China delivered no subsonic aircraft to African countries, while the major Western European nations delivered 20. This suggests some growth in deliveries for China for subsonic aircraft during 2008–2011 relative to 2004–2007, but the major Western European shrunk in deliveries of subsonic combat aircraft to African countries.

Russia has been an important supplier to African countries in helicopters and surface-to-air missiles. During 2008–2011, Russia was the lead supplier of helicopters (70) and surface-to-air missiles (120), while the US and China did not supply any. The major Western European countries had reduced their supply of helicopters, providing 30 during 2008–2011 and 40 during 2004–2007. During 2004–2007, Russia supplied 40 helicopters to African countries, which indicates some growth in deliveries during the subsequent period. Russia experienced substantive growth in surface-to-air missiles, delivering none during 2004–2007 and delivering 120 during 2008–2011. Neither the US nor China delivered surface-to-air missiles in 2008–2011 or in 2004–2007, while the major Western European countries delivered only 10 during 2004–2007 and delivered none during 2008–2011.

Sub-Saharan African countries have expanded their purchases of combat helicopters and multirole fighter aircraft; indeed, African purchases of aviation have doubled in volume during 2008–2013 relative to the prior period. Although Egypt, Morocco, and Algeria in North Africa have traditionally been the largest acquirers of aircraft, sub-Saharan African countries are expanding their purchases. Many African countries purchased new or upgraded Sukhoi and MiG multirole aircraft, made by Russia. Indeed, Ethiopia ordered 18 upgraded older Su-30K fighter jets in July 2013, while in May 2013, Uganda ordered six Su-30MK Russian-made jets (Uganda has acquired 12 Su-30s since 2011). Angola operates six Su-30MKs, while Chad ordered three Russian MiG-29s in 2009. As of the summer of 2014, Angola was likely to purchase more aircraft due to concerns about the Congo. Uganda needed to protect its oil fields, as well as supported the government of South Sudan in the Sudanese civil war.[60]

Conclusion

This chapter has explored the challenges and opportunities faced by the global defense industrial base by focusing on the trends in global arms deliveries by supplier nations to various regions – Asia, the Near East, Latin America, and Africa. The evolution in the demand for various countries for particular categories of equipment has been the result of shifting defense priorities, expanding defense budgets, and the development of their indigenous defense industrial bases. The increasing ambitions of China in the Asia-Pacific region, as well as the threat posed by Iran in the Middle East, have particularly led to increased demand for defense equipment. This chapter has also examined the historical development of supplier nations for various categories of defense equipment to various importing nations. The importance of supplier nations has changed and is likely to continue to change as importing countries develop their own defense industrial bases. Indeed, while many of the Asian countries have held competitions for equipment from international defense contractors, they have also formed alliances with foreign defense contractors, engaged in "offset" agreements, and expanded their defense industrial bases. Moreover, sanctions against supplier countries, such as Russia in the Ukrainian crisis (discussed in Chapter 7) may reduce the role of supplier countries for importing nations.

Many developing countries are dealing with fiscal constraints and slower economic growth which results in the difficulties faced by them concerning whether to purchase more expensive, modern equipment, or cheaper, less innovative equipment. These are challenges similar to those faced by the US and European countries. Moreover, shifts in defense priorities require greater diversity in the types of equipment needed and the quantities. The trends and challenges in growing domestic defense industrial bases versus importing equipment, as well as the evolving demands for certain types of equipment, can be seen in much of the analysis in this chapter on global trade and highlights the importance of flexibility in production and supply by various nations to each other. In short, the expansion of globalization in the defense sector in recent years has been and will continue to be key in maintaining stability and prosperity in the global environment in the coming years.

References

Agence France-Presse, 2013a. "China, Malaysia to Hold Joint Military Drills." *Defense News*, October 30.

60 Nikala, 2014

——. 2013b. "EADS Still Hopes to Sell Eurofighter to South Korea." *Defense News*, September 26.

——. 2013c. "India Plays Down Prospect of Early Rafale Fighter Deal." *Defense News*, October 30.

——. 2013d. "Sri Lanka Raises Defense Budget Despite Foreign Pressure." *Defense News*, October 21.

——. 2013e. "Taiwan Gets First Batch of US-Made Attack Helicopters." *Defense News*, November 5.

——. 2013f. "Taiwan Receives First US Anti-submarine Aircraft." *Defense News*, September 25.

——. 2013g. "Taiwan Military: China Able to Invade by 2020." *Defense News*, October 8.

Bekdil, Burak Ege, 2013. "Turkey Signs Contract for First National Drone." *Defense News*, October 30.

Blumenthal, Dan, 2010. "India Prepares for a Two Front War." *Wall Street Journal Online*, March 1.

Bradsher, Keith, 2013. "Bolstering Military Ties, US Gives Philippines Aid." *New York Times*, December 18, p. A8.

Cameron, Doug, 2013. "Clipped by U.S., Lockheed CEO Aims Abroad." *Wall Street Journal*, December 7–8, pp. B1 and B3.

"Chinese Small Arms Sales May Fan Conflicts, Equip Terrorists." 2013. *National Defense*, October 11.

Erwin, Sandra I., 2013. "Missile Defense Market Getting Tougher for U.S. Firms." *National Defense*, October 29.

Gulati, Nikhil, 2010. "Boeing Gets Initial India Request for Refueling Planes." *Wall Street Journal Online*, January 18.

Hayashi, Yuka and Pasztor, Andy, 2013. "Japan, US at Odds over China's Air Zone." *Wall Street Journal*, December 2, p. A 16.

Higuera, Jose, 2014. "South America's Growing Interest in Antarctica Drives Polar Ship Buys." *Defense News*, August 4.

"Israel Aerospace Targets the Warship Market," 2013. *Globes*, October 9.

Kallender-Umezu, Paul, 2014. "With New Radar, Japan Sends Message to China." *Defense News*, April 28.

Krishna, R. Jai, 2010. "Larsen Eyes Revenue from Defense Nuclear Power." *Wall Street Journal Online*, February 16.

Lerner, Jeremy, 2010. "Cuts and Bureaucracy Stall Fleet Replacement." *Financial Times*, July 19.

Mehta, Aarton and Fryer-Biggs, Zachary, 2012. "Mideast Buying Lifts Stagnant Global Market." *Defense News*, November 12.

Minnick, Wendell, 2012. "Still Vigorous Asian Budgets Focus on Naval, Air Forces." *Defense News*, October 24.

——. 2013. "China Establishes UAV Industrial Base." *Defense News*, November 5.

——. 2014a. "Russian Fighters for China Still on Hold." *Defense News*, June 2.

——. 2014b. "Beijing Continues S. China Sea Expansion." *Defense News*, June 16.

——. 2014c. "Experts Say China Planning a Much Larger Carrier." *Defense News*, June 30.

——. 2014d. "China Develops Mature, Broad-Based UAV Sector." *Defense News*, July 14.

Misquitta, Sonya, 2009. "Defense Contractors Target Big Jump in India's Military Spending." *Wall Street Journal*, July 17, p. B1.

Mustafa, Awad, 2014a. "Center of Vehicle Market Shifts to Mideast, North Africa." *Defense News*, June 16.

——. 2014b. "Mideast, US Lead Sharp Rise in Munitions Market." *Defense News*, July 28.

Nikala, Oscar, 2014. "Multirole Combat Jets Propagate Across Sub-Saharan Africa." *Defense News*, July 14.

Page, Jeremy, 2013. "China Takes Territorial Dispute to New Depths." *Wall Street Journal*, December 2, pp. A1, A16.

Perlo-Freeman, Sam and Carina Solmirano, 2014. *SIPRI Fact Sheet: Trends in World Military Expenditure, 2013*. April.

Pfeiffer, Sylvia, 2010. "Defense Sector Watches to See Where the Axe Will Fall." *Financial Times*, July 19, p. 6.

Raghuvanshi, Vivek, 2013a. "India Extends Relations With China, Russia." *Defense News*, October 25.

——. 2013b. "India to Bolster UAV Fleet for Border Surveillance." *Defense News*, October 29.

——. 2012. "India to Focus Resources on Naval Operations." *Defense News*, October 24.

——. 2014a. "India Proposes $2.25B Tender for ASW Shallow Water Craft." *Defense News*, June 9.

——. 2014b. "India to Restart $10B Vehicle Program." *Defense News*, June 9.

——. 2014c. "India's Fighter Jet Negotiations Stall over Delivery Commitments." *Defense News*, June 16.

——. 2014d. "India's Promise of Defense Spending Boost Fizzles." *Defense News*, July 14.

Sanger, David E., 2013. "In the East China Sea, a Far Bigger Test of Power Looms." *New York Times*, December 2, p. A3.

Shanker, Thom. 2013. "Hagel Criticizes Chinese Navy, Citing Near Miss." *New York Times*, December 20, p. A4.

Siboro, Tiarma, 2013. "Indonesia, US Deepen Defense Ties Amid Exercises and Arms Deals." *Defense News*, September 30.

Spitzer, Kirk. 2013. "Japan Unveils Sweeping Plans to Bolster Defense." *USA Today*, December 18, p. 7A.

Conclusion

The global defense sector is faced with a variety of challenges as it continues to maintain global security and stability. Slow economic growth in a variety of countries has contributed to rising federal debts and deficits. The resulting fiscal constraints designed to reduce the substantial debt levels in the US and other countries have limited defense spending and will continue to do so in future years. In addition, the defense sectors have also faced shifting defense priorities which impact the size and scope of military forces, as well as the demand for certain types of equipment from the defense industrial base. For example, as the US withdraws its forces from Iraq and Afghanistan and focuses more on the Asia-Pacific region due to China's expanding regional ambitions, its need for the size of the Army relative to the Navy and the Air Force has changed, as well as its demand for ships and aircraft relative to tanks and ground vehicles. The impact of the evolution in demand for equipment and services due to fiscal constraints and shifting defense priorities from governments affects the sustainability of defense contractors in the defense industrial base. Their profitability and productivity, in turn, impacts economic growth in local regions where their manufacturing facilities and employees are located. As a result, Congressional representatives from those areas face tradeoffs between reducing government spending in the defense sector in order to help reduce federal debt versus potentially negatively impacting production facilities and employment in their local areas. This book has discussed the challenges facing governments in the US and other countries, as well as defense contractors, and has provided perspectives on potential strategies to mitigate risks and to sustain the global defense industrial base.

The US Department of Defense faces a variety of tradeoffs due to these fiscal constraints and shifting defense priorities. Chapter 1 emphasized the importance of the defense sector as a share of GDP and its volatility over the past 50 years as defense priorities shifted and the demand for military personnel and equipment varied. As of FY 2015, the proposed DoD base budget at $495.6 billion is almost half of the $1.014 trillion in discretionary spending. Moreover, non-discretionary spending, such as Social Security and Medicare, has risen over the past 50 years. Due to the budget constraints, the services face tradeoffs between purchasing quantities of new models of equipment based on the latest technologies, purchasing quantities of older models of equipment, sustaining and operating existing equipment, and investing in R&D to develop the next generation of equipment. This involves tradeoffs between procurement, operations and maintenance, and RDT&E. Sequestration in 2013 led to significant delays in operating and maintaining equipment, and the uncertainty of whether full sequestration will commence again in 2016 has made future procurement decisions difficult. Moreover, due to the shift in defense priorities away from Iraq and Afghanistan and toward the Asia-Pacific, the Army will be significantly reduced, which will further reduce military personnel spending (the second largest category of DoD's budget, following operations and maintenance). Consequently, DoD faces tradeoffs between defense priorities – the equipment, personnel, and training needed in anticipation of a variety of potential types of conflicts in certain regions – and fiscal constraints (the size and growth of the defense budget), as well as tradeoffs between various categories of the budget.

In an effort to balance these tradeoffs, DoD has endeavored to find solutions to handle the complexities of rising costs, limited financial resources, and demand for the current generation of equipment, as well as the next generation of equipment. As discussed in Chapters 1 and 2, one strategy is for older equipment to be eliminated and the funds to be channeled into high-priority modernization programs. For example, the Air Force could save money by eliminating older equipment, such as the A-10, and funneling the money into critical modernization programs, such as the KC-46a tanker, the future long-range bomber, and the F-35. A second related strategy is to substitute older equipment with less expensive equipment; the Air Force has been considering replacing F-16s and A-10s with UAVs and unmanned aircraft. A third option is to replace older equipment with more established models. Indeed, as discussed in Chapter 2, the Navy plans to replace older equipment with established models, such as replacing the older EA-6B Prowler with the EA-18G Growler, the older P-3C Orion aircraft with the P-8A Poseidon, and the older helicopters with the MH-60R and the MH-60S. As discussed in Chapter 7, they have also considered replacing the aging C-2s with V-22 Ospreys, which would serve as the next-generation carrier onboard delivery system, despite various tradeoffs in capabilities between the different types of equipment. A fourth strategy would be to standardize and limit the types of models used in an effort to reduce operations and maintenance costs. For example, the Navy plans to have carrier strike groups deploy five models of aircraft with five types of engines by 2023, rather than 10 models of aircraft and eight different types of engines, as was the case in 2005. Another approach would be to eliminate unneeded equipment or convert it for another similar use and thereby reduce costs. The sale of unneeded equipment to foreign military establishments is also an option. Indeed, with the shifting defense priorities and the repositioning of the role of the Army, the Army has less need for MRAPs. As a result, it is considering converting MRAP vehicles into trucks, using them for other functions, such as training, and/or re-selling them to foreign military establishments. Foreign countries have shown interest in acquiring cheaper, used equipment; for example, countries such as Libya have shown interest in acquiring older US Army Chinook helicopters.

A number of other countries are facing the same tradeoffs as the US between fiscal constraints and shifting defense priorities, as discussed in Chapters 7 and 8. For example, as discussed in Chapter 7, although France continues to have the highest defense expenditures, followed by the UK and Germany, many of the

European countries have experienced significant declines in defense spending between 2008 and 2013, including France, the UK, Italy, Spain, the Netherlands, and Greece. Many of these countries have followed strategies of reducing their purchases of existing equipment, as the US has done. Indeed, France announced in 2013 that it would halve its purchases of Dassault Rafale fighter planes over the next six years, while Greece has reduced its plans to upgrade its Mirage 2000 fighter jets and Italy cut its F-35 order from 131 aircraft to 90 aircraft. The German Army, like the US Army, will shrink, which will result in significant military and civilian cutbacks, as well as the reduction of purchases of equipment, such as NH90 helicopters and Tiger support helicopters. A key strategy among European countries has been collaboration in training and in assessing key equipment functions. As discussed in Chapter 7, the UK and Norway considered collaborating for training and support for the F-35, while Lithuania, Latvia, and Estonia discussed collaboration on joint arms purchases.

The conflict between the role of fiscal constraints in reducing the defense budget and limiting purchases of defense equipment, and the need for buying more defense equipment with perceived military threats is highlighted by the impact of the Ukrainian crisis on the European defense industrial base. The crisis may counter the tendencies of many European countries to reduce military expenditures in the wake of slower economic growth; indeed, Latvia, Lithuania, and Estonia are planning to increase their defense spending, and Poland and the Czech Republic are expanding their defense budgets. Moreover, sanctions against Russia imposed in 2014 may reduce the equipment purchased by the Russian military from other countries, which could negatively impact the effectiveness of the Russian military, as well as those defense equipment manufacturers that export equipment to Russia. For example, in early September 2014, France announced that it would not deliver a Mistral warship to Russia, which was part of a contract negotiated in 2011 for four Mistral warships. In addition, the sanctions could also limit countries from purchasing Russian equipment, which could damage the Russian defense industrial base, as well as limit the capabilities of the military forces in the countries which had planned to import Russian defense equipment.

The challenges faced by the defense forces in countries can significantly impact the sustainability of their defense industrial bases. The US tradeoffs in the types and quantities of equipment purchased by the services – the Army, the Navy and Marine Corps, and the Air Force – have powerful implications for the US defense industrial base. As discussed in Chapter 2, some of the areas of the defense industrial base have exhibited significant growth, such as cybersecurity and UAVs. On the other hand, other areas of the defense industrial base have experienced closures of production facilities, such as the shipyards. Indeed, one of the six major shipyards in the US, the Avondale shipyard, has been slated for closure. Moreover, other newer shipyards, such as Marinette Marine in Wisconsin and the Austal shipyard in Mobile, which are dependent on production of littoral combat ships, will face the uncertainty of whether DoD will reduce the purchase of littoral combat ships from 52 down to 32. This reflects the tradeoffs faced by

DoD between fiscal constraints (the need to reduce procurement costs) and defense priorities (the need for littoral combat ships in the Asia-Pacific region). The impact of sequestration on defense spending, as well as uncertainties regarding its timing, have also resulted in layoffs and concern about investing in R&D by defense contractors.

Similarly, the significant defense spending cuts in many overseas countries have led to the closure of facilities and loss of key jobs. For example, as discussed in Chapter 7, the UK's reduction in defense spending has led to plans for the closure of significant BAE shipbuilding facilities in Portsmouth, which has operated for 500 years. Nevertheless, many countries continue to expand their defense industrial bases in weapons sectors which they consider to be critical. Indeed, as discussed in Chapter 8, in addition to purchasing equipment from overseas, India, Indonesia, Israel, Saudi Arabia, Turkey, and the UAE have further developed their defense industrial bases. Offset agreements also provide opportunities to expand their domestic defense industrial bases, despite the purchase of materials from overseas countries, since a certain percentage of the contract has to be reinvested in the local economy by the recipient of the contract.

Atrophy of the US defense industrial base due to a lack of demand for certain types of equipment can have a significant impact on economic growth in areas where the production facilities are located, and this can, in turn, exert a powerful influence on the perspectives of Congressional representatives. As discussed in Chapter 2, the top states with substantive aerospace and defense industrial employment include: California, Washington, Texas, Florida, Arizona, Connecticut, Virginia, Kansas, New York, and Pennsylvania. In states such as Kansas, Washington, Arizona, Connecticut, Alabama, New York, Vermont, and Maine, the defense industry makes up 4% or more of the state's GDP. As a result, Congressional representatives from affected districts are faced with the challenge of lowering defense spending to reduce federal debt, but potentially damaging economic growth in their districts due to closure of production facilities.

Chapter 5 illustrated the incentives of Congressional representatives to maintain defense production in their states by analyzing the tanker competition between Boeing and Northrop Grumman/EADS. The proposed Northrop Grumman/EADS tanker would have created 48,000 jobs in the US, especially in Alabama, since, although some parts would be manufactured overseas, the assembly of the tanker would occur in Mobile. As a result, the Alabama Congressional delegation strongly supported it. The Boeing tanker, on the other hand, would have involved 44,000 jobs across 40 states and 300 suppliers, and Everett, Washington and Wichita, Kansas would have been the major locations of tanker production. As a result, Congressional representatives from Washington and Kansas strongly supported the Boeing proposal. Similarly, examples of conflicts between Congressional representatives supporting the continuation of production of a weapons system in their districts and the perspectives of various services in eliminating a particular weapons system due to shifting defense priorities and/or fiscal constraints include debates regarding the C-17 and the C-27J, discussed in Chapter 2. Moreover, the conflicts between attempting to preserve the U2 at the expense of the Global Hawk and the elimination of the A-10 reflect the differences in the perspectives of the services on the importance of particular types of weapons systems in the wake of fiscal constraints relative to the perspectives of Congressional representatives.

Atrophy of particular sectors in the US defense industrial base due to a lack of demand for certain types of equipment can also lead to difficulties in regenerating those sectors, if defense priorities shift in future years and there is a significant increase in demand for those weapons systems. This is because current reductions in demand provide defense contractors with little incentive to hire younger workers and, since a significant share of workers are likely to retire, there is little opportunity for an intergenerational transfer of skillsets. Consequently, if there is an increase in demand for a given weapons system in the future, it may be difficult for production levels to meet demand. This risk has been partially mitigated, but not eliminated completely, in situations where low demand for a given weapons system has significantly slowed production, such that some employment continues.

The uncertainty of defense spending and its impact on production has not only limited investments in additional skilled employees and resulted in volatility in production, but it also has contributed to greater concern about investing in R&D. Innovation and R&D are important for DoD in terms of developing the next generation of military equipment. Nevertheless, defense contractors fear that the sunk costs may not be recovered if demand for the new products diminishes. Unfortunately, RDT&E has only been about 13% of the enacted FY 2013, the enacted FY 2014, and the proposed FY 2015 defense base budgets, so there is significant burden on defense contractors to invest in new technologies. The services have attempted, however, to stimulate R&D for products in demand. For example, as discussed in Chapter 2, the Army awarded technology investment agreements to aircraft companies which match investments from the companies over six years with funding from the Army's Aviation and Missile Research, Development, and Engineering Center (AMRDEC) in order that the Army can ultimately acquire new vertical lift aircraft. The strategy of defense contractors to sustain innovation includes expanding in growing areas of the defense industrial base, such as cybersecurity, electronic network equipment, and UAVs. R&D is more likely to be recouped in growing sectors; moreover, many of the designs can be also be used in the civilian sector. Similarly, many of the other countries are facing comparable difficulties in sustaining innovation in their defense industrial base. Several European countries have collaborated in these efforts. For example, as discussed in Chapter 7, the UK and France signed an MOU in July, 2014 to study an unmanned combat aerial system (UCAS) with the goal of replacing Rafale and Typhoon fourth generation fighters in future years.

As the military establishments in the US and in other countries struggle with budget constraints and demand for equipment, while governmental representatives struggle with the challenges of reducing federal debt at the cost of hurting regional growth in some areas, defense contractors also face challenges. These include: uncertainty regarding demand for current and existing products; current and potential plant closures and employment reductions; attenuation of skillsets of employees; decreased incentives to invest in R&D; and potential ripple effects into other product lines (for example, a reduction in the number of aircraft carriers can impact the demand for the number of planes, etc.). Defense contractors, however, have developed various strategies to mitigate risk and to sustain their production capabilities. These include:

a. engaging in mergers and/or alliances with other firms;
b. expanding into the civilian markets; and
c. expanding into the overseas markets.

Mergers and alliances among defense contractors in order to sustain the defense industrial base can assist in pooling knowledge, jointly investing in R&D, providing stronger bids in competitions, and improving productivity and cost efficiencies through economies of scale. In the wake of the end of the Cold War, large defense firms merged to improve economies of scale and generate cost efficiencies, as well as to integrate skillsets in order to expand into other defense equipment markets. Chapter 3 discussed the mixed results in efficiencies in various equipment categories due to mergers between various defense contractors, as well as the impact on market concentration in various equipment categories. While it is unlikely that mergers between large defense contractors will occur in the upcoming years, it is likely that mergers between larger firms and smaller firms will occur in growing markets, such as UAVs, cybersecurity, etc. This would provide larger firms with the innovation and skillsets of the smaller firms. Smaller firms would be more likely to receive research funding from the larger firms through mergers; indeed, they would potentially have difficulties in accessing capital markets if they did not merge. As discussed in Chapter 2, mergers between larger and smaller firms in the UAV market have led to creation of new products through skillset transfer. For example, Northrop Grumman acquired Ryan Aeronautics which had expertise in target drone production and design, as well as Swift Engineering which had expertise in designing blended wing UAVs. Moreover, Boeing's acquisition of the Institu Group and Frontier assisted Boeing in developing UAVs. Similarly, larger companies, such as Cisco, Intel, and Google, have acquired smaller companies in order to develop their role in the cybersecurity market: Cisco purchased Sourcefire, Google bought VirusTotal, etc.

As discussed in Chapter 4, alliances provide defense contractors with many of the benefits of mergers, in terms of sharing knowledge, developing new products, and providing stronger bids in international competitions. Since larger defense contractors are less likely to merge due to market concentration issues, alliances provide them with the ability to work with other defense contractors to produce better products. For example, the F-35 Joint Strike Fighter involves nine large contractors from various countries, led by Lockheed Martin. Similarly, Boeing and Lockheed Martin have formed an alliance to create a strong proposal for the Air Force's Long Range Strike Program, which helps to integrate Boeing's 50 years of knowledge of B-52s, as well as its knowledge as the primary contractor of the KC-46 tanker, and Lockheed Martin's knowledge as the lead contractor on the F-35 program.

Expansion of defense contractors into civilian markets is a key strategy in mitigating the risks of reduced defense spending and demand for various products. As discussed earlier, investing in research in growing areas, such as cybersecurity and UAVs which have both military and civilian uses, enables the development of common designs and reduces the potential of financial losses associated with innovative research. Chapters 6 and 7 highlight the expansion of defense contractors into civilian markets, including the aerospace industry. Nevertheless, while slow economic growth increases the likelihood of reductions in defense spending due to growth in federal debt, it can also increase the likelihood of reduced purchases of related equipment in civilian markets. For example, as discussed in Chapter 6, the degree to which airlines require new civilian aircraft is often based on their projections of passenger travel, which may be limited due to the risk of job loss and slow economic growth. As discussed in Chapter 2, a number of the top 20 global defense companies receive a significant portion of their revenues from civilian markets; examples include: Northrop Grumman (79.1%), General Dynamics (60.3%), Boeing (60%), Thales (56.3%), and Finmeccanica (49.6%).

Expansion of defense contractors into overseas markets, both for civilian and military equipment, is another key strategy that can mitigate risk and improve profitability. As discussed throughout the book, defense contractors in a number of countries have significant sales in overseas markets which enable them to diversify uncertainty and risk in their own countries. Chapter 2 illustrates that, although 46 of the top 100 global defense firms are in the US, the remainder are from other countries, such as the UK, Russia, Japan, France, Israel, South Korea, and Germany. Chapter 8 examined the global arms deliveries in the worldwide defense market in detail and the analysis indicates that between 2004 and 2011, the US was the leading supplier of global arms deliveries, followed by Russia, France, the UK, China, Germany, Italy, etc. Within developing countries, Saudi Arabia, India, China, Egypt, Pakistan, Israel, the UAE, South Korea, Taiwan, and Algeria were the leading recipients of arms deliveries between 2004 and 2011. The focus of Chapter 8 on the evolution and expansion of defense markets in Asia, the Middle East/Near East, Latin America, and Africa emphasized the growth of overseas markets and the reaction of various countries to shifting defense priorities and expanding or contracting defense budgets. Indeed, the top five defense spenders in Asia – China, India, Japan, South Korea, and Taiwan – captured 87% of Asian defense spending in 2011 and jointly spent twice the amount in 2011 that they had spent in 2000. Similarly, the Middle East is a growing market. Countries such as Kuwait, Saudi Arabia, and the UAE are estimated to spend $175 billion over the next five years in defense. Saudi Arabia's defense spending in 2011 was more than the combined defense budgets of Bahrain, Egypt, Israel, Iraq, Jordan, Kuwait, Syria, Oman, and Lebanon. Nevertheless, although Saudi Arabia's defense spending increased by 14% between 2012 and 2013, countries with smaller budgets, such as Iraq and

Bahrain, surpassed it in growth at 27% and 26%, respectively. Chapter 7 provided case studies of the evolution in recent years of the leading suppliers and recipients of global imports and exports in aircraft and ships.

Nevertheless, as discussed in Chapter 7, expansion of defense contractors into overseas countries can also have limitations. Many of the countries which are significant importers of defense equipment experienced an increase in debt as a share of GDP in 2007 and this trend continued following the financial crisis in 2008 and 2009. While some of these countries subsequently reduced their debt as a share of GDP, other countries continued to experience increasing debt. Both rising debt and slow GDP growth for importing countries limits their abilities to import defense equipment, which, in turn, limits the profitability of overseas sales for defense contractors. Similarly, as discussed in Chapter 7, since many of the exporting countries also had rising debt as a share of GDP and slow GDP growth, there were increased incentives for defense contractors in these countries to expand overseas to sustain their profitability.

The desire for some countries to preserve their domestic defense industrial bases limits the perception of their markets as "open markets" for foreign defense contractors, which can limit their strategies to expand overseas sales. Chapter 5 discusses the tendencies in the US to prevent foreign entrants from merging with domestic competitors often due to perceptions that the acquisition could impact national security. It also discusses the role of tariffs and quotas designed to limit imports into the US, as well as the strengthening of the Buy America legislation. Congressional representatives often play a key role in preventing the loss of jobs in their home states and districts by protecting domestic jobs from foreign competition. The 2008 tanker competition between Boeing and an alliance between Northrop Grumman and EADS highlighted the conflicts of opening US markets to foreign suppliers, even if the foreign suppliers are creating jobs and building production facilities in the US, rather than producing and assembling the equipment overseas. Indeed, the tanker competition highlighted the divide in the US between a desire for open markets and for the best product at the lowest costs, versus the potential for loss of jobs in certain areas of the country. Moreover, Chapter 6 suggests that only 6-7% of the DoD procurement budget is spent on purchases from foreign suppliers and that, depending on the defense equipment category, the share of purchases in that equipment category by DoD from foreign suppliers is often under 2.5% and never over 11%. Nevertheless, between 2009 and 2012, DoD purchased about one-third of its petroleum from foreign suppliers and purchased over 40% in 2013. It purchased between 10% and 17% of its subsistence and construction materials from foreign suppliers. Consequently, although the US is an important market for the global defense industrial base, it still attempts to protect its own defense industrial base in military equipment from foreign suppliers. Similarly, in the future, as overseas countries further develop their defense industrial bases, they may be less likely to import military equipment due to concerns that they will lose jobs domestically.

In conclusion, this book discusses the current and future challenges facing the global defense industrial base and provides perspectives on potential strategies for governments and companies to sustain global security and stability. While the integration of global markets beyond the defense sector impacts economic growth, inflation / deflation, and employment levels, due to their interconnections in financial markets and trade, participants in the global economy, especially manufacturers in the defense sector, can mitigate risks through collaboration as they face common challenges, both in the economic arena and the national security arena. Although strategies to sustain the defense industrial base involve risks for defense equipment manufacturers and governments, their continued collaborative efforts across borders to promote innovation and cost efficiencies by sharing resources, as well as diversifying across product types and borders, will assist in protecting the safety and security of current and future generations throughout the global community.

Index

Note: Page numbers in **bold** indicate figures, *italic* numbers indicate tables.

Printed in the United States
by Baker & Taylor Publisher Services

Printed in the United States
by Baker & Taylor Publisher Services